A PRACTICAL GUIDE TO MAGNETIC CIRCULAR DICHROISM SPECTROSCOPY

BICENTENNIAL 1807 WILEY 2007 BICENTENNIAL

THE WILEY BICENTENNIAL–KNOWLEDGE FOR GENERATIONS

Each generation has its unique needs and aspirations. When Charles Wiley first opened his small printing shop in lower Manhattan in 1807, it was a generation of boundless potential searching for an identity. And we were there, helping to define a new American literary tradition. Over half a century later, in the midst of the Second Industrial Revolution, it was a generation focused on building the future. Once again, we were there, supplying the critical scientific, technical, and engineering knowledge that helped frame the world. Throughout the 20th Century, and into the new millennium, nations began to reach out beyond their own borders and a new international community was born. Wiley was there, expanding its operations around the world to enable a global exchange of ideas, opinions, and know-how.

For 200 years, Wiley has been an integral part of each generation's journey, enabling the flow of information and understanding necessary to meet their needs and fulfill their aspirations. Today, bold new technologies are changing the way we live and learn. Wiley will be there, providing you the must-have knowledge you need to imagine new worlds, new possibilities, and new opportunities.

Generations come and go, but you can always count on Wiley to provide you the knowledge you need, when and where you need it!

William J. Pesce
WILLIAM J. PESCE
PRESIDENT AND CHIEF EXECUTIVE OFFICER

Peter Booth Wiley
PETER BOOTH WILEY
CHAIRMAN OF THE BOARD

A PRACTICAL GUIDE TO MAGNETIC CIRCULAR DICHROISM SPECTROSCOPY

W. Roy Mason
Department of Chemistry and Biochemistry
Northern Illinois University
DeKalb, Illinois 60115

BICENTENNIAL 1807 WILEY 2007 BICENTENNIAL

WILEY-INTERSCIENCE
A JOHN WILEY & SONS, INC., PUBLICATION

Published by John Wiley & Sons, Inc., Hoboken, New Jersey
Published simultaneously in Canada

Wiley Bicentennial Logo: Richard J. Pacifico

***Library of Congress Cataloging-in-Publication Data*:**

Mason, W. Roy.
A practical guide to magnetic circular dichroism spectroscopy/by W. Roy Mason.
p. cm.
Includes bibliographical references and index.
ISBN 978-0-470-06978-3 (cloth)
1. Molecular spectroscopy. 2. Molecular spectra. 3. Magnetic circular dichroism. I. Title.
QC454.M6M37 2008
543'.54--dc22

2007011030

10 9 8 7 6 5 4 3 2 1

CONTENTS

APPENDIXES

PREFACE

This practical guide is intended to present a concise description of magnetic circular dichroism (MCD) spectroscopy and to illustrate how it can be applied to the interpretation of molecular electronic spectra. The presentation here is intended to be descriptive and thereby to help the reader visualize the optical spectroscopic effects presented by MCD measurements. An important purpose of this text is to call attention to the added dimension that experimental MCD spectra contribute to conventional absorption spectroscopy: The sign of the differential absorption ΔA, which, together with its magnitude, has the capability in many cases of supporting specific transition assignments and eliminating other hypothetical possibilities, often based upon simple, elegant symmetry arguments. The wave description of polarized light and the development of MCD theory assumes that the reader has familiarity with wave properties of light, quantum chemistry, electronic states, and molecular electronic structure. Advanced symmetry arguments are described which are based on the irreducible tensor methods outlined by S. B. Piepho and P. N. Schatz in their book *Group Theory and Spectroscopy with Applications to Magnetic Circular Dichroism* (Wiley-Interscience, New York, 1983). This book, including its symmetry tables (Appendixes) and precise specification of standard conventions, is invaluable for interpretive MCD spectroscopy, and all investigators in the field owe the authors a great debt of gratitude. Some practical considerations for experimental MCD measurements based on the present author's experience are included, but the discussion here is perhaps more general than any specific case may require. The illustrative examples discussed in the case studies were chosen to show the breadth of application of MCD measurements to the interpretation of electronic spectra in the vis–UV region and to the formulation of electronic structure models; the case studies included here are not intended to represent an exhaustive review of MCD spectroscopy. Furthermore, some of the details of the interpretive arguments in the examples are, for reasons of complexity, left for the reader to consult the original literature. The text concludes with a descriptive chapter on vibrational and rotation–vibrational MCD in the IR or near-IR region (MVCD) and on MCD in the X-ray region (XMCD), followed by a chapter that introduces magnetic linear dichroism (MLD) spectroscopy, a related and complementary magneto-optical technique. Although every effort has been made to eliminate errors in the text, the author takes sole responsibility for any that remain.

W. Roy Mason

DeKalb, Illinois
October 2006

1 Introduction

Magnetic circular dichroism (MCD) spectroscopy is based upon the measurement of the difference in absorption between left circularly polarized (lcp) light and right circularly polarized (rcp) light, induced in a sample by a strong magnetic field oriented parallel to the direction of light propagation. The difference absorption, or dichroism, is defined by convention as $\Delta A = A_- - A_+$, where A_- is lcp absorption and A_+ is rcp absorption. The measured ΔA is the same quantity measured for natural circular dichroism (CD) which is observed for chiral (optically active) molecules in regions of absorption. However, the origin of CD and MCD is quite different. Natural CD requires a molecular environment where molecular structure features distribute electric charge in a spatial array that has helical "handedness." In contrast, MCD is due to electromagnetic interaction of the external field with electronic charge within the sample, no matter how it is distributed, and is a universal property of light absorption for all matter when placed in a magnetic field. A chiral molecular structure is not a requirement for MCD. In either case, however, the quantity ΔA is analogous to conventional light absorption, where $A = (A_- + A_+)/2$, in that it obeys the Beer–Lambert law and thus is proportional to the molar concentration of the absorbing species, c, and the path length through the sample, ℓ (in centimeters); in addition, magnetically induced ΔA for MCD is proportional to the magnetic field **B** Eq. (1.1), where

$$\Delta A = A_- - A_+ = \Delta\varepsilon_M c\ell \mathbf{B} \tag{1.1}$$

$\Delta\varepsilon_M$ is the differential molar absorptivity per Tesla of field, analogous to ε, the molar absorptivity. It should be apparent that ΔA differs from conventional absorption in that the measured quantity has a sign associated with it that is dependent upon whether A_- or A_+ is larger. The positive or negative sign of ΔA for MCD adds another dimension to absorption spectra, which present only positive values of A versus energy or wavelength.

The ΔA for MCD results from a magnetic perturbation (the Zeeman effect) of the states involved in optical transition(s) responsible for light absorption. The phenomenon is related to the Faraday effect, discovered by Michael Faraday in the 1840s when it was observed that the plane of polarized light passing through a sample placed in a magnetic field was rotated in transparent regions of the spectrum (magnetic optical rotation or MOR). The optical rotation results from

---

A Practical Guide to Magnetic Circular Dichroism Spectroscopy, by W. Roy Mason

magnetically induced refractive index differences (birefringence) for plane polarized light in different directions. In regions of absorption, a magnetic field can induce not only differences in index of refraction but also differences in absorption of lcp and rcp light, which differences give rise to elliptically polarized light emerging from the sample. The optical rotation is related to the rotation of the major axis of the ellipse, while ΔA is related to the ratio of the minor and major axes of the ellipse as will be explained in Chapter 2. The wavelength dependence of MOR (magnetic optical rotatory dispersion or MORD) and MCD are related to each other. In principle, one can be calculated from the other by means of integral transforms (Kramers–Kronig type) over all wavelengths. However, MCD is almost exclusively used today because of experimental considerations. Perhaps one of the most important is that ΔA, like A, is ≈ 0 in transparent regions of the spectrum and is not strongly affected by birefringent imperfections in optical elements (windows, lenses, cells, etc.) that are part of spectrometers. In contrast, optical rotation can have significant magnitude in transparent regions of the spectrum and can be strongly affected by birefringence due to strains or other imperfections in optical elements.

The development of MCD really began in the 1930s when a quantum mechanical theory of MOR in regions outside absorption bands was formulated. The expansion of the theory to include MCD and MOR effects in the region of absorptions, which were referred to as "anomalous dispersions," was developed soon thereafter. There was, however, little effort made to refine MCD as a modern spectroscopic technique until the early 1960s. The publication of important reviews by Buckingham and Stephens [review 1] in 1966 and Schatz and McCaffery [review 2] in 1969 called attention to the potential of MCD, with the added dimension of the sign of ΔA, as an interpretative tool for electronic spectra. The ability of MCD to interrogate and characterize electronic excited states in a way that was unavailable from other spectroscopic methods at the time was one of the exciting aspects noted by these reviews. Since that time there have been numerous studies of MCD spectra for a very large variety of samples, including stable molecules in solutions, in isotropic solids, and in the gas phase, as well as unstable molecules entrapped in noble gas matrices. More recently, MCD has found useful application in the study of biologically important systems including metalloenzymes and proteins containing metal centers. The application of MCD spectroscopy has been broad and diverse; a single volume cannot hope to touch on all aspects, but the goal here is to concisely describe the technique and then present some examples that illustrate not only the breadth of application, but also some of the more subtle aspects of electronic structure that can be experimentally characterized. MCD is an experimental spectroscopic tool that can help support or reject various proposed models for bonding and electronic structure that are the stock and trade of the chemical scientist.

The phenomenon of polarized light is first described in Chapter 2, followed by the theoretical framework for MCD spectroscopy in Chapter 3, which includes the Rigid Shift model and the definition and properties of MCD A, B, and C terms. The presentation follows closely the important book by S. B. Piepho and P. N.

Schatz [ref. 1], which describes the origin of MCD in considerable detail and then uses MCD as the basis for the application of advanced symmetry methods which employ powerful irreducible tensor techniques. The symbolism, definitions, and standard basis conventions of Piepho and Schatz are used here, for the most part, so that if further elaboration of the mathematical development is required, the reader can easily consult the relevant sections of their book. Chapter 4 presents a brief description of practical MCD spectral measurements, which is followed in Chapter 5 by a discussion of how MCD spectra are interpreted. In Chapters 6 and 7 a number of case studies are presented in order to illustrate the utility and breadth of MCD spectroscopy as applied to the study of the properties of electronic states and their electronic transitions in the vis–UV region. Chapter 6 is concerned with diamagnetic systems and A and B terms, while Chapter 7 discusses examples of paramagnetic systems and C terms. Chapter 8 presents brief descriptions of MCD for vibrations and rotation-vibrations in the IR or near-IR region (MVCD), and it also discusses MCD in the X-ray region (XMCD). Chapter 9 introduces magnetic linear dichroism (MLD) spectroscopy, a related and complementary magneto-optical technique involving an external field that is transverse to the light propagation direction.

A list of reviews of MCD spectroscopy is included at the end, which may provide useful historical perspectives on the development of the technique and standard practice used today. It should be noted that the MCD term definitions, which feature so prominently in the interpretation of MCD spectra, originally differed from the standard conventions based on the Stephens definitions of 1976 [review 3] in common use today. The differences between the older definitions and those in standard use today are noted in Appendix A of Piepho and Schatz [ref. 1]. Also some earlier literature report MCD spectra in terms of molar ellipticity angles per Gauss $[\theta]_M$ rather than ΔA or molar absorptivities per Tesla, $\Delta\varepsilon_M$. The relationships between these older quantities and those in use today are also described in the Appendix A of ref. 1. It should be noted that the shape of the MCD spectra will appear the same because of the proportionality of $[\theta]_M$ and $\Delta\varepsilon_M$, but the quantitative magnitudes will certainly be different due to field and path unit differences.

2 Polarized Light

An understanding of the nature of polarized light and the transitions it can induce in atoms and molecules in the presence of a magnetic field will provide useful background for an understanding of MCD. In this chapter a wave model description of polarized light and its interaction with matter will be given. The description here is brief; for further clarification of the electromagnetic wave properties of light, standard reference works such as Born and Wolf [ref. 2, especially Section 1.4 on vector waves] can be consulted.

2.1. LINEAR POLARIZATION AND PLANE POLARIZED WAVES

A linearly or plane polarized light wave can be conveniently described by the real part (Re) of the time-dependent vector potential $\mathbf{A} = \mathbf{A^0} \exp(i2\pi\nu(t - nz/c))$, Eq. (2.1), where $\mathbf{A^0}$ is the vector amplitude, ν is the wave frequency, n is the index of refraction which is inversely related to the wave velocity in the medium, t is time, z is the propagation direction, and c is the speed of light.

$$\mathrm{Re}\,\mathbf{A} = \mathrm{Re}\,\mathbf{A^0} \exp(i2\pi\nu(t - nz/c)) = \mathbf{A^0} \cos 2\pi\nu(t - nz/c) \tag{2.1}$$

The electric **E** and magnetic **B** field vectors of such a wave are perpendicular to each other, and they each oscillate in a single plane containing the direction of propagation. The oscillating fields can each be described by vectors perpendicular to the propagation direction. Thus if we take the propagation to be in the positive z direction in a right-handed coordinate system, the electric and magnetic fields will be in the xy plane. The electric and magnetic field vectors in Gaussian units (electrostatic units for **E** and electromagnetic units for **B**) are given by

$$\mathbf{E} = -(1/c)\partial\mathbf{A}/\partial t = -(i2\pi\nu/c)\mathbf{A} \tag{2.2}$$

$$\mathbf{B} = \nabla \times \mathbf{A} = \mu\mathbf{H} \approx \mathbf{H} \tag{2.3}$$

within the Coulomb gauge (the so-called “transverse” or “radiation” gauge) defined by the relation $\nabla \cdot \mathbf{A} = 0$ and where the magnetic permeability $\mu \approx \mu_0 = 1$ (the permeability in a vacuum) for nonmagnetic environments (the usual case). It may be remarked here that **H** is the magnetic field strength and **B** is the

A Practical Guide to Magnetic Circular Dichroism Spectroscopy, by W. Roy Mason

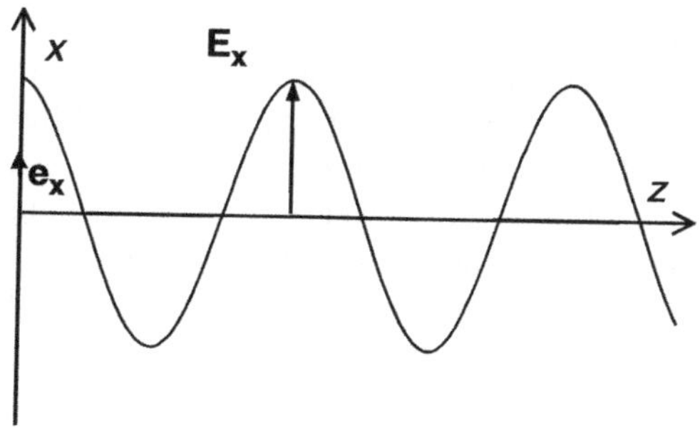

Figure 2.1. A plane-polarized wave in the x direction with propagation in the z direction.

magnetic field flux density (magnetic induction) within a sample but $\mathbf{B} = \mathbf{H}$ is assumed here for samples in common nonmagnetic solution or matrix environments. Furthermore, $\nabla \cdot A = 0$ because **E** and **B** are both perpendicular to the direction of propagation; they are also perpendicular to each other. The vector amplitude $\mathbf{A^0}$ contains polarization information and can be written as $\mathbf{A^0} = \boldsymbol{\pi}\mathrm{A}^0$, where $\boldsymbol{\pi}$ is a unit polarization vector. Thus, for the electric field vector amplitude from Eq. (2.2) we have

$$\mathbf{E^0} = -(i2\pi\nu/c)\mathbf{A^0} = -(i2\pi\nu/c)\boldsymbol{\pi}\mathrm{A}^0 = \boldsymbol{\pi}\mathrm{E}^0 \tag{2.4}$$

The unit vectors in the coordinate directions are symbolized by $\mathbf{e_x}$, $\mathbf{e_y}$, and $\mathbf{e_z}$. Thus a plane-polarized wave in the xz plane would be represented by

$$\mathbf{E_x} = \mathrm{Re}\ \mathbf{e_x}E^0 \exp(i2\pi\nu(t - nz/c)) = \mathbf{e_x}E^0 \cos 2\pi\nu(t - nz/c) \tag{2.5}$$

The $\mathbf{E_x}$ vector would always point in the x direction as it is propagated along the z direction. Such a wave is said to be linearly polarized in the x direction. Such a wave is illustrated in Figure 2.1. The **B** vector for this wave would, of course, point in the y direction, $\mathbf{B_y}$.

2.2. CIRCULAR POLARIZATION AND CIRCULARLY POLARIZED WAVES

A circularly polarized wave can be visualized as the sum of two plane-polarized waves $\mathbf{E_x}$ and $\mathbf{E_y}$ that differ in phase by $\pi/2$, or one-quarter of a wavelength. Thus circularly polarized waves are given by

$$\begin{aligned}\mathbf{E}_\pm &= \mathrm{Re}[(-i2\pi\nu A^0/c)(1/\sqrt{2})(\mathbf{e_x} \pm i\mathbf{e_y})\exp(i2\pi\nu(t - n_\pm z/c))] \\ &= (E^0/\sqrt{2})[\mathbf{e_x}\cos 2\pi\nu(t - n_\pm z/c) \mp \mathbf{e_y}\sin 2\pi\nu(t - n_\pm z/c)]\end{aligned} \tag{2.6}$$

where the + subscript indicates a rcp wave and the − subscript indicates an lcp wave, respectively, and the phase difference between the two plane waves is $\pm\pi/2$ [the term $\pm i\mathbf{e}_y$ in Eq. (2.6) results from the phase difference $\exp(\pm i\pi/2)$ for $\mathbf{E}_y$ relative to $\mathbf{E}_x$]. The resultant vector executes a clockwise (+) or counterclockwise (−) helical motion when viewed face-on or looking toward the source ($-z$ direction) as they precess about the propagation direction ($+z$ direction). Now if an rcp and an lcp wave are both affected by the medium in exactly the same way (an isotropic medium), then they will have equal index of refraction ($n_+ = n_-$) and we have for $n = (n_+ + n_-)/2$

$$\mathbf{E}_x = (1/\sqrt{2})(\mathbf{E}_+ + \mathbf{E}_-) = \mathbf{e}_x E^0 \cos 2\pi\nu(t - nz/c) \tag{2.7}$$

which is a plane-polarized wave in the x direction (see Figure 2.2). Thus two plane-polarized waves with $\pm\pi/2$ phase difference can be combined to give an lcp or an rcp circularly polarized wave [by Eq. (2.6)]; likewise, two circularly polarized waves, one lcp and the other rcp, traveling through the medium with the same velocity ($n_+ = n_-$) can be combined to give a plane-polarized wave [by Eq. (2.7)].

For an anisotropic (birefringent) medium where $n_+ \neq n_-$, the two circularly polarized components of Eq. (2.6) will travel through the medium with different

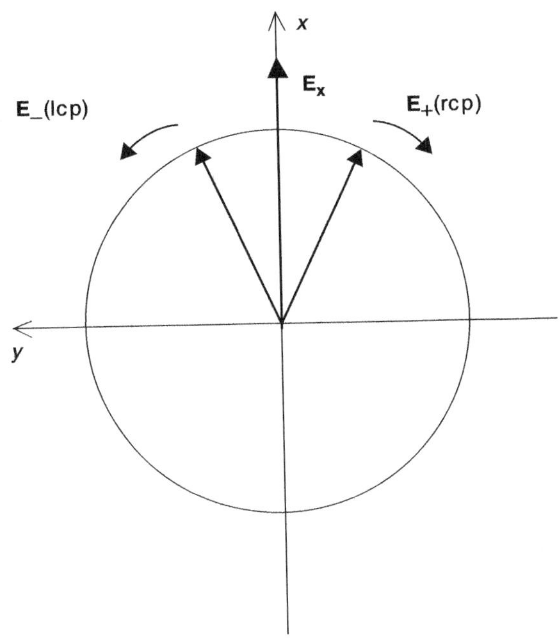

Figure 2.2. As viewed in the $-z$ direction (toward the source), two circularly polarized waves $\mathbf{E}_-$ and $\mathbf{E}_+$ combine to give a linearly polarized wave $\mathbf{E}_x$ as given by Eq. (2.7).

velocities and will differ in phase such that $\psi_\pm = 2\pi\nu(t - n_\pm z/c)$. If we write $\psi = (\psi_+ + \psi_-)/2 = 2\pi\nu(t - nz/c)$, where $n = (n_+ + n_-)/2$, the mean index of refraction, and write $\delta = (\psi_+ - \psi_-)/2 = 2\pi\nu z(n_- - n_+)/2c = \pi\nu z\Delta n/c$, where Δn is the index of refraction difference between lcp and rcp waves, then the **E** vector from Eq. (2.6) is given by [ref. 3]

$$\begin{aligned}\mathbf{E} = (1/\sqrt{2})(\mathbf{E}_+ + \mathbf{E}_-) &= (E^0/2)[\mathbf{e_x}\cos\psi_+ - \mathbf{e_y}\sin\psi_+ + \mathbf{e_x}\cos\psi_- + \mathbf{e_y}\sin\psi_-]\\ &= (E^0/2)[\mathbf{e_x}(\cos\psi_+ + \cos\psi_-) - \mathbf{e_y}(\sin\psi_+ - \sin\psi_-)]\\ &= (E^0/2)[2\mathbf{e_x}\cos\psi\cos\delta - 2\mathbf{e_y}\sin\delta\cos\psi]\\ &= E^0\cos\psi[\mathbf{e_x}\cos\delta - \mathbf{e_y}\sin\delta] \qquad (2.8)\end{aligned}$$

which describes a plane-polarized wave propagating in the z direction, but rotated by an angle of δ relative to the x direction. As shown in Figure 2.3, a birefringent medium causes the two circularly polarized components of a plane polarized wave to travel at different velocities, and the emergent plane polarized wave will be rotated about z relative to the incident plane of polarization (assumed here to be the xz plane). Anisotropic or birefringent media thus cause rotation of polarized light which is proportional to the distance traveled through the medium and to the difference in index of refraction for the two circularly polarized components Δn. The angle of rotation, conventionally symbolized by ϕ, for a wave

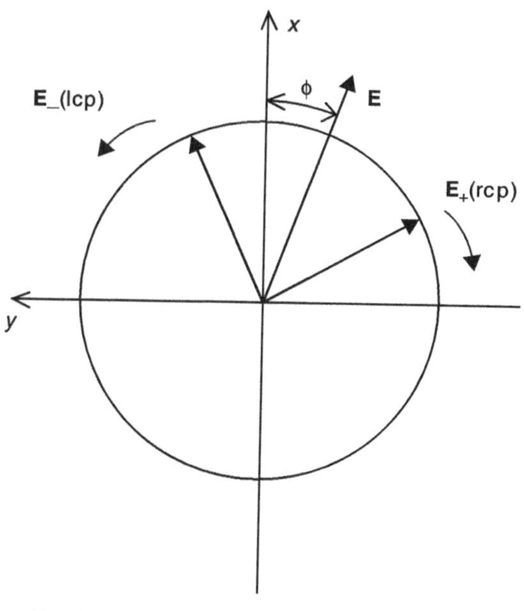

Figure 2.3. The combination of two circularly polarized waves that travel at differing velocities through the medium ($n_- > n_+$) gives rise to a linearly polarized wave that is rotated relative to the x axis. Viewed in the $-z$ direction (looking toward the source).

that has traveled a distance ℓ along the z direction is given by

$$\phi = \arctan(-E_y/E_x) = \arctan[-E^0\mathbf{e_y}\sin(\pi\nu\ell\Delta n/c)/(E^0\mathbf{e_x}\cos(\pi\nu\ell\Delta n/c))] \quad (2.9)$$

The above description of light passing through a medium holds for transparent regions of the spectrum. If the medium is not transparent but absorbs light with a frequency dependence, then not only will plane-polarized light be rotated on passage through the anisotropic absorbing medium, but also the vector amplitude of the circularly polarized components may no longer be the same. In this case, eliptically polarized light will emerge from the medium. The intensity in the z direction for a distance ℓ through the medium is given by

$$\mathbf{e_z}I(\nu) = \mathbf{e_z}I_0\exp(-4\pi k\nu\ell/c) \quad (2.10)$$

where k is an absorption coefficient. If $k_+ = k_-$, then the circularly polarized vectors have the same absorption coefficients in the medium, but if $k_+ \neq k_-$, then the medium is said to be dichroic. The total absorption can be expressed by the sum of the mean absorption of the circularly polarized vectors and their difference: $k \pm k'$, where $k = (k_+ + k_-)/2$ and $k' = (k_- - k_+)/2 = \Delta k/2$. At this point it is useful to define the complex index of refraction:

$$n_\pm = n_\pm - ik_\pm \quad (2.11)$$

Then upon substitution into Eq. (2.6) we have

$$\begin{aligned}\mathbf{E}_\pm &= \mathrm{Re}[(E^0/\sqrt{2})(\mathbf{e_x} \pm i\mathbf{e_y})\exp(i2\pi\nu(t - n_\pm\ell/c))] \\ &= \mathrm{Re}[(E^0/\sqrt{2})(\mathbf{e_x} \pm i\mathbf{e_y})\exp(2\pi\nu k_\pm\ell/c)\exp(i2\pi\nu(t - n_\pm\ell/c))] \quad (2.12)\end{aligned}$$

The $\mathbf{E}$ vector of the emergent light from the medium can then be written in terms of k, Δk, $\psi = 2\pi\nu(t - n\ell/c)$, (where $n = (n_+ + n_-)/2$), Δn, and $\phi = \pi\nu\ell\Delta n/c$ (δ in Eq. (2.8) for $z = \ell$):

$$\begin{aligned}\mathbf{E} &= (1/\sqrt{2})(\mathbf{E}_+ + \mathbf{E}_-) \\ &= \mathrm{Re}\{(E^0/2)[(\mathbf{e_x} + i\mathbf{e_y})\exp(-(k - \Delta k/2)2\pi\nu\ell/c)\exp(i2\pi\nu(t - n_+\ell/c)) \\ &\quad + (\mathbf{e_x} - i\mathbf{e_y})\exp(-(k + \Delta k/2)2\pi\nu\ell/c)\exp(i2\pi\nu(t - n_-\ell/c))]\} \\ &= \mathrm{Re}\{(E^0/2)\exp(-2\pi\nu\ell k/c)\exp(i\psi)[(\mathbf{e_x} + i\mathbf{e_y})\exp(\Delta k\pi\nu\ell/c)\exp(i\phi) \\ &\quad + (\mathbf{e_x} - i\mathbf{e_y})\exp(-\Delta k\pi\nu\ell/c)\exp(-i\phi)]\} \quad (2.13)\end{aligned}$$

Thus the rcp and lcp vectors are both subjected to the mean absorption and mean index of refraction by the common factors $\exp(-2\pi\nu\ell k/c)$ and $\exp(i\psi)$, respectively, in Eq. (2.13), but the rcp and lcp vectors are further modified by the absorption difference factors $\exp(+\Delta k\pi\nu\ell/c)$ and $\exp(-\Delta k\pi\nu\ell/c)$, respectively. It can be shown that the expansion of the $\exp(\pm i\phi)$ terms can be described by unit vectors $\mathbf{e}'_x$ and $\mathbf{e}'_y$ that have been rotated through an angle ϕ with respect to the x axis, which, with some rearrangement, gives Eq. (2.14) [ref. 3]:

$$\begin{aligned}\mathbf{E} &= \mathrm{Re}\{(E^0/2)\exp(-2\pi\nu\ell k/c)\exp(i\psi)\,[\mathbf{e}'_x[\exp(\Delta k\pi\nu\ell/c)+\exp(-\Delta k\pi\nu\ell/c]\\ &\quad + i\mathbf{e}'_y[\exp(\Delta k\pi\nu\ell/c)-\exp(-\Delta k\pi\nu\ell/c]\}\\ &= \mathrm{Re}\{(E^0)\exp(-2\pi\nu\ell k/c)\exp(i\psi)[\mathbf{e}'_x\cosh(\Delta k\pi\nu\ell/c)+i\mathbf{e}'_y\sinh(\Delta k\pi\nu\ell/c)]\}\end{aligned} \tag{2.14}$$

The emergent electric field vector $\mathbf{E}$ is elliptically polarized and traces out an ellipse as shown in Figure 2.4. The ellipticity, conventionally symbolized as Ψ, for a wave that has traveled a distance ℓ along the z direction represents the eccentricity of the ellipse, and the tangent of Ψ is the ratio of the minor axis (b) to major axis (a) of the ellipse:

$$\begin{aligned}\tan\Psi &= b/a = (|\mathbf{E}_-|-|\mathbf{E}_+|)/(|\mathbf{E}_-|+|\mathbf{E}_+|)\\ &= \tanh(\Delta k\pi\nu\ell/c) = \sinh(\Delta k\pi\nu\ell/c)/(\cosh(\Delta k\pi\nu\ell/c))\\ &\approx \Delta k\pi\nu\ell c \qquad \text{for } \Delta k\pi\nu\ell/c \le 0.1 \quad \text{(the usual case)}\end{aligned} \tag{2.15}$$

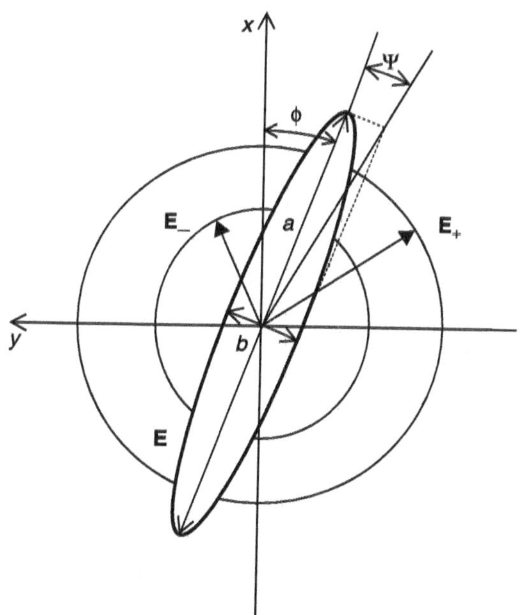

Figure 2.4. Elliptical polarization of the **E** vector which emerges from a medium in which $k_- > k_+$ and $n_- > n_+$. Viewed in the $-z$ direction (looking toward the source).

The important point here is that as light passes through the medium, a differential index of refraction Δn for two circularly polarized waves leads to optical rotation, while differential absorption Δk causes dichroism. It is the latter phenomenon, the differential absorption of circularly polarized light by matter, caused by the presence of an external magnetic field oriented along the propagation direction, that provides the origin of MCD.

2.3. ABSORPTION PROBABILITIES

The absorption of light passing through a distance ℓ in the medium is given by

$$dI/I = -\kappa d\ell \tag{2.16}$$

where I is the light intensity, ℓ is the path through the absorbing medium, and κ is the Lambert's law absorption coefficient. From Eq. (2.10) we note that κ is related to the absorption coefficient k by $\kappa = 4\pi\nu k/c$. Therefore

$$\kappa = -dI/Id\ell = 4\pi\nu k/c \tag{2.17}$$

In terms of transition probabilities for a transition from state a to state j, the differential $-dI/d\ell$ becomes

$$-dI/d\ell = h\nu(N_a P_{a\to j} - N_j P_{j\to a}) \tag{2.18}$$

It is assumed that $P_{a\to j} = P_{j\to a}$, and therefore $\kappa(\nu)$ can be written

$$\kappa(\nu) = [h\nu/I(\nu)](N_a - N_j)P_{a\to j} \tag{2.19}$$

From electromagnetic theory, the radiation intensity $I(\nu)$ is the time average ($\langle\ \rangle$) of the Poynting vector [ref. 2] and is given by

$$I(\nu) = |(c/4\pi)\langle \mathrm{Re}\,\mathbf{E}(\nu) \times \mathrm{Re}\,\mathbf{B}(\nu)\rangle| = (c/4\pi)|(-2\pi i\nu \mathbf{A^0}/c)^2\langle \cos^2 2\pi\nu(t - nz/c)\rangle|$$
$$= (c/4\pi)|\mathbf{E^0}|^2(1/2) = (c/8\pi)|\mathbf{E^0}|^2 \tag{2.20}$$

but also since $|i|^2 = 1$ we have

$$I(\nu) = (c/4\pi)(-2\pi|i|\nu A^0/c)^2(1/2) = \pi(A^0)^2\nu^2/2c \tag{2.21}$$

The transition probability $P_{a\to j}$ is given by

$$P_{a\to j} = (\pi^2/h^2)\left|\left\langle j\left|\sum (q_k/m_k c)\mathbf{A}^{0*}\cdot\mathbf{P}_k \exp(i2\pi\nu z_k/c)\right|a\right\rangle\right|^2 \rho_{aj}(\nu) \tag{2.22}$$

where $\rho_{aj}(\nu)$ is the absorption lineshape function such that $\int \rho_{aj}(\nu)d\nu = 1$ and the sum is over k charges q_k with masses m_k and momenta $\mathbf{P}_k$; the $*$ for $\mathbf{A}^{0*}$ indicates the complex conjugate. Transition probabilities are derived by means of time-dependent perturbation theory, and derivations can be found in quantum chemistry texts such as Eyring, Walter and Kimball [ref. 4, Chapter VIII]. The term $\exp(i2\pi\nu z_k/c)$ in Eq. (2.20) can be expanded as

$$\exp(i2\pi\nu z_k/c) = 1 + i2\pi\nu z_k/c + \cdots \tag{2.23}$$

and when we use a molecule fixed coordinate system, Eq. (2.22) becomes

$$\begin{aligned} P_{a\to j} &= (\pi^2/h^2)\left|\left\langle j\left|\sum(e_k/m_k c)\boldsymbol{\pi}^* A^0 \cdot \mathbf{p}_k \exp(i2\pi\nu z_k/c)\right|a\right\rangle\right|^2 \rho_{aj}(\nu) \\ &= [\pi^2(A^0)^2/(c^2h^2)]\left|\left\langle j\right|\sum e_k/m_k\boldsymbol{\pi}^* \right. \\ &\quad \left. \times\, i2\pi m_k(E_j - E_a)/h\mathbf{r_k}(1 + i2\pi\nu z_k/c)\left|a\right\rangle\right|^2 \rho_{aj}(\nu) \\ &= [\pi^2(A^0)^2\nu^2|i|^2/(c^2h^2)]|\langle j|\mathbf{m}\cdot\boldsymbol{\pi}^*|a\rangle + \langle j|(\mathbf{e_z}\times\boldsymbol{\pi}^*\boldsymbol{\mu})|a\rangle \\ &\quad + (i\pi\nu/c)\langle j|(\boldsymbol{\pi}_1^* Q_{xz} + \boldsymbol{\pi_2} * Q_{yz})|a\rangle|^2 \rho_{aj}(\nu) \end{aligned} \tag{2.24}$$

where $\nu = \nu_{aj} = (E_j - E_a)/h$ and the unit vector $\boldsymbol{\pi} = (\mathbf{e_x}\pi_1 + \mathbf{e_y}\pi_2)$ describes the polarization properties for light propagating along the z direction. The first term in Eq. (2.24) contains the electric dipole operator $\mathbf{m} = \sum e_k\mathbf{r_k}$; the second contains the magnetic dipole operator $\boldsymbol{\mu} = -e/2m_e c\sum(\mathbf{l_k} + 2\mathbf{s_k}) = -\mu_B\sum(\mathbf{l_k} + 2\mathbf{s_k})$, where μ_B = Bohr magneton; $\mathbf{l_k}$ and $\mathbf{s_k}$ are orbital and spin angular momentum operators, respectively, in units of $h/2\pi$; and the third term contains the electric quadrupole operators $Q_{\alpha\beta} = \sum q_k(\mathbf{r}_{k\alpha}\mathbf{r}_{k\beta} - (\mathbf{r_k}\cdot\mathbf{r_k}/3)\delta_{\alpha\beta})$. Of these, the electric dipole term is the largest contributor and therefore is the most important. Using Eq. (2.19) and substituting the expression for $I(\nu)$ from Eq. (2.21), $\kappa(\nu)$ can then be written

$$\begin{aligned} \kappa(\nu) &= (8\pi^3\nu/hc)(N_a - N_j)|\langle j|\mathbf{m}\cdot\boldsymbol{\pi}^*|a\rangle + \langle j|(\mathbf{e_z}\times\boldsymbol{\pi}^*\mu)|a\rangle \\ &\quad + (i\pi\nu/c)\langle j|(\boldsymbol{\pi}_1^* Q_{xz} + \boldsymbol{\pi}_2^* Q_{yz})|a\rangle|^2 \rho_{aj}(\nu) \end{aligned} \tag{2.25}$$

Finally, the Lorentz effective field approximation is assumed: $\mathbf{E}^0_{\text{microscopic}} = \alpha\mathbf{E}^0_{\text{macroscopic}}$, where $\alpha = (n^2+2)/3$, and $\mathbf{B}^0_{\text{microscopic}} = \alpha'\mathbf{B}^0_{\text{macroscopic}}$, where $\alpha' \approx 1$ for a nonmagnetic matrix (the usual case). Also in most instances the index of refraction n is $\approx$ constant for a nonabsorbing solvent or an isotropic matrix. Equation (2.25) becomes

$$\begin{aligned} \kappa(\nu) &= (8\pi^3\nu/hc)(N_a - N_j)|(\alpha/\sqrt{n})\langle j|\mathbf{m}\cdot\boldsymbol{\pi}^*|a\rangle + \sqrt{n}[\langle j|(\mathbf{e_z}\times\boldsymbol{\pi}^*\mu)|a\rangle \\ &\quad + (i\pi\nu/c)\langle j|(\boldsymbol{\pi}_1^* Q_{xz} + \boldsymbol{\pi}_2^* Q_{yz})|a\rangle|^2]\rho_{aj}(\nu) \end{aligned} \tag{2.26}$$

The absorption coefficients for rcp (+) and lcp (−) light are then given by

$$\kappa_{\pm}(\nu) = (8\pi^3\nu/hc)(N_a - N_j)|(\alpha/\sqrt{n})\langle j|m_{\mp}|a\rangle \pm \sqrt{n}[\langle j|(\mu_{\mp}|a\rangle \pm (i\pi\nu/c)\langle j|Q \pm |a\rangle]|^2\rho_{aj}(\nu) \tag{2.27}$$

where $m\pm = (m_x \pm im_y)/\sqrt{2}$; $\mu\pm = (\mu_x \pm i\mu_y)/\sqrt{2}$; and $Q\pm = (Q_{yz} \pm iQ_{xz})/\sqrt{2}$. Upon expanding Eq. (2.27) and making use of $\langle j|Op_{\mp}|a\rangle = \langle a|Op \pm |j\rangle^*$, we have

$$\kappa_{\pm}(\nu) = (8\pi^3\nu/hc)(N_a - N_j)\{\underbrace{\alpha^2/n|\langle a|m \pm |j\rangle|^2}_{\text{electric dipole}} + \underbrace{n|\langle a|\mu \pm |j\rangle|^2}_{\text{magnetic dipole}}$$
$$+ \underbrace{(n\pi\nu/c)|\langle a|Q \mp |j\rangle|^2}_{\text{electric quadrupole}} \mp 2\alpha\mathrm{Im}[\langle a|m \pm |j\rangle(\langle j|\mu_{\mp}|a\rangle$$
$$- (i\pi\nu/c)\langle j|Q \pm |a\rangle)] + 2n\mathrm{Im}[\langle a|\mu \pm |j\rangle(\pi\nu/c)\langle j|Q \pm |a\rangle]\}\rho_{aj}(\nu) \tag{2.28}$$

The first three terms in Eq. (2.28) are dependent upon the external field and are zero when $B = 0$, whereas the last three terms are nonzero for chiral molecules, even when $B = 0$, and thus are responsible for natural CD. Therefore only the first three terms are included for a description of MCD. It is also necessary to sum over all transitions that are part of the $a \to j$ band. Combining all field dependent quantities we have

$$\Delta\kappa(\nu) = \kappa_- - \kappa_+ = (8\pi^3\nu/hc)\sum_{aj}(N_a - N_j)[\underbrace{\alpha^2/n(|\langle a|m_-|j\rangle|^2 - |\langle a|m_+|j\rangle|^2)}_{\text{electric dipole terms}}$$
$$+\underbrace{n(|\langle a|\mu_-|j\rangle|^2 - |\langle a|\mu_+|j\rangle|^2)}_{\text{magnetic dipole terms}}$$
$$+ \underbrace{(n\pi\nu/c)(|\langle a|Q_+|j\rangle|^2 - |\langle a|Q_-|j\rangle|^2)}_{\text{electric quadrupole terms}}]\rho_{aj}(\nu) \tag{2.29}$$

The electric dipole terms are usually much larger than the magnetic dipole and electric quadrupole terms, and therefore the latter are neglected here. Finally the differential absorption coefficient is given by

$$\Delta\kappa(\nu) = (8\pi^3\nu\alpha^2/(hcn))\sum_{aj}(N_a - N_j)[|\langle a|m_-|j\rangle|^2 - |\langle a|m_+|j\rangle|^2]\rho_{aj}(\nu) \tag{2.30}$$

The sum of the lcp and rcp components $(\kappa_- + \kappa_+)/2$ also gives the normal absorption coefficient $\kappa(\nu)$:

$$\kappa(\nu) = (4\pi^3\nu\alpha^2/(hcn))\sum_{aj}(N_a - N_j)[|\langle a|m_-|j\rangle|^2 + |\langle a|m_+|j\rangle|^2]\rho_{aj}(\nu) \tag{2.31}$$

These absorption coefficients can be converted to conventional chemical units where A and ΔA are the absorbance and differential absorbance, respectively:

$$A/(c\ell) = \kappa(\nu)\log_{10}e/c = \varepsilon, \quad \text{where } c = \text{mol/L and } \ell = \text{path (in cm)}$$

$$(N_a - N_j)/c = (N_a - N_j)/N \times N_0 \times 10^{-3}, \quad \text{where } N_0 = \text{Avogadro's number; } N = \text{total number of absorbing particles}$$

Therefore, $\Delta A = (\Delta\varepsilon c\ell) = \Delta\kappa(\nu)\log_{10}e\ell(N_a - N_j)/N \times N_0 \times 10^{-3}$ and we can then write Eqs. (2.30) and (2.31) in terms of ΔA and A expressed in units of energy E:

$$\Delta A/E = (\Delta\varepsilon c\ell)/E = \gamma\sum(N_a - N_j)/N(|\langle a|m_-|j\rangle|^2 - |\langle a|m_+|j\rangle|^2)\rho_{aj}(E) \quad (2.32)$$

$$A/E = (\varepsilon_- + \varepsilon_+)c\ell/2E = (\gamma/2)\sum(N_a - N_j)/N(|\langle a|m_-|j\rangle|^2 + |\langle a|m_+|j\rangle|^2)\rho_{aj}(E) \quad (2.33)$$

where $E = h\nu$ and the constant γ is given by the constants

$$\gamma = 2N_0\pi^3\alpha^2 c\ell\log_{10}e/(250hcn) = 3.266 \times 10^{38}c\ell(\alpha^2/n)$$

where cgs units are used for h and c. If the electric dipole matrix elements are expressed in Debye units (10^{-18} esu cm), consistent with the use of Gaussian units, then the expression for γ simplifies to

$$\gamma = 326.6c\ell(\alpha^2/n)$$

Normally the effective field correction, α^2/n, can be ignored since MCD to absorption ratios are used and these ratios are independent of this factor (see Chapter 3).

Equations (2.32) and (2.33) are the basic light absorption equations of MCD spectroscopy. It should be emphasized here that the electric dipole matrix elements $m_\pm$ and line shape $\rho_{aj}(E)$ are magnetic field-dependent quantities, and in the next chapter the effects of a perturbation from an external field along z will be developed. The coordinates for both equations are space-fixed axes fixed at the center of gravity of the molecule, but parallel to laboratory axes.

3 Theoretical Framework: Definition of MCD Terms

As Eqs. (2.32) and (2.33) in Chapter 2 show, the electric dipole expressions for ΔA (MCD) and for A (absorption) are very similar, and both depend upon the electric dipole matrix elements $\langle a|m_{\pm}|j\rangle$ and the band-shape function $\rho_{aj}(E)$. The matrix elements are expanded by assuming the Born–Oppenheimer/Franck–Condon approximations, and then a Zeeman magnetic field perturbation is applied and simplified to first-order in the magnetic field by using the Rigid-Shift approximation to the band shape. The resulting mathematical expression for ΔA is rather cumbersome and is conveniently divided into "terms," each of which is concerned with different aspects of the magnetic effects upon the dichroism. The definition and properties of the MCD terms (symbolized by A, B, and C) and the parameters that describe them is the main purpose of this chapter. The expression for the absorption A also leads to parameterization (symbolized by D) and is often combined with the MCD parameters to give ratios that are approximately independent of medium effects and the precise form of the band-shape functions.

3.1. BORN–OPPENHEIMER/FRANCK–CONDON APPROXIMATION

Born–Oppenheimer (BO) states within the Franck–Condon (FC) approximation for the ground state A and excited states J, and involving vibrational functions $|g\rangle$ and $|j\rangle$, respectively, can be symbolized as the following:

$$|A_{\alpha g}\rangle = \phi_{A\alpha}(q, Q)\chi_g(Q) = |A\alpha\rangle|g\rangle$$

$$|J_{\lambda j}\rangle = \phi_{J\lambda}(q, Q)\chi_j(Q) = |J\lambda\rangle|j\rangle$$

Then the summations for ΔA and A [Eqs. (2.32) and (2.33)] will be over $\alpha\lambda gj$ and we can write

$$\Delta A/E = \gamma \sum (N_{A\alpha g} - N_{J\lambda j})/N(|\langle A_{\alpha g}|m_-|J_{\lambda j}\rangle|^2 - |\langle A_{\alpha g}|m_+|J_{\lambda j}\rangle|^2)\rho_{AJ}(E) \quad (3.1)$$

$$A/E = (\gamma/2) \sum (N_{A\alpha g} - N_{J\lambda j})/N(|\langle A_{\alpha g}|m_-|J_{\lambda j}\rangle|^2 + |\langle A_{\alpha g}|m_+|J_{\lambda j}\rangle|^2)\rho_{AJ}(E) \quad (3.2)$$

A Practical Guide to Magnetic Circular Dichroism Spectroscopy, by W. Roy Mason

In the presence of an external magnetic field along the z light propagation direction, we then assume a Zeeman perturbation of $-\mu_z B$, where $\mu_z = -\mu_B(L_z + 2S_z)B$, μ_B is the Bohr magneton, and L_z and S_z are orbital and spin angular momentum operators, respectively, with orbital and spin angular momentum in units of $h/(2\pi)$. The perturbation is assumed to be small compared with kT and the bandwidth (this is the so-called linear limit). The (diagonal) perturbations of states A and J will be given by (δ's are Kronecker deltas that equal 1 if subscripts are the same, otherwise 0)

$$\begin{aligned} \langle A_{\alpha g}| -\mu_z B|A_{\alpha' g'}\rangle &= -\langle A_\alpha|\mu_z|A_\alpha\rangle B\delta_{\alpha\alpha'}\delta_{gg'} \\ \langle J_{\lambda j}| -\mu_z B|J_{\lambda' j'}\rangle &= -\langle J_\lambda|\mu_z|J_\lambda\rangle B\delta_{\lambda\lambda'}\delta_{jj'} \end{aligned} \tag{3.3}$$

Using first-order perturbation theory, the perturbed wavefunctions A and J (primes) are

$$\begin{aligned} |A_{\alpha g}\rangle' &= |A_{\alpha g}\rangle - \sum_{K \neq A, \kappa} \langle K_\kappa|\mu_z|A_\alpha\rangle^\circ B/(W_A{}^\circ - W_K{}^\circ)|K_{\kappa g}\rangle \\ |J_{\lambda j}\rangle' &= |J_{\lambda j}\rangle - \sum_{K \neq J, \kappa} \langle K_\kappa|\mu_z|J_\lambda\rangle^\circ B/(W_J{}^\circ - W_K{}^\circ)|K_{\kappa j}\rangle \end{aligned} \tag{3.4}$$

where the superscript $^\circ$ indicates field free quantities. The electric dipole matrix elements to first-order in the field become

$$\begin{aligned} \langle A_{\alpha g}|m_\pm|J_{\lambda j}\rangle' = \Bigg\{\Bigg[&\langle A_\alpha|m_\pm|J_\lambda\rangle^\circ \\ &+ \sum_{K \neq A, \kappa} \langle A_\alpha|\mu_z|K_\kappa\rangle^\circ \langle K_\kappa|m_\pm|J_\lambda\rangle^\circ B/(W_K{}^\circ - W_A{}^\circ) \\ &+ \sum_{K \neq J, \kappa} \langle A_\alpha|m_\pm|K_\kappa\rangle^\circ \langle K_\kappa|\mu_z|J_\lambda\rangle^\circ B/(W_K{}^\circ - W_J{}^\circ)\Bigg]\Bigg\}\langle g|j\rangle \end{aligned} \tag{3.5}$$

where the $\sum$'s over k for $\langle k|g\rangle$ and $\langle k|j\rangle = |g\rangle$ and $|j\rangle$, respectively. The vibronic energies to first-order in the field are

$$\begin{aligned} E'_{A\alpha g} &= E_{A\alpha g} - \langle A_\alpha|\mu_z|A_\alpha\rangle^\circ B \\ E'_{J\lambda j} &= E_{J\lambda j} - \langle J_\lambda|\mu_z|J_\lambda\rangle^\circ B \end{aligned} \tag{3.6}$$

The energy $E'_{A\alpha g}$ can be used to evaluate the population factor for $|A_{\alpha g}\rangle$. We assume $N_{J\lambda j} = 0$ because the excited-state population is near zero if A and J are well-separated in energy and a conventional light source is used. Then we have

$$\begin{aligned} (N_{A\alpha g} - N_{J\lambda j})/N \approx N_{A\alpha g}/N &= N'_{A\alpha g}\Big/\Big(\sum_{\alpha g} N'_{A\alpha g}\Big) \\ &= \exp(-E'_{A\alpha g}/(kT))\Big/\sum_{\alpha g}\exp(-E'_{A\alpha g}/(kT)) \end{aligned} \tag{3.7}$$

so that

$$N'_{A\alpha g}\Big/\left(\sum_{\alpha g} N'_{A\alpha g}\right) = \frac{\exp(-E'_{A\alpha g}/(kT))\exp(\langle A_\alpha|\mu_z|A_\alpha\rangle^\circ B/(kT))}{\sum_{\alpha g}\exp(-E'_{A\alpha g}/(kT))\exp(\langle A_\alpha|\mu_z|A_\alpha\rangle^\circ B/(kT))} \tag{3.8}$$

The Zeeman splitting of the ground state will always sum to zero: $\sum\langle A_\alpha|\mu_z|A_\alpha\rangle = 0$. Let the electronic degeneracy of the ground state A be $|A|$ so that

$$\sum_{\alpha g} N'_{A\alpha g} = |A|\sum_{g} N'_{A\alpha g}$$

By means of the expansion $\exp(\langle A_\alpha|\mu_z|A_\alpha\rangle^\circ B/(kT)) \approx 1 + \langle A_\alpha|\mu_z|A_\alpha\rangle^\circ B/(kT) + \cdots$, we have

$$N'_{A\alpha g}\Big/\left(\sum_{\alpha g} N'_{A\alpha g}\right) \approx \left(N_g\Big/\left(|A|\sum_{g} N_g\right)\right)(1 + \langle A_\alpha|\mu_z|A_\alpha\rangle^\circ B/(kT) \tag{3.9}$$

3.2. RIGID-SHIFT APPROXIMATION

The Rigid-Shift (RS) approximation assumes the band shifts by the Zeeman effect, but does not change shape as shown in Figure 3.1. Thus we have

$$\rho_{AJ}(E) = \rho_{AJ}(E - a_{\alpha\lambda}B) \tag{3.10}$$

where

$$a_{\alpha\lambda}B = (E'_{J\lambda j} - E'_{A\alpha g}) - (E_{J\lambda j} - E_{A\alpha g}) = -(\langle J_\lambda|\mu_z|J_\lambda\rangle^\circ - \langle A_\alpha|\mu_z|A_\alpha\rangle^\circ)B \tag{3.11}$$

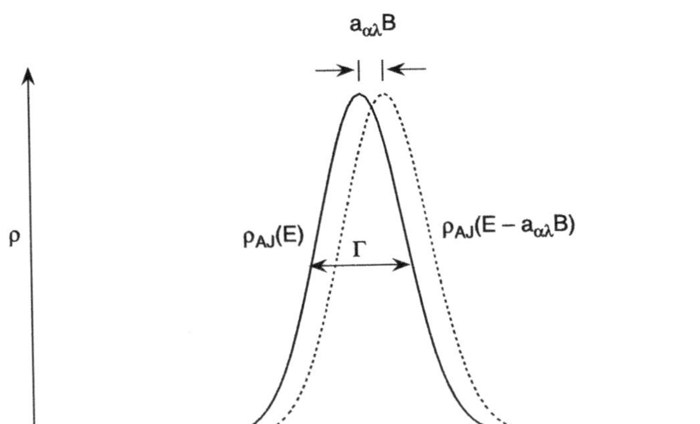

Figure 3.1. Rigid-Shift approximation: The band shifts due to the Zeeman perturbation but does not change shape. (Adapted from ref. 1.)

Equation (3.10) can be expanded in a Taylor series $f(x+h) = f(x) + h\partial f(x)/\partial x + h^2\partial^2 f(x)/\partial^2 x + \cdots$

$$\begin{aligned}\rho_{AJ}(E - a_{\alpha\lambda}B) &= \rho_{AJ}(E) + \partial\rho_{AJ}(E)/\partial E(-a_{\alpha\lambda}B) + \cdots \\ &= \rho_{AJ}(E) + (\langle J_\lambda|\mu_z|J_\lambda\rangle^\circ - \langle A_\alpha|\mu_z|A_\alpha\rangle^\circ)B\partial\rho_{AJ}(E)/\partial E \qquad (3.12)\end{aligned}$$

This expansion assumes the Zeeman shift $a_{\alpha\lambda}B$ is small compared to the $A \rightarrow J$ linewidth Γ of the composite band. The zero-field lineshape function is assumed to be independent of α and λ: $\rho_{A\alpha g,\,J\lambda j}(E) = \rho_{gi}(E)$. Substitution of Eqs. (3.5), (3.9), and (3.12) into Eq. (3.1) gives

$$\begin{aligned}\Delta A/E &= \gamma \sum (N_{A\alpha g} - N_{J\lambda j})/N(|\langle A_{\alpha g}|m_-|J_{\lambda j}\rangle|^2 - |\langle A_{\alpha g}|m_+|J_{\lambda j}\rangle|^2)\rho_{AJ}(E) \\ &= \gamma N_g \Big/ \left(|A| \sum N_g\right)(1 + \langle A_\alpha|\mu_z|A_\alpha\rangle^\circ B/(kT))\ \langle g|j\rangle^2 \\ &\times \Bigg\{\Bigg[\langle A_\alpha|m_-|J_\lambda\rangle^\circ + \Bigg[\sum_{K\neq A,\,\kappa} 1/(W_k{}^\circ - W_A{}^\circ)\langle A_\alpha|\mu_z|K_\kappa\rangle^\circ\langle K_\kappa|m_-|J_\lambda\rangle^\circ \\ &+ \sum_{K\neq J,\,\kappa} 1/(W_K{}^\circ - W_J{}^\circ)\langle A_\alpha|m_-|K_\kappa\rangle^\circ\langle K_\kappa|\mu_z|J_\lambda\rangle^\circ\Bigg]B\Bigg]^2 \\ &- \Bigg[\langle A_\alpha|m_+|J_\lambda\rangle^\circ + \Bigg[\sum_{K\neq A,\kappa} 1/(W_K{}^\circ - W_A{}^\circ)\langle A_\alpha|\mu_z|K_\kappa\rangle^\circ\langle K_\kappa|m_+|J_\lambda\rangle^\circ \\ &+ \sum_{K\neq J,\kappa} 1/(W_K{}^\circ - W_J{}^\circ)\langle A_\alpha|m_+|K_\kappa\rangle^\circ\langle K_\kappa|\mu_z|J_\lambda\rangle^\circ\Bigg]B\Bigg]^2\Bigg\} \\ &\times [\rho_{AJ}(E) + (\langle J_\lambda|\mu_z|J_\lambda\rangle^\circ - \langle A_\alpha|\mu_z|A_\alpha\rangle^\circ)B\partial\rho_{AJ}(E)/\partial E]\end{aligned}$$

This expression for ΔA is both complicated and cumbersome. At this point it is convenient to organize the expression according to the type of magnetic interaction with the ground and excited states and also between states to first order in B. This is best accomplished by introducing some simplifying definitions and relations. First, the lineshape function is defined as

$$f(E) = (1/N_G)\sum_g N_g \sum_j |\langle g|j\rangle|^2 \rho_{gj}(E) \qquad \text{such that} \int f(E)\ dE = 1$$

where $N_G = \sum_g N_{A\alpha g}$ and $N_g = \sum_g N_g$. Then the following definitions are also used:

$$m_{\pm 1} = \mp m_\pm = \mp 1/\sqrt{2}\ (m_x \pm im_y)$$

$$\langle A_\alpha|m_{\pm 1}|J_\lambda\rangle^* = -\langle J_\lambda|m_{\mp 1}|A_\alpha\rangle$$

$$\sum_{\alpha\lambda} (|\langle A_\alpha|m_{-1}|J_\lambda\rangle|^2 - |\langle A_\alpha|m_{+1}|J_\lambda\rangle|^2 = 0$$

$$\mu_z = -\mu_B(L_z + 2S_z) \qquad \text{where } \mu_B = \text{Bohr magneton}$$

Upon substitution of these definitions, collecting terms, dropping the superscript $^\circ$ notation, and performing some rearrangement, the BO–FC–RS approximation, after discarding those terms in B^2, finally becomes

$$\Delta A/E = \gamma\mu_B B[A_1(-\partial f(E)/\partial E) + (B_0 + C_0/(kT))f(E)] \tag{3.13}$$

where the A_1, B_0, and C_0 parameters define the MCD A, B, and C terms, respectively. The subscripts on the parameter labels refer to the order of the spectral moments associated with their respective band shapes, which will be discussed below (spectral moments will be discussed in Section 3.14). They are given by Eqs. (3.14)–(3.16) (Stephens 1976 conventions, [ref. 1, Appendix A; review 3]:

$$\begin{aligned} A_1 = (1/|A|)\sum_{\alpha\lambda}(\langle J_\lambda|L_z + 2S_z|J_\lambda\rangle - \langle A_\alpha|L_z + 2S_z|A_\alpha\rangle) \\ \times (|\langle A_\alpha|m_{-1}|J_\lambda\rangle|^2 - |\langle A_\alpha|m_{+1}|J_\lambda\rangle|^2) \end{aligned} \tag{3.14}$$

$$\begin{aligned} B_0 = (2/|A|)\mathrm{Re}\sum_{\alpha\lambda}\Bigg[\sum_{K\neq J,\kappa} 1/(W_K - W_J)\langle J_\lambda|L_z + 2S_z|K_\kappa\rangle \\ \times (\langle A_\alpha|m_{-1}|J_\lambda\rangle\langle K_\kappa|m_{+1}|A_\alpha\rangle - \langle A_\alpha|m_{+1}|J_\lambda\rangle\langle K_\kappa|m_{-1}|A_\alpha\rangle) \\ + \sum_{K\neq A,\kappa} 1/(W_K - W_A)\langle K_\kappa|L_z + 2S_z|A_\alpha\rangle \\ \times (\langle A_\alpha|m_{-1}|J_\lambda\rangle\langle J_\lambda|m_{+1}|K_\kappa\rangle - \langle A_\alpha|m_{+1}|J_\lambda\rangle\langle J_\lambda|m_{-1}|K_\kappa\rangle)\Bigg] \end{aligned} \tag{3.15}$$

$$\begin{aligned} C_0 = -(1/|A|)\sum_{\alpha\lambda}\langle A_\alpha|L_z + 2S_z|A_\alpha\rangle \\ \times (|\langle A_\alpha|m_{-1}|J_\lambda\rangle|^2 - |\langle A_\alpha|m_{+1}|J_\lambda\rangle|^2) \end{aligned} \tag{3.16}$$

Finally, D_0, the dipole strength parameter, is defined so that

$$A/E = \gamma D_0 f(E) \tag{3.17}$$

where

$$D_0 = (1/(2|A|))\sum_{\alpha\lambda}(|\langle A_\alpha|m_{-1}|J_\lambda\rangle|^2 + |\langle A_\alpha|m_{+1}|J_\lambda\rangle|^2) \tag{3.18}$$

It should be noted that the operators in Eqs. (3.14)–(3.16) and (3.18) refer to space-fixed coordinates (fixed at the center of mass of the molecule and with axes parallel to laboratory axes) and must be converted to molecule-fixed coordinates (which rotate with the molecule); degenerate states for $|A\alpha\rangle$ and $|J\lambda\rangle$ are diagonal in the space-fixed operator μ_z. To remove this latter restriction, the expressions for A_1 and C_0 can be rewritten in basis invariant form. This may be accomplished easily if we assume that only unitary transformations are used for the conversion. This is based on the principle of "spectroscopic stability" (Piepho and Schatz [ref. 1], Section 4.6). The changes affect only Eqs. (3.14) and (3.16); B_0 and D_0 [Eqs. (3.15) and (3.18)] are unchanged:

$$A_1 = -(1/|A|)\sum_{\alpha\alpha'\lambda\lambda'} (\langle J_\lambda|L_z + 2S_z|J_{\lambda'}\rangle\delta_{\alpha\alpha'} - \langle A_\alpha|L_z + 2S_z|A_\alpha\rangle\delta_{\lambda\lambda'}) \times (\langle A_\alpha|m_{-1}|J_\lambda\rangle\langle J_{\lambda'}|m_{+1}|A_{\alpha'}\rangle - \langle A_\alpha|m_{+1}|J_\lambda\rangle\langle J_{\lambda'}|m_{-1}|A_{\alpha'}\rangle) \tag{3.19}$$

and

$$C_0 = (1/|A|)\sum_{\alpha\alpha'\lambda\lambda'} \langle A_{\alpha'}|L_z + 2S_z|A_\alpha\rangle \times (\langle A_\alpha|m_{-1}|J_\lambda\rangle\langle J_{\lambda'}|m_{+1}|A_{\alpha'}\rangle - \langle A_\alpha|m_{+1}|J_\lambda\rangle\langle J_{\lambda'}|m_{-1}|A_{\alpha'}\rangle) \tag{3.20}$$

Finally, since molecular properties are based on molecule-fixed operators, the space-fixed coordinates must be converted to molecule-fixed coordinates. This requires spatial averaging of the molecule-fixed coordinates, which involves integration over all possible orientations of the molecules relative to the laboratory axes (this procedure is described in Piepho and Schatz [ref. 1, Section 4.6, pp. 86–88]): Two sets of coordinates with the same origin are related by

$$(x, y, z)_{\text{space-fixed}} = (A)(x', y', z')_{\text{molecule-fixed}}$$

where an orthogonal transformation matrix (A) is needed for the conversion. It follows that

$$\begin{aligned} x &= A_{11}x' + A_{12}y' + A_{13}z' \\ y &= A_{21}x' + A_{22}y' + A_{23}z' \\ z &= A_{31}x' + A_{32}y' + A_{33}z' \end{aligned}$$

The spatial average of a function $F(\theta, \phi, \psi)$ is given by the integration

$$\overline{F} = \frac{\int\int\int F(\theta, \phi, \psi)d\tau}{\int\int\int d\tau} \qquad \text{where } d\tau = \sin\theta\, d\theta\, d\phi\, d\psi \tag{3.21}$$

It has been shown that if θ, ϕ, and ψ are Eulerian angles and the integration is θ from 0 to 2π, ϕ from 0 to 2π, and ψ from 0 to π, then $\int d\tau = 8\pi^2$ and

$$\begin{aligned} &1/(8\pi^2)\int\int\int A_{i\lambda}A_{j\mu}\sin\theta\, d\theta\, d\phi\, d\psi = (1/3)\delta_{ij}\delta_{\lambda\mu} \\ &1/(8\pi^2)\int\int\int A_{i\lambda}A_{j\mu}A_{k\nu}\sin\theta\, d\theta\, d\phi\, d\psi = (1/6)\varepsilon_{ijk}\varepsilon_{\lambda\mu\nu} \end{aligned} \tag{3.22}$$

where the alternating tensor ε_{ijk} equals 0 unless i, j, k are all different and then $\varepsilon_{ijk} = 1$ for $i, j, k = 1, 2, 3$ or any even permutation of i, j, k and $\varepsilon_{ijk} = -1$ for any odd permutation. The MCD terms can be written in a vector form $(\mathbf{L} + 2\mathbf{S})\cdot \mathbf{m} \times \mathbf{m}$, where the spatially averaged quatities are written with a bar over the parameters in Eqs. (3.23)–(3.26):

$$\bar{A}_1 = (i/(3|A|)) \sum_{\alpha\alpha'\lambda\lambda'} (\langle J_\lambda|\mathbf{L}+2\mathbf{S}|J_{\lambda'}\rangle\delta_{\alpha\alpha'} - \langle A_{\alpha'}|\mathbf{L}+2\mathbf{S}|A_\alpha\rangle\delta_{\lambda\lambda'})$$
$$\cdot (\langle A_\alpha|\mathbf{m}|J_\lambda\rangle \times \langle J_{\lambda'}|\mathbf{m}|A_{\alpha'}\rangle) \qquad (3.23)$$

$$\bar{B}_0 = (2/(3|A|)) \,\mathrm{Im} \sum_{K\neq J,\,\kappa,\,\alpha\lambda} \Big[\sum 1/(W_K - W_J)\langle J_\lambda|\mathbf{L}+2\mathbf{S}|K_\kappa\rangle$$
$$\cdot (\langle A_\alpha|\mathbf{m}|J_\lambda\rangle \times \langle K_\kappa|\mathbf{m}|A_\alpha\rangle) + \sum_{K\neq A,\,\kappa,\,\alpha\lambda} 1/(W_K - W_A)\langle K_\kappa|\mathbf{L}+2\mathbf{S}|A_\alpha\rangle$$
$$\cdot (\langle A_\alpha|\mathbf{m}|J_\lambda\rangle \times \langle J_\lambda|\mathbf{m}|K_\kappa\rangle)\Big] \qquad (3.24)$$

$$\bar{C}_0 = (i/(3|A|)) \sum_{\alpha\alpha'\lambda} \langle A_{\alpha'}|\mathbf{L}+2\mathbf{S}|A_\alpha\rangle \cdot (\langle A_\alpha|\mathbf{m}|J_\lambda\rangle \times \langle J_{\lambda'}|\mathbf{m}|A_{\alpha'}\rangle) \qquad (3.25)$$

$$\bar{D}_0 = (1/(3|A|)) \sum_{\alpha\lambda} |\langle A_\alpha|\mathbf{m}|J_\lambda\rangle|^2 \qquad (3.26)$$

If the molecule is oriented or isotropic (belongs to the cubic point groups O_h, O, or T_d), the basis functions can be chosen diagonal in molecule-fixed μ_z and the molecule-fixed z' axis is chosen to coincide with the space-fixed z axis. Then the spatially averaged parameters are equivalent to the space-fixed forms (3.14)–(3.16) and (3.18) so that

$$\bar{A}_1 = A_1, \qquad \bar{B}_0 = B_0, \qquad \bar{C}_0 = C_0, \qquad \bar{D}_0 = D_0$$

For molecules freely rotating in solution, randomly oriented in an isotropic medium, or for noncubic symmetries, space averaging is necessary and the space-averaged forms of the MCD parameters must be used.

3.3. A_1-, B_0-, AND C_0-TERM PARAMETERS

The MCD term parameters conveniently separate the important contributions to the observed MCD spectra. In the following sections the parameters will be discussed in terms of the type of information that can extracted from their qualitative and quantitative determination from experimental spectra.

3.4. *A* TERMS

The A_1 parameter describes the derivative-shaped A term. In terms of the magnetic moment operator $\mu_z = -\mu_B(L_z + 2S_z)$, Eq. (3.14) can be rewritten

$$A_1 = -(1/(\mu_B|A|)) \sum\nolimits_{\alpha\lambda} (\langle J_\lambda|\mu_z|J_\lambda\rangle - \langle A_\alpha|\mu_z|A_\alpha\rangle)$$
$$\times (|\langle A_\alpha|m_{-1}|J_\lambda\rangle|^2 - |\langle A_\alpha|m_{+1}|J_\lambda\rangle|^2) \qquad (3.27)$$

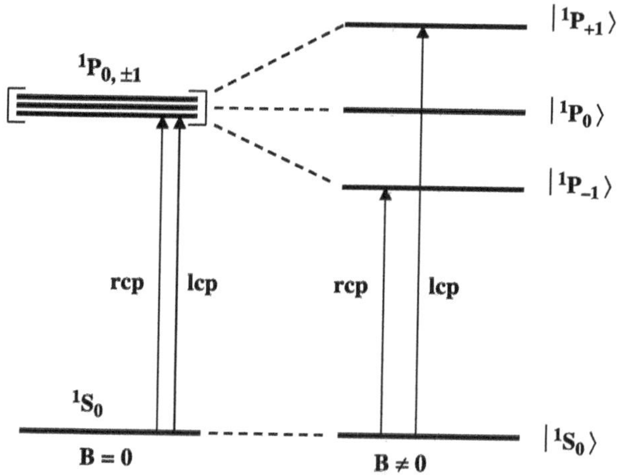

Figure 3.2. Energy levels for the atomic $s^2 \rightarrow sp$ transition $^1S_0 \rightarrow {}^1P_{1,0,-1}$.

Consider the atomic $s^2 \rightarrow sp$ transition $^1S_0 \rightarrow {}^1P_{1,0,-1}$. The ground state $|^1S_0\rangle$ is nondegenerate, but the excited state $|^1P_{0,\pm1}\rangle$ is threefold degenerate and therefore subject to Zeeman splitting $H_{\text{Zeeman}} = -\mu_z B$ in the presence of a magnetic field (Figure 3.2). In this case where $|\psi\rangle = |\text{SLJM}_J\rangle$ the excited-state g factor is given by the Landè formula with $J = L = 1$, and $S = 0$ (singlet states)

$$g = 1 + \frac{J(J+1) + S(S+1) - L(L+1)}{2J(J+1)} = +1 \tag{3.28}$$

and the Zeeman energies are given by

$$H_{\text{Zeeman}} = g\mu_B M_J B \tag{3.29}$$

Therefore, $H_{\text{Zeeman}} = \pm g\mu_B B$ for $|^1P_{\pm1}\rangle$ and $H_{\text{Zeeman}} = 0$ for $|^1P_0\rangle$.

The A term defined by the A_1 parameter for this case is given by

$$A_1 = -(1/(\mu_B|A|)) \sum_{\alpha\lambda, a_{\alpha\lambda}} (|\langle A_\alpha|m_{-1}|J_\lambda\rangle|^2 - |\langle A_\alpha|m_{+1}|J_\lambda\rangle|^2) \tag{3.30}$$

where $a_{\alpha\lambda} = -(\langle J_\lambda|\mu_z|J_\lambda\rangle - \langle A_\alpha|\mu_z|A_\alpha\rangle)$, the difference per Tesla of B in Zeeman energies between $|A\rangle$ and $|J\rangle$, the Zeeman shift in the RS approximation. So we have

$$\begin{aligned} a_{0,-1} &= -\langle^1P_{-1}| - \mu_B(L_z + 2S_z)|^1P_{-1}\rangle = (-1)(-\mu_B)(-1)\langle^1P_{-1}|^1P_{-1}\rangle \\ &= -\mu_B = -g\mu_B \qquad \text{for } g = +1 \end{aligned}$$

$$a_{0,0} = \langle^1P_0| - \mu_B(L_z + 2S_z)|^1P_0\rangle = 0$$

$$\begin{aligned} a_{0,+1} &= -\langle^1P_{+1}| - \mu_B(L_z + 2S_z)|^1P_{+1}\rangle = (-1)(-\mu_B)(+1)\langle^1P_{+1}|^1P_{+1}\rangle \\ &= +\mu_B = +g\mu_B \qquad \text{for } g = +1 \end{aligned}$$

Therefore we have

$$A_1 = (1/\mu_B(1))[-g\mu_B(|\langle {}^1S_0|m_{-1}|{}^1P_{-1}\rangle|^2 - |\langle {}^1S_0|m_{+1}|{}^1P_{-1}\rangle|^2) + 0$$
$$= 0\,(\text{lcp}) \qquad \neq 0\,(\text{rcp})$$
$$+ g\mu_B(|\langle {}^1S_0|m_{-1}|{}^1P_{+1}\rangle|^2 - |\langle {}^1S_0|m_{+1}|{}^1P_{+1}\rangle|^2)]$$
$$\neq 0\,(\text{lcp}) \qquad = 0\,(\text{rcp})$$
$$= 2g|\langle {}^1S_0|m_{\mp1}|{}^1P_{\pm1}\rangle|^2 \tag{3.31}$$

Note that lcp (m_{-1}) is allowed to $|{}^1P_{+1}\rangle$ but forbidden to $|{}^1P_{-1}\rangle$. Similarly, rcp (m_{+1}) is allowed to $|{}^1P_{-1}\rangle$ but forbidden to $|{}^1P_{+1}\rangle$. Both lcp and rcp are forbidden to $|{}^1P_0\rangle$. These strong selection rules for polarized light provide the origin of the MCD *A* term. Finally, the expression for D_0 given by

$$D_0 = (1/(2|A|))\sum_{\alpha\lambda}(|\langle A_\alpha|m_{-1}|J_\lambda\rangle|^2 + |\langle A_\alpha|m_{+1}|J_\lambda\rangle|^2) \tag{3.32}$$

can be written for the $|{}^1S_0\rangle \rightarrow |{}^1P_{0,\pm1}\rangle$ transition

$$= (1/2(1))(|\langle {}^1S_0|m_{-1}|{}^1P_{+1}\rangle|^2 + |\langle {}^1S_0|m_{+1}|{}^1P_{-1}\rangle|^2)$$
$$= |\langle {}^1S_0|m_{\pm1}|{}^1P_{\mp1}\rangle|^2 \tag{3.33}$$

Therefore, the ratio of A_1 from the MCD and D_0 from the absorption gives

$$A_1/D_0 = 2g \tag{3.34}$$

the excited-state *g* factor (proportional to the excited state magnetic moment). In cases where the ground state is also degenerate, the *A* term provides the difference between the excited-state and the ground-state *g* factors.

Figure 3.3 illustrates the formation of the *A* term for this example. The left side of the figure shows that the absorption of lcp and rcp light are identical so that ΔA is zero when $B = 0$. In the presence of the field, the right side of the figure shows the small Zeeman shift of A_- to higher energy, which gives rise to a nonzero derivative-shaped ΔA—in this case a positive *A* term, assuming *g* is positive. Normally the Zeeman splitting of the absorption band (exaggerated in the right side of the figure for clarity) would be so small that it would be difficult to measure, especially if the bandwidths were typical for molecular species (100–1000 cm^{-1}). It should be emphasized that *A* terms possess a sign, depending upon whether A_1/D_0 is positive or negative. Figure 3.4 illustrates positive and negative *A* terms. The positive *A* term, as in the above example (Figure 3.3), has, by convention [ref. 1], ΔA negative on the low-energy side (low-frequency or higher wavelength) of the crossing point and ΔA positive on the high-energy side (high-frequency or lower wavelength) of the crossing point. The crossing point would, of course, correspond with the maximum in the absorption *A*. The negative *A*

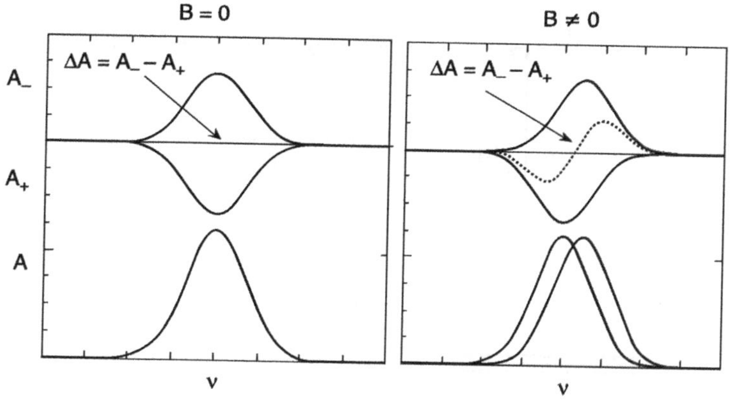

Figure 3.3. Origin of a positive *A* term.

term would be just the reverse as shown in Figure 3.4. A negative *A* term would result if the Zeeman multiplet in Figure 3.2 were inverted, for example.

Now consider the transition $|^1P_{0,\pm1}\rangle \rightarrow |^1S_0\rangle$ (Figure 3.5). Using the above approach, we find $a_{\alpha\lambda} = -(\langle J_\lambda|\mu_z|J_\lambda\rangle - \langle A_\alpha|\mu_z|A_\alpha\rangle) = \langle A_\alpha|\mu_z|A_\alpha\rangle$, and therefore

$$\begin{aligned} a_{-1,0} &= \langle ^1P_{-1}|-\mu_B(L_z + 2S_z|^1P_{-1}\rangle = (-\mu_B)(-1)\langle ^1P_{-1}|^1P_{-1}\rangle \\ &= \mu_B = +g\mu_B \qquad \text{for } g = +1 \end{aligned}$$

$$a_{0,0} = \langle ^1P_0|-\mu_B(L_z + 2S_z)|^1P_0\rangle = 0$$

$$\begin{aligned} a_{+1,0} &= \langle ^1P_{+1}|-\mu_B(L_z + 2S_z)|^1P_{+1}\rangle = (-\mu_B)(+1)\langle ^1P_{+1}|^1P_{+1}\rangle \\ &= -\mu_B = -g\mu_B \qquad \text{for } g = +1 \end{aligned}$$

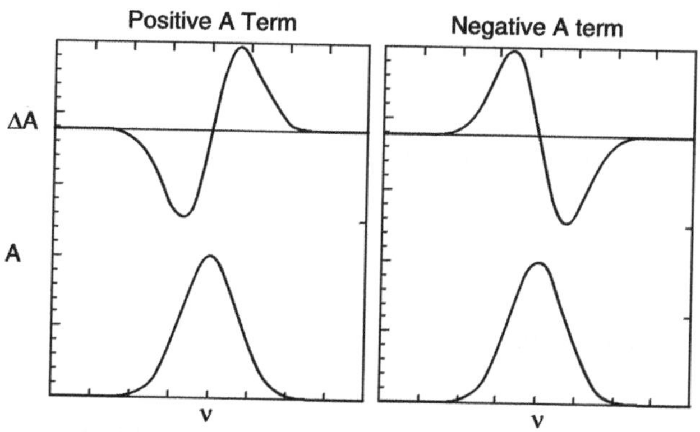

Figure 3.4. Positive and negative *A* terms.

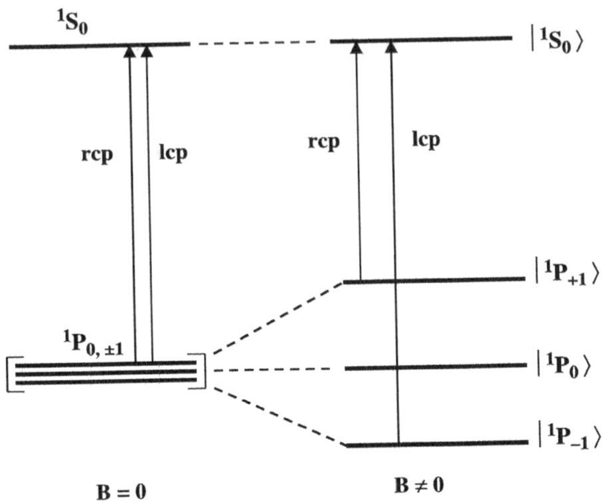

Figure 3.5. Energy levels for the transition $|^1P_{0,\pm1}\rangle \rightarrow |^1S_0\rangle$.

Thus we have just the opposite sign for the Zeeman energies. Therefore we have

$$A_1 = (1/\mu_B(3))[+g\mu_B(\underbrace{|\langle^1P_{-1}|m_{-1}|^1S_0\rangle|^2}_{\neq 0\,(\mathrm{lcp})} - \underbrace{|\langle^1P_{-1}|m_{+1}|^1S_0\rangle|^2}_{=0\,(\mathrm{rcp})}) + 0$$

$$- g\mu_B(\underbrace{|\langle^1P_{+1}|m_{-1}|^1S_0\rangle|^2}_{=0\,(\mathrm{lcp})} - \underbrace{|\langle^1P_{+1}|m_{+1}|^1S_0\rangle|^2}_{\neq 0\,(\mathrm{rcp})})]$$

$$= (2/3)g|\langle^1P_{\pm1}|m_{\pm1}|^1S_0\rangle|^2 = (2/3)g|\langle^1S_0|m_{\pm1}|^1P_{\mp1}\rangle|^2 \tag{3.35}$$

This time, however, rcp is allowed for $|^1P_{+1}\rangle$ and lcp is allowed for $|^1P_{-1}\rangle$, just the reverse of the previous case. As above, we have for D_0

$$D_0 = (1/3)|\langle^1P_{\mp1}|m_{\pm1}|^1S_0\rangle|^2 = (1/3)|\langle^1S_0|m_{\pm1}|^1P_{\mp1}\rangle|^2 \tag{3.36}$$

and therefore the ratio has the same sign as before

$$A_1/D_0 = ((2/3)g)/(1/3) = 2g \tag{3.37}$$

and gives a measure of the ground-state magnetic moment.

3.5. *C* TERMS

The C_0 parameter describes the MCD C term, which is temperature-dependent and due to ground-state degeneracy.

$$C_0/(kT) = (1/(\mu_B|A|))\sum_{\alpha}\langle A_\alpha|\mu_z|A_\alpha\rangle/(kT)\Delta\alpha \tag{3.38}$$

where

$$\Delta\alpha = \sum_{\lambda}(|\langle A_\alpha|m_{-1}|J_\lambda\rangle|^2 - |\langle A_\alpha|m_{+1}|J_\lambda\rangle|^2) \tag{3.39}$$

Since $(N'_{A\alpha g} - N_{A\alpha g})/N = \Delta N'_{A\alpha g}/N$, the difference in population of the ground state upon application of the magnetic field, we can use Eq. (3.9) to give

$$\begin{aligned}\Delta N'_{A\alpha g}/N &= (N_g/(|A|N_G))(1 + \langle A_\alpha|\mu_z|A_\alpha\rangle B/(kT)) - N_g/(|A|N_G)\\ &= \langle A_\alpha|\mu_z|A_\alpha\rangle/(|A|kT)BN_g/N_G \approx \langle A_\alpha|\mu_z|A_\alpha\rangle B/(|A|kT)\end{aligned} \tag{3.40}$$

Thus

$$C_0/(kT) = (1/(\mu_B|A|))\sum_{\alpha}(\Delta N'_{A\alpha}/NB)\,\Delta\alpha \tag{3.41}$$

which emphasizes the population change ($\Delta N'_{A\alpha} \neq 0$) for state $|A_\alpha\rangle$ when $B \neq 0$. The transition $|A_\alpha\rangle \rightarrow |J_\lambda\rangle$ must also absorb lcp and rcp light to a different extent ($|A_\alpha\rangle$ must be degenerate so that $\Delta\alpha$ will $\neq 0$).

The Zeeman sublevels will be divided into Kramer's pairs $|A_\alpha\rangle$ and $|A_{\alpha'}\rangle$ such that

$$\langle A_\alpha|\mu_z|A_\alpha\rangle = -\langle A_{\alpha'}|\mu_z|A_{\alpha'}\rangle$$

For example, in the $|^1P_{0,\pm1}\rangle \rightarrow |^1S_0\rangle$ example above we have

$$\begin{aligned}&\langle ^1P_{+1}|\mu_z|^1P_{+1}\rangle = -\langle ^1P_{-1}|\mu_z|^1P_{-1}\rangle\\ &-g\mu_B = -(+g\mu_B) \quad \text{and} \quad \Delta\alpha = -\Delta\alpha'\\ &(|\langle ^1P_{-1}|m_{-1}|^1S_0\rangle|^2 - |\langle ^1P_{-1}|m_{+1}|^1S_0\rangle|^2)\\ &\qquad \neq 0\,(\text{lcp}) \qquad\qquad 0\,(\text{rcp})\\ &\qquad\qquad = (|\langle ^1P_{+1}|m_{-1}|^1S_0\rangle|^2 - |\langle ^1P_{+1}|m_{+1}|^1S_0\rangle|^2)\\ &\qquad\qquad\quad = 0\,(\text{lcp}) \qquad\qquad \neq 0\,(\text{rcp})\end{aligned}$$

Therefore

$$C_0/(kT) = (2/(\mu_B|A|))\sum_{i\text{ pairs}}(\Delta N'_{A\alpha}/NB)\Delta\alpha = \sum\langle A_\alpha|\mu_z|A_\alpha\rangle/(|A|kT)\Delta\alpha_i$$

For $|^1P_{0,\pm1}\rangle \rightarrow |^1S_0\rangle$ we have

$$C_0/(kT) = (2g/(3kT))|\langle ^1P_{\mp1}|m_{\pm1}|^1S_0\rangle|^2$$

so that the term parameters $C_0 = A_1$ in this case [see Eq. (3.35)], and finally

$$C_0/D_0 = 2g$$

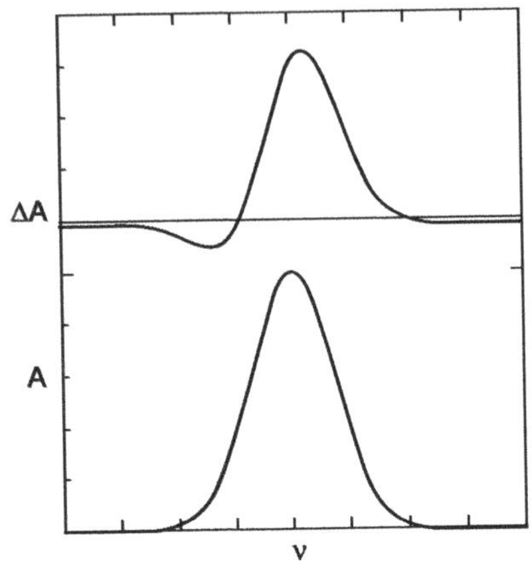

Figure 3.6. Positive C term.

just as for the A term in this case. The main difference between A and C terms is the strong temperature dependence and therefore the field-induced population difference in the ground state for the C term. Figure 3.6 shows the shape of the positive C term assumed for this case if g is positive. The plot shows that the upper Zeeman sublevel has some population ($T > 0\,\text{K}$), and therefore there is a small negative ΔA on the low-frequency side of the term (exaggerated in the figure perhaps). As the temperature is lowered and the population devolves into the lower-energy Zeeman sublevel, the shape of the MCD spectrum parallels that of the absorption giving a single-signed band that maximizes (or minimizes) at the frequency of the absorption maximum.

3.6. *B* TERMS

B terms arise as a result of field-induced mixing of the zero-field wavefunctions. Both ground-state and excited-state functions can be affected, but the inverse energy differences between the states that mix tend to make the effects larger for closely spaced excited states if they are energetically distant from the ground state. Thus from Eq. (3.2) we have

$$|A_{\alpha g}\rangle' = |A_{\alpha g}\rangle - \sum_{K \neq A, \kappa} \langle K_\kappa | \mu_z | A_\alpha \rangle^\circ B/(W_A{}^\circ - W_K{}^\circ)|K_{\kappa g}\rangle$$

Ground-state mixing with $|K_{\kappa g}\rangle$

$$|J_{\lambda j}\rangle' = |J_{\lambda j}\rangle - \sum_{K \neq J, \kappa} \langle K_\kappa | \mu_z | J_\lambda \rangle^\circ B/(W_J{}^\circ - W_K{}^\circ)|K_{\kappa j}\rangle$$

Excited-state mixing with $|K_{\kappa j}\rangle$

As Eq. (3.15) shows, a necessary condition for the mixing of state $|K_\kappa\rangle$ is simultaneous nonzero μ_z and $m_{\pm 1}$ matrix elements. Furthermore, the B term for the state $|J_\lambda\rangle$ interacting with state $|K_\kappa\rangle$ will have opposite sign from the B term for the state $|K_\kappa\rangle$ interacting with the state $|J_\lambda\rangle$, whether or not $|J_\lambda\rangle$ is higher or lower in energy than $|K_\kappa\rangle$. The extent of the mixing, and therefore the magnitude of the *B* term, is inversely proportional to the energy difference between states. Excited-state mixing is usually more common because the energy difference between ground and excited states, $W_A{}^\circ - W_K{}^\circ$, is normally larger. Equation (3.15) for excited-state mixing, neglecting ground-state mixing, is given as

$$B_0 = -(2/(\mu_B|A|))\mathrm{Re}\sum_{\alpha\lambda}\left[\sum_{K\neq J,\kappa} 1/(W_K - W_J)\langle J_\lambda|L_z + 2S_z|K_\kappa\rangle \right.$$
$$\left. \times\, (\langle A_\alpha|m_{-1}|J_\lambda\rangle\langle K_\kappa|m_{+1}|A_\alpha\rangle - \langle A_\alpha|m_{+1}|J_\lambda\rangle\langle K_\kappa|m_{-1}|A_\alpha\rangle)\right] \quad (3.42)$$

In fact, if $W_J{}^\circ - W_K{}^\circ$ is small, we can have overlapping *B* terms of opposite sign which give the appearance of an *A* term. Such a situation is referred to as a "pseudo" *A* term and will be discussed below. Similarly, if $W_A{}^\circ - W_K{}^\circ$ were small, we would have a "pseudo" *C* term, which would be temperature-dependent because of population differences.

B_0 and B_0/D_0 are more difficult to calculate than A_1 or C_0 because of the sum over all mixing states $|K_\kappa\rangle$ and the fact that transition moments do not cancel in the ratios as in the case of A_1/D_0 or C_0/D_0 as illustrated above. B_0 calculations are feasible if only a few states $|K_\kappa\rangle$ are involved and they are close in energy to the state $|J_\lambda\rangle$ of interest.

As an example, consider a case of two atomic transitions: $s \rightarrow p_x$ and $s \rightarrow p_y$, which might be at different energies in a complex where a different "crystal field" environment is present along the *x* and *y* axes as shown in Figure 3.7. Two

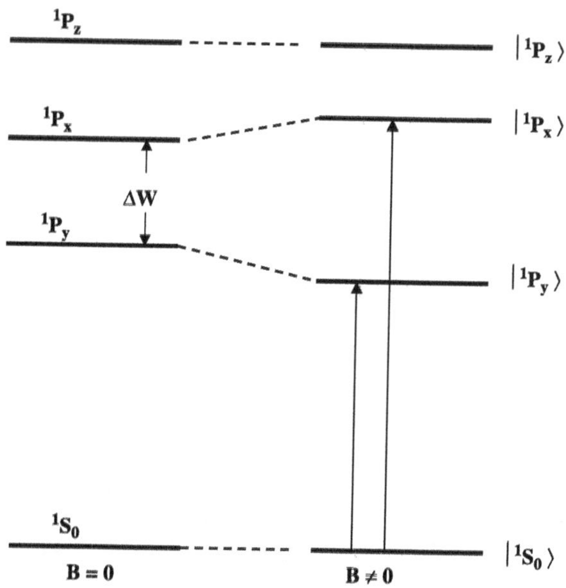

Figure 3.7. Energy levels for two atomic transitions $s \rightarrow p_x$ and $s \rightarrow p_y$, with a different "crystal field" environment along the *x* and *y* axes.

transitions separated by $\Delta W = W_x - W_y$ would then be expected. If we limit the intermixing of states to just the two $|^1P_x\rangle$ and $|^1P_y\rangle$ states, then we can calculate the B terms for each of them by using the following relations:

$$|^1P_x\rangle = (1/\sqrt{2})(|^1P_{+1}\rangle - |^1P_{-1}\rangle) \quad \text{and} \quad |^1P_y\rangle = (i/\sqrt{2})(|^1P_{+1}\rangle + |^1P_{-1}\rangle)$$

then

$$\langle^1S|m_{\pm 1}|^1P_x\rangle = \pm(1/\sqrt{2})\langle^1S|m_{\pm 1}|^1P_{\mp 1}\rangle \quad \text{and}$$
$$\langle^1S|m_{\pm 1}|^1P_y\rangle = (i/\sqrt{2})\langle^1S|m_{\pm 1}|^1P_{\mp 1}\rangle \quad \text{and also}$$
$$\langle^1P_x|L_z|^1P_y\rangle = -i \quad \text{and} \quad \langle^1P_y|L_z|^1P_x\rangle = +i$$

Upon substituting into Eq. (3.42), we obtain

$$\begin{aligned} B_0(^1P_x) &= -(2/(\mu_B \Delta W))\text{Re}\{\langle^1P_x|\mu_z|^1P_y\rangle[\langle^1S|m_{-1}|^1P_x\rangle\langle^1P_y|m_{+1}|^1S\rangle \\ &\quad - \langle^1S|m_{+1}|^1P_x\rangle\langle^1P_y|m_{-1}|^1S\rangle]\} \\ &= (2(-i)/\Delta W)[1/\sqrt{2}\langle^1S|m_{-1}|^1P_{+1}\rangle(i^*/\sqrt{2})\langle^1S|m_{-1}|^1P_{+1}\rangle^* \\ &\quad - (1/\sqrt{2})\langle^1S|m_{+1}|^1P_{-1}\rangle(-i^*/\sqrt{2})\langle^1S|m_{+1}|^1P_{-1}\rangle^*] \\ &= (2(-i)/\Delta W)[(-i/2)|\langle^1S|m_{-1}|^1P_{+1}\rangle|^2 - (i/2)|\langle^1S|m_{+1}|^1P_{-1}\rangle|^2] \\ &= -(2/\Delta W)|\langle^1S|m_{\pm 1}|^1P\mp 1\rangle|^2 \end{aligned}$$

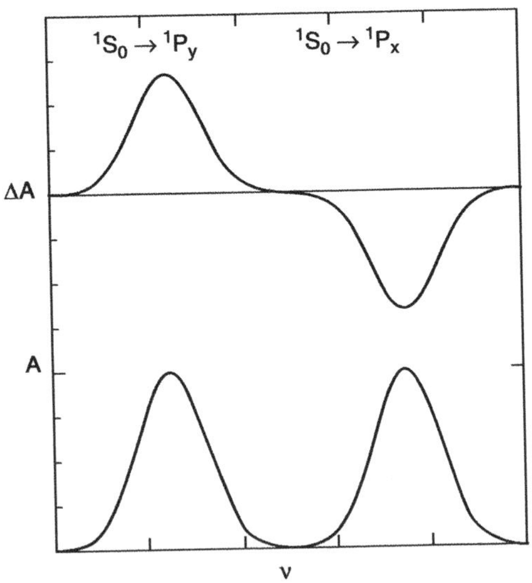

Figure 3.8. Two B terms of opposite sign resulting from Zeeman interaction between the ^1P_y and ^1P_x states.

and similarly

$$B_0(^1P_y) = -(2/(\mu_B \Delta W))\mathrm{Re}\{\langle ^1P_y|\mu_z|^1P_x\rangle[\langle ^1S|m_{-1}|^1P_y\rangle\langle ^1P_x|m_{+1}|^1S\rangle - \langle ^1S|m_{+1}|^1P_y\rangle\langle ^1P_x|m_{-1}|^1S\rangle]\} = +(2/\Delta W)|\langle ^1S|m_{\pm 1}|^1P_{\mp 1}\rangle|^2$$

Therefore the two *B* terms have opposite signs as expected:

$$B_0(^1P_x) = -B_0(^1P_y)$$

Figure 3.8 shows the two *B* terms for this case, assuming intermixing only between the ^1P_y and ^1P_x states; the lower-energy transition to ^1P_y has a *B* term of opposite sign than the higher-energy transition to ^1P_x.

3.7. PSEUDO *A* TERMS: OVERLAPPING *B* TERMS

As noted above, if two states are close in energy, say within their bandwidths, and undergo field-induced mixing giving rise to opposite signed *B* terms, the MCD will have the appearance of an *A* term. Consider two transitions $A \rightarrow J_1$ and $A \rightarrow J_2$, where $W_{J2} - W_{J1} = \Delta W$. Assume the Zeeman interaction between J_1 and J_2 is small compared to ΔW so that $\langle J_1\lambda|\mu_z|J_2\lambda'\rangle B/\Delta W \ll 1$, then the contributions to the MCD will be additive:

$$\Delta A/E = \gamma\mu_B B \sum_{i=1,2}[A_1(J_i)(-\partial f_i(E)/\partial E) + (B_0(J_i) + C_0(J_i)/(kT))f_i(E)]$$

Also, $f_2(E) = f_1(E - \Delta W)$ if J_1 and J_2 have identical potential surfaces. Similarly,

$$A/E = \gamma \sum_{i=1,2} D_0(J_i)f_i(E)$$

If the field-induced mixing is only between J_1 and J_2, then by means of the properties of the $m_{\pm 1}$ matrix elements it can be shown that

$$B_0(J_1) \equiv B_0(J_1, J_2) = -B_0(J_2) \equiv B_0(J_2, J_1)$$

Then if $\Delta A/E$ is due only to these *B* terms (i.e., $A_1(J_i) = C_0(J_i) = 0$), we have

$$\Delta A/E = \gamma\mu_B B[B_0(J_1)f_1(E) + B_0(J_2)f_2(E)] = \gamma\mu_B B[B_0(J_1)f_1(E) - B_0(J_1)f_2(E)]$$
$$= \gamma\mu_B B[B_0(J_1)f_1(E) - B_0(J_1)f_1(E - \Delta W)]$$

Then using a Taylor series expansion $f(x+h) = f(x) + h\partial f(x)/\partial x + h^2\partial^2 f(x)/\partial^2 x + \cdots$ for $f_1(E - \Delta W)$, we can write

$$\Delta A/E \approx \gamma\mu_B B[-B_0(J_1)\Delta W(-\partial f_1(E)/\partial E)] = \gamma\mu_B B[A_1'(J_1, J_2)\Delta W(-\partial f_1(E)/\partial E)]$$

where $A_1'(J_1, J_2) = -B_0(J_1)\Delta W$. In the limit $\Delta W \rightarrow 0, A_1'(J_1, J_2)$ becomes equal to $A_1(J)$, when J_1 and J_2 in fact become degenerate.

In the case of the $s \rightarrow p_x$ and $s \rightarrow p_y$ transitions in the example above, as $\Delta W \rightarrow 0$ and the $|^1P_x\rangle$ and $|^1P_y\rangle$ states become degenerate, the A term will be given by Eq. (3.27):

$$\begin{aligned} A_1 &= (2/\mu_B)\{\langle ^1P_x|\mu_z|^1P_y\rangle[\langle ^1S|m_{-1}|^1P_x\rangle\langle ^1P_y|m_{+1}|^1S\rangle \\ &\quad - \langle ^1S|m_{+1}|^1P_x\rangle\langle ^1P_y|m_{-1}|^1S\rangle]\} = +2|\langle ^1S|m_{\pm 1}|^1P_{\mp 1}\rangle|^2 \\ &= -B_0(^1P_x)\Delta W \qquad \text{(in the limit } \Delta W \rightarrow 0) \end{aligned}$$

Note that this result is the same as found for the degenerate excited state $|^1S_0\rangle \rightarrow |^1P_{0,\pm 1}\rangle$ example above. The A term is determined entirely by the $|^1P_{x,y}\rangle$ states and their $|^1P_{\pm 1}\rangle$ counterparts, with no contribution from the $|^1P_0\rangle = |^1P_z\rangle$ state.

3.8. OVERLAPPING C TERMS

Using the relationship for a small ΔW and a Taylor series expansion of $f_1(E - \Delta W)$, we obtain

$$f_2(E) \approx f_1(E) - \Delta W(\partial f_1(E)/\partial E)$$

Then if $A_1 = B_0 = 0$ (only C terms present) for J_1 and J_2, we have

$$\Delta A/E = (\gamma\mu_B B/(kT))\{[C_0(J_1) + C_0(J_2)]f_1(E) - C_0(J_2)\Delta W(\partial f_1(E)/\partial E)\}$$

If $C_0(J_1)$ and $C_0(J_2)$ are of the same sign, then the first term dominates for small ΔW. However, if $C_0(J_1) = -C_0(J_2)$, then

$$\Delta A/E = (\gamma\mu_B B/(kT))[C_0(J_2)\Delta W(-\partial f_1(E)/\partial E)]$$

which has the appearance of an A term, but it is temperature-dependent and decreases to 0 as $\Delta W \rightarrow 0$.

3.9. GROUND-STATE NEAR DEGENERACY

If we have a case where $\langle A_{1\alpha}|\mu_z|A_{2\alpha'}\rangle B/\Delta W \ll 1$ for small ΔW, the MCD and absorbance are additive, but they are weighted by fractional populations, δ_i:

$$\Delta A/E = \gamma\mu_B B \sum_{i=1,2} \delta_i[A_1(A_i)(-\partial f_i(E)/\partial E) + (B_0(A_i) + C_0(A_i)/(kT)f_i(E)]$$

and

$$A/E = \gamma \sum_{i=1,2} \delta_i D_0(A_i)\, f_i(E)$$

where

$$\delta_1 = \frac{|A_1|}{[|A_1| + |A_1|\exp(-\Delta W/(kT))]} \quad \text{and} \quad \delta_2 = 1 - \delta_1$$

If A_1, A_2, and J are all nondegenerate, then

$$\Delta A/E = (\gamma\mu_B B\Delta W/2)[B_0'(A_1)(-\partial f_1(E)/\partial E) + B_0'(A_1)/(kT)f_1(E)]$$
$$A/E = (\gamma/2)[D_0(A_1) + D_0(A_2))]f_1(E)$$

The MCD is independent of ΔW since $\Delta WB_0'(A_1)$ is independent of ΔW, and we thus have a pseudo A term and a pseudo C term in this case.

3.10. DEVIATION FROM THE LINEAR LIMIT

In the case of very low temperatures or very high fields when the Zeeman splitting is $>kT$, we cannot make the approximation

$$\exp(\langle A_\alpha|\mu_z|A_\alpha\rangle^\circ B/(kT)) \approx 1 + \langle A_\alpha|\mu_z|A_\alpha\rangle^\circ B/(kT) + \cdots$$

Then

$$N'_{A\alpha g}/N = \frac{\exp(\langle A_\alpha|\mu_z|A_\alpha\rangle^\circ B/(kT))}{\sum_{\alpha g}\exp(\langle A_\alpha|\mu_z|A_\alpha\rangle^\circ B/(kT))}$$

and in the BO–FC–RS approximation, Eq. (3.13) becomes

$$\Delta A/E = \gamma\mu_B B\left\{\frac{\sum_{\alpha\lambda}\exp(\langle A_\alpha|\mu_z|A_\alpha\rangle^\circ B/(kT))}{\sum_{\alpha\lambda}\exp(\langle A_\alpha|\mu_z|A_\alpha\rangle^\circ B/(kT))} \times [A_1(\alpha\lambda)(-\partial f(E)/\partial E)\right.$$
$$\left. + B_0(\alpha\lambda)f(E) + (|\langle A_\alpha|m_{-1}|J_\lambda\rangle|^2 - |\langle A_\alpha|m_{+1}|J_\lambda\rangle|^2)f(E)/(\mu_B B)]\right\}$$

Then for C terms at low temperature ($A_1(\alpha\lambda)$ and $B_0(\alpha\lambda)\approx 0$), with

$$\Delta\alpha_i = (|\langle A_\alpha|m_{-1}|J_\lambda\rangle|^2 - |\langle A_\alpha|m_{+1}|J_\lambda\rangle|^2)$$

we write

$$(\Delta A/E)_{\text{sat'd}}$$
$$=\gamma f(E)\frac{\{\sum_{i\text{pairs}}\Delta_{\alpha i}[\exp(\langle A_{\alpha i}|\mu_z|A_{\alpha i}\rangle B/(kT)) - \exp(-\langle A_{\alpha i}|\mu_z|A_{\alpha i}\rangle B/(kT))]\}}{\sum_{i\text{pairs}}[\exp(\langle A_{\alpha i}|\mu_z|A_{\alpha i}\rangle B/(kT)) + \exp(\langle A_{\alpha i}|\mu_z|A_{\alpha i}\rangle B/(kT))]}$$

If one of the Zeeman pairs is the only one populated at the very low temperature, then the saturated limit is given by

$$(\Delta A/E)_{\text{sat'd}} = \gamma\Delta\alpha_1 f(E)$$

At higher temperatures than the saturated limit where only one of a Kramer's pair is populated and by using the definition of the $\tanh(u) = (e^u - e^{-u})/(e^u + e^{-u})$, the

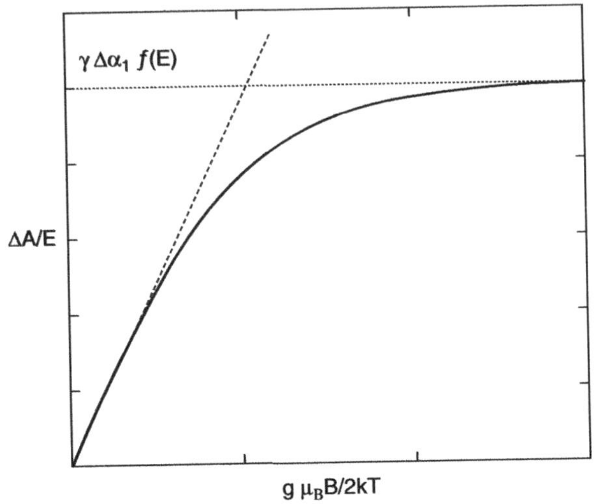

Figure 3.9. Deviation from the linear limit at low temperature or high field for a Kramer's doublet. (Adapted from ref. 1.)

expression for $(\Delta A/E)_{\text{sat'd}}$ becomes

$$\begin{aligned}(\Delta A/E)_{\text{sat'd}} &= \gamma\Delta\alpha f(E)\tanh(\langle A_{\alpha i}|\mu_z|A_{\alpha i}\rangle B/(kT)) \\ &= \gamma\Delta\alpha f(E)\tanh(g\mu_B B/(2kT))\end{aligned}$$

This is illustrated in Figure 3.9. There are complications if the ground-state basis is subject to orientational averaging. The matrix elements $\langle A_{\alpha i}|\mu_z|A_{\alpha i}\rangle$ are, in general, orientation-dependent, and molecules with different orientations with respect to the field direction will have different Zeeman energies and wavefunctions because of the exponentials. The problem is simplified for isotropic Zeeman splitting. Then the exponentials become orientation-independent. Finally, it may be remarked that saturation effects are much less important for *A* and *B* terms.

3.11. ZEEMAN SPLITTING ≥ BANDWIDTH

In the case of very sharp bands where the Zeeman splitting is ≥ bandwidth but still $< kT$, we can no longer make the approximation $\rho_{AJ}(E - a_{\alpha\lambda}B) = \rho_{AJ}(E) + \partial\rho_{AJ}(E)/\partial E(-a_{\alpha\lambda}B) + \cdots$ in writing Eq. (3.12). Therefore, the BO–FC–RS approximation becomes

$$\Delta A/E = (\gamma\mu_B B/|A|)\sum_{\alpha\lambda}[B_0(\alpha\lambda) + C_0(\alpha\lambda)/(kT)) + R_0(\alpha\lambda)]f'_{A\alpha,J\lambda}(E)$$

where

$$R_0(\alpha\lambda) = (1/(\mu_B B))(|\langle A_\alpha|m_{-1}|J_\lambda\rangle|^2 - |\langle A_\alpha|m_{+1}|J_\lambda\rangle|^2)$$

and

$$f'_{A\alpha,J\lambda}(E) = (1/N_G)\sum_g N_G \sum_j |\langle g|j\rangle|^2 \rho'_{A\alpha g,J\lambda j}(E)$$

This term $R_0(\alpha\lambda)$ reflects the fact that Zeeman transitions are circularly polarized. There is a complicated orientation dependence for Zeeman transitions, and care must be taken to ensure isotropic conditions.

3.12. RELATIVE MAGNITUDE OF *A*, *B*, AND *C* TERMS

It is clear from Eqs. (3.13)–(3.16) that the magnitude of the *B* term varies as $1/(\Delta W)$, the energy difference between the intermixing states, and the magnitude of the *C* term varies as $1/(kT)$. The maximum amplitude of *B* and *C* terms also depend upon the inverse bandwidth, $1/\Delta$. However, as a consequence of the derivative in Eq. (3.13), the *A* term depends upon $1/\Delta^2$. These characteristics can be demonstrated by assuming a specific band model such as the Gaussian band. For absorption and *B* or *C* terms, an isolated Gaussian band $f(E)$ is given by

$$f(E) = (1/\Delta\sqrt{\pi}) \exp(-(E - E^0)^2/\Delta^2)$$

where 2Δ is the bandwidth at f_{max}/e. The quantity f_{max} is given by $f(E)$ at $E = E^0$:

$$f_{max} = (1/(\Delta\sqrt{\pi})) \exp(-(E^0 - E^0)^2/\Delta^2) = 1/(\Delta\sqrt{\pi})$$

Therefore *B*- and *C*-term magnitudes will vary as $1/\Delta$. For the A term, $\partial f(E)/\partial E$ is required:

$$\begin{aligned}\partial f(E)/\partial E &= (1/(\Delta\sqrt{\pi})) \exp(-(E - E^0)^2/\Delta^2)\partial/\partial E[-E^2 + 2EE^0 - (E^0)^2] \\ &= -(2(E-E^0)/(\Delta^3\sqrt{\pi})) \exp(-(E - E^0)^2/\Delta^2)\end{aligned}$$

The minimum (f'_{min}) and maximum (f'_{max}) of $\partial f(E)/\partial E$ occur at $E^0 \pm \Delta/\sqrt{2}$, respectively:

$$\begin{aligned}f'_{max} &= -(2(\Delta/\sqrt{2} + E^0 - E^0)/(\Delta^3\sqrt{\pi})) \exp(-(\Delta/\sqrt{2} + E^0 - E^0)^2/\Delta^2) \\ &= -2/(\Delta^2\sqrt{2}\sqrt{e}\sqrt{\pi})\end{aligned}$$

Similarly,

$$f'_{min} = +2/(\Delta^2\sqrt{2}\sqrt{e}\sqrt{\pi})$$

Thus the magnitude of the *A* term as given by f'_{max} or f'_{min} depends upon $1/\Delta^2$; consequently, *A* terms are more sensitive to transition bandwidths than *B* or *C* terms. Finally, the ratio of the A-term maximum and the absorption maximum is given by

$$f'_{max}/f_{max} = -2(\Delta\sqrt{\pi})/(\Delta^2\sqrt{2}\sqrt{e}\sqrt{\pi}) = -2/(\Delta\sqrt{2}\sqrt{e}) = -1/(1.17\Delta)$$

The relative magnitudes of the MCD to absorption ratio $(\Delta A/A)_{\max}$ can be estimated from the ratio of Eqs. (3.13) and (3.17) and is approximately $g\mu_B B/\Delta : g\mu_B B/(\Delta W) : g\mu_B B/(kT)$ for A, B, and C terms, repectively. For a typical broad-band case at room temperature ($g = 2$, $B = 1\,\mathrm{T}$, $\Delta = 10^3\,\mathrm{cm}^{-1}$, $\Delta W = 10^4\,\mathrm{cm}^{-1}$, and $kT = 200\,\mathrm{cm}^{-1}$) the approximate relative values are $10:1:50$ for A, B, and C terms. For a narrow band at low temperature ($\Delta = 10\,\mathrm{cm}^{-1}$ and $kT = 7\,\mathrm{cm}^{-1}$), the approximate relative values become $1000:1:1000$ [ref. 1, Section 5.3].

3.13. PARAMETER EVALUATION. LINESHAPE FUNCTION

One approach to evaluation of A_1, B_0, C_0, and D_0 parameters is to assume a specific lineshape function for an isolated band (not the usual case, unfortunately—most experimental spectra consist of overlapping bands, sometimes only partially resolved and closely spaced in energy). For example, if a Gaussian fit of an isolated band is assumed, we have

$$(\Delta A/E)_{E=E^0} = \gamma\mu_B B(B_0 + C_0/(kT))(1/(\Delta\sqrt{\pi}))$$

$$(\Delta A/E)_{E=\Delta/\sqrt{2}} = \gamma\mu_B B A_1[-2/(\Delta^2\sqrt{2}\sqrt{e}\sqrt{\pi})] + (\Delta A/E)_{E=E^0}[\exp(-(\Delta/\sqrt{2} - E^0)^2/\Delta^2)]$$

And for the absorption, the expression is

$$(A/E)_{E=E^0} = \gamma D_0 1/(\Delta\sqrt{\pi})$$

This approach is not very satisfactory because bands are frequently not exactly Gaussian or consist of several overlapping bands that must be fitted by some procedure. Such a fitting procedure usually involves guessing parameters and then comparing a calculated band shape with the experimental spectrum. Parametric fits suffer from the fact that any curve can be fit if enough parameters are arbitrarily chosen. The significance of the parameters is often not meaningful.

3.14. PARAMETER EVALUATION. METHOD OF MOMENTS

Another approach that does not require a specific lineshape function to be assumed is the Method of Spectral Moments. We define the nth moment as

$$\langle\alpha(E)\rangle_n^{E0} = \int \alpha(E)(E - E^0)^n dE$$

where the integration over the band is from zero $\alpha(E)$ on the low-energy side to zero $\alpha(E)$ on the high-energy side of the band. $E^0 \approx \overline{E}$, where $\overline{E}$ is defined so that $\langle\alpha(E)\rangle_1^{E0} = 0$:

$$\overline{E} = \int \alpha(E)E\,dE \Big/ \left(\int \alpha(E)\,dE\right)$$

and then the n moments are calculated about $\overline{E}$:

$$\langle\alpha(E)\rangle_n^{E0} = \int \alpha(E)(E - \overline{E})^n dE$$

The method assumes the BO–FC approximation to be valid. MCD and absorption parameters are then given by moments determined about $\overline{E}$ in the following expressions (c = molar concentration and ℓ = path length in cm); the zeroth and first MCD moments are given by ($\gamma = 326.6\ \mathrm{D}^2$ and $\mu_B = 0.4669\,\mathrm{cm}^{-1}\mathrm{T}^{-1}$):

$$\langle\Delta A/E\rangle_0^{E0} = \langle\Delta\varepsilon c\ell/E\rangle_0^{E0} = \gamma\mu_B B(B_0 + C_0/kT) = 152.5Bc\ell(B_0 + C_0/(kT))$$

$$\langle\Delta A/E\rangle_1^{E0} = \langle\Delta\varepsilon c\ell/E\rangle_1^{E0} = \gamma\mu_B BA_1 = 152.5Bc\ell A_1$$

and for zeroth absorption moment:

$$\langle A/E\rangle_0^{E0} = \langle\varepsilon c\ell/E\rangle_0^{E0} = \gamma D_0 = 326.6c\ell D_0$$

4 Measurement of MCD Spectra

The measurement of MCD spectra requires that the sample be placed in a strong magnetic field oriented along the light path of a sensitive CD spectrometer. The field direction (the orientation of the **B** vector along the $+z$ coordinate direction, the light propagation direction), by convention, should have the $N \rightarrow S$ field vector pointing toward the detector. In this chapter a description of a typical MCD spectrometer will be given, together with some points concerning data acquisition, optical elements, and stray polarization, calibration, and magnet systems. The accurate and precise (reproducible) measurement of MCD spectra together with the corresponding absorption spectra is fundamental to the use of MCD as a valid spectroscopic tool. The spectroscopist must be vigilant against the measurement of sample or, more seriously, spectrometer artifacts, some of which can be orders of magnitude larger than the intended MCD measurement.

4.1. MCD SPECTROMETER

Figure 4.1 shows a sketch of a spectrometer, designed and built in the author's laboratory, which will measure not only the MCD (or CD alone in the absence of the magnetic field) but also the absorbance of the sample [ref. 5]. The simultaneous measurement of absorbance A and the MCD (or CD) $\Delta A = A_- - A_+$ along the same light path is highly desirable in order to avoid errors that could develop by using two separate instruments for measuring A and ΔA. The interpretation of MCD spectra requires comparison with the corresponding absorption spectra, and MCD parameters are often derived from ratios obtained by integration of the respective spectra. In Figure 4.1, the light from a powerful source (xenon arc lamp) is dispersed by means of a monochromator and then passed through a Rochon prism linear polarizer to form two beams of plane-polarized light. One beam (the extraordinary beam) is deviated slightly from the other (the ordinary beam), and the two beams differ in polarization by 90°. The deviated beam is detected by means of a photomultiplier (PMT_2), which provides a signal proportional to the input light intensity; the PMT current is conventionaly converted to a voltage signal proportional to the light intensity by means of a current to voltage amplifier. The undeviated beam is passed through a photoelastic modulator (PEM) which acts as a transparent, dynamic quarter-wave plate. When

---

A Practical Guide to Magnetic Circular Dichroism Spectroscopy, by W. Roy Mason

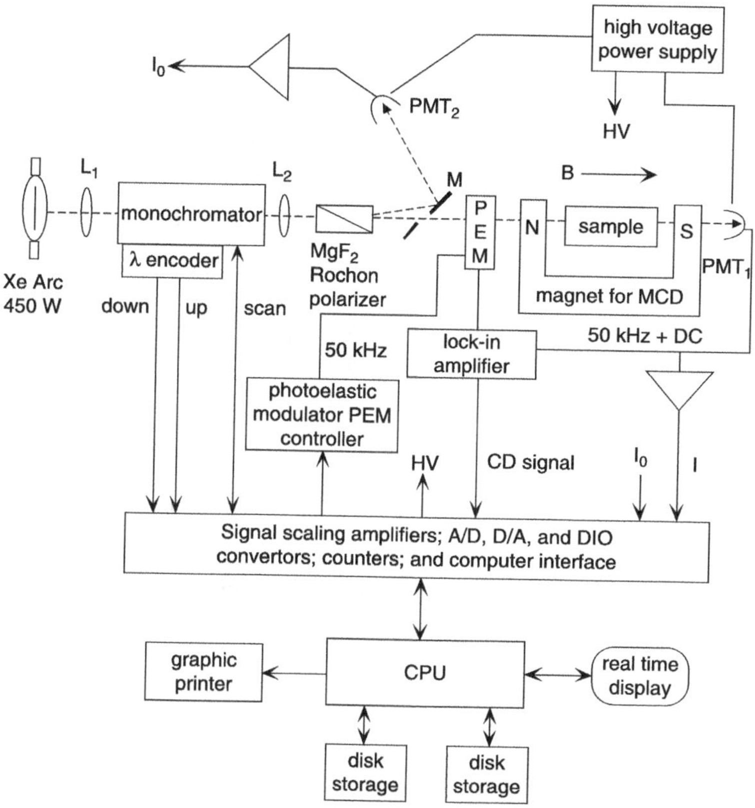

Figure 4.1. Block diagram of a computer-controlled MCD spectrometer.

linearly polarized light passes through a quarter-wave plate, it is converted to circularly polarized light (see Chapter 2, Section 2.2). The linearly polarized light with intensity I_0 can be described as the sum of two circularly polarized components that are in phase with each other:

$$I_0 = (1/2)(I_-^0 + I_+^0) \tag{4.1}$$

The PEM will retard one of the components (or advance the other) by a quarter wavelength, $\pm\lambda/4$, with a time-dependent retardation, δ, proportional to the modulator driving voltage:

$$\delta = \delta_0 \sin \omega t \tag{4.2}$$

where $\delta_0 =$ peak retardation proportional to the driving voltage of the PEM and adjusted to give $\pm\lambda/4$ for the current wavelength λ; $\omega =$ modulator frequency, which is typically a high acoustic frequency such as 50 kHz; and $t =$ time. The modulator thus produces alternately lcp and rcp light with a sine-wave time dependence. This is illustrated in Figure 4.2, where the emergent light polarization over

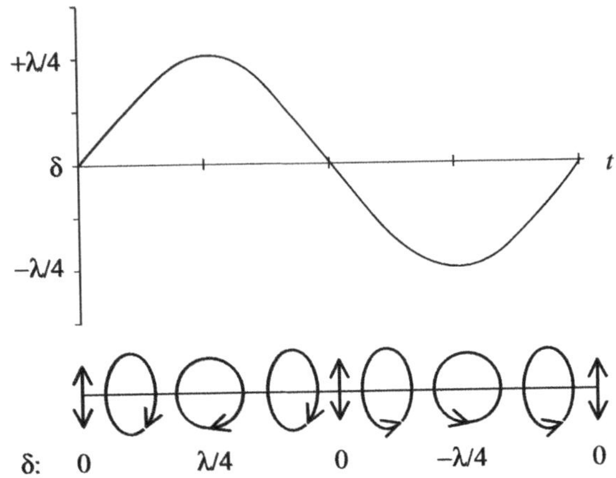

Figure 4.2. Polarization of light emergent from the PEM over the course of the sine-wave modulation. rcp (λ/4) and lcp (− λ/4) are produced during a single period.

the course of a single modulation frequency period is shown. The modulated circularly polarized light is then passed through the sample situated in the bore of a magnet and is finally detected by another photomultiplier, PMT_1.

4.2. DETECTOR SIGNALS: MEASUREMENT OF ΔA AND A

The light intensity at the PMT_1 detector is given by

$$I_{det} = (I_0/2)[(1 - \sin(\delta_0 \sin \omega t))10^{-A_-} + (1 + \sin(\delta_0 \sin \omega t))10^{-A_+}] \tag{4.3}$$

where $A_\pm$ is the absorbance of the lcp (−) or rcp (+) component of the polarized light. The signal from this detector (the photocurrent again is converted to a voltage by means of a current to voltage amplifier) consists of a dc component, V_{dc}, proportional to the light intensity passed through the sample and a small ac component, V_{ac}, at the modulation frequency ω that develops if the sample absorbs lcp or rcp to a different extent:

$$V_{det} = V_{dc} + V_{ac} = eQGI_{det} \tag{4.4}$$

where e is the electronic charge, Q is the photocathode efficiency of the PMT, and G includes the detector gain and current to voltage amplification factor. The voltage then is given by

$$V_{det} = (eQGI_0/2)[(10^{-A_-} + 10^{-A_+} + (10^{-A_+} - 10^{-A_-})\sin(\delta_0 \sin \omega t)] \tag{4.5}$$

with $V_{dc} = (eQGI_0/2)(10^{-A_-} + 10^{-A_+})$ and $V_{ac} = (eQGI_0/2)[(10^{-A_+} - 10^{-A_-}) \sin(\delta_0 \sin \omega t)]$. The ratio of these two components is independent of $eQGI_0/2$ and can be shown to be proportional to the MCD (or CD) $\Delta A = A_- - A_+$ when the $10^{-A_\pm}$ terms are expressed exponentials of e:

$$\begin{aligned}\frac{V_{ac}}{V_{dc}} &= \frac{[\exp(2.303(\Delta A)/2) - \exp(-2.303(\Delta A)/2)]\sin(\delta_0 \sin\omega t)}{\exp(2.303(\Delta A)/2) + \exp(-2.303(\Delta A)/2)} \\ &= \tanh(2.303\Delta A/2)\sin(\delta_0 \sin\omega t) = \tanh(1.1515\Delta A)\sin(\delta_0 \sin\omega t)\end{aligned} \tag{4.6}$$

If ΔA is <0.1, the usual case, then the ratio can be approximated by

$$V_{ac}/V_{dc} \approx 1.1515\Delta A \sin(\delta_0 \sin\omega t) \tag{4.7}$$

The function $\sin(\delta_0 \sin\omega t)$ can be further expanded in a Fourier series:

$$\sin(\delta_0 \sin\omega t) = 2J_1(\delta_0)\sin\omega t + 2J_3(\delta_0)\sin 3\omega t + \cdots \tag{4.8}$$

where J_i are Bessel functions of order i. The first term of the series in Eq. (4.8) is given by $J_1(\delta_0) = \delta_0/2$ is the largest so that

$$\sin(\delta_0 \sin\omega t) \approx 2\delta_0/2 \sin\omega t = \delta_0 \sin\omega t \tag{4.9}$$

which is an approximate sine wave with respect to t. If the measurement is made under conditions where V_{dc} is held constant, and since δ_0 will be constant at a given wavelength, then

$$V_{ac} = 1.1515 V_{dc}\delta_0 \Delta A \sin\omega t \tag{4.10a}$$

or in terms of ΔA we obtain

$$\Delta A = V_{ac}/(1.1515 V_{dc}\delta_0 \sin\omega t) \tag{4.10b}$$

Thus, ΔA is proportional to the V_{ac} signal with frequency ω. This signal can be detected with high sensitivity by means of a phase-sensitive lock-in amplifier (LIA) tuned to the modulator reference frequency ω_{ref}. The LIA employs phase-sensitive detection (psd) by taking the product of V_{ac} and a constant reference ac signal, V_{ref}, at the ω_{ref} reference frequency from the PEM modulator controller:

$$V_{psd} = V_{ac} V_{ref} \sin(\omega t + \theta)\sin(\omega_{ref} t + \theta_{ref}) \tag{4.11}$$

$$\begin{aligned} &= (1/2)V_{ac}V_{ref}\cos[(\omega - \omega_{ref})t + \theta - \theta_{ref}] \\ &\quad - (1/2)V_{ac}V_{ref}\cos[(\omega + \omega_{ref})t + \theta + \theta_{ref}]\end{aligned} \tag{4.12}$$

where θ and θ_{ref} are the variable phase of V_{ac} and the constant phase of V_{ref}, respectively. If ω and ω_{ref} are equal, then a low-pass filter in the LIA will reject the sum frequency term in Eq. (4.12) and the LIA output will be

$$V_{LIA} = V_{ac}V_{ref}\cos(\theta - \theta_{ref}) \tag{4.13}$$

The LIA output, V_{LIA}, is essentially a DC voltage that is proportional to V_{ac} times the cosine of the phase difference $(\theta - \theta_{ref})$ between V_{ac} and V_{ref}. This phase difference of the ac signal relative to the PEM reference voltage is determined by whether the absorption of lcp or rcp light is greater and therefore the sign of the LIA output, and thus the sign of ΔA; it will be positive if A_- is greater than A_+. The proportionality of ΔA to V_{LIA}, together with the phase difference, must be determined experimentally by a calibration, for example by using a standard sample with a known sign and magnitude of ΔA (see Section 4.5).

Finally, in Figure 4.1 the PMTs are carefully selected to have very nearly the same response to a given intensity of light. The gains of the two PMTs are then changed simultaneously to keep V_{dc} from PMT_1 at a constant value. As the sample absorbs light, the gain of PMT_1 (and therefore PMT_2 also) must be increased to maintain a constant V_{dc}. The dc signal from PMT_2 will therefore increase as the absorption increases. By taking the log ratio of the constant dc signal (V_{dc}) from PMT_1 and the dc signal from $PMT_2(V_2)$ which will vary as the detector gains are changed, the sample absorbance can be determined: $A = -\log(V_{dc}/V_2)$. By plotting the A from the log ratio of the two dc signals and ΔA from the LIA output as a function of wavelength, two spectra are obtained simultaneously and synchronously from the same light path.

Spectral results are often reported as molar absorptivities $\varepsilon = A/(c\ell)$, where c is the sample molar concentration and ℓ is the path length (in centimeters) through the sample; and for differential absorptivities we have $\Delta\varepsilon = \Delta A/(c\ell)$. For MCD, work $\Delta\varepsilon$ is usually normalized for the magnetic field B ($\approx H$) in Tesla: $\Delta\varepsilon_M = \Delta A/(c\ell B)$. However, it should be noted here that older MCD data may be reported as ellipticity, θ (degrees or millidegrees), or molar ellipticities, $[\theta]_M$ (degrees or millidegrees deciliter $\text{mol}^{-1}\text{decimeter}^{-1}$); this practice followed older units for natural CD results. Furthermore, for MCD results, $[\theta]_M$ values were usually normalized for the magnetic field in Gauss. Presenting MCD data with these older units, of course, does not change the shape of the spectral features or affect the qualitative interpretation of A, B, or C terms, but when quantitative comparisons are made with ΔA or $\Delta\varepsilon_M$ values from more modern spectra, care must be taken to ensure compatible units of field and path length. The relations between θ and $[\theta]_M$ and ΔA and $\Delta\varepsilon_M$, respectively, are given by Eqs. (4.14) and (4.15) (see ref. 1, Appendix A).

$$\theta = 32{,}982\ \Delta A \qquad (\theta \text{ in mdeg}) \tag{4.14}$$

$$\frac{[\theta]_M}{10^4\text{Gauss}} = \frac{3298.2\,\Delta\varepsilon_M}{\text{Tesla}} \tag{4.15}$$

4.3. COMPUTER CONTROL AND DATA ACQUISITION

Typical MCD (or CD) instrumentation, such as shown in Figure 4.1, is computer-controlled, and data are acquired, stored, and plotted in digital form. The quality of the measured spectra depends upon the signal-to-noise ratio of the detectors, the spectral bandwidth that determines the spectral resolution, and acquisition data density. The signal-to-noise ratio in a measurement limits the ultimate sensitivity of the measurement and is often limited by the light flux that passes through the sample and reaches the PMT detector. The greater the flux, the better will be the signal-to-noise ratio. Light flux is limited by such experimental parameters as source brightness, the cross-sectional area of the light beam, and the various apertures, including the entrance and exit slits, which limit light passage through the spectrometer. PMT detectors are remarkably sensitive, but signal-to-noise ratio is greatly improved if the light flux reaching the photocathode is large so that the detector can be operated at lower gain (lower high voltage applied to the dynodes). This reduces the contributions from shot noise and thermal noise in the PMT. The computer's analog-to-digital (A/D) signal acquisition must have sufficient dynamic range to detect small changes in signal. Thus a 16-bit A/D converter is better than a 12-bit converter. Ideally, the noise level should be less than the least significant bit of the converter. For example, a 16-bit A/D converter with a full-scale range of 10 volts can distinguish between signals that differ by 0.153 mV (1 part per 65536), but if the converter is only 12 bits, it can only distinguish between 2.44 mV (1 part per 4096). With an intense (bright) light source, a monochromator with high throughput, and low-noise preamplifiers and LIA, it is possible to achieve a sensitivity for ΔA in the range of 10^{-6}–10^{-7}. Spectral bandwidth is largely determined by the dispersion characteristics of the monochromator, which in turn is dependent upon the dispersing element (grating or prism or combination) and the slit width. If the slit width is too narrow (good spectral bandwidth), then there will be greatly reduced light flux (poor signal-to-noise ratio). It is often a compromise between bandwidth concerns and signal-to-noise matters. The compromise is usually determined by the nature of the sample and the objective of the measurements; it is not sensible to use narrow bandwidth for a sample that only exhibit broad spectra, because the signal-to-noise ratio will suffer, and information will be lost on the signal measurement end of the experiment. Finally, the density of the acquired data must similarly be considered against the spectral bandwidth. If the measurement resolution is relatively low, good results might be had with a data density of only 10 points per nanometer of wavelength in the vis and UV region. On the other hand, if the resolution warrants, data resolution can be 1000 points per nanometer or higher. The data resolution is, of course, limited by the smallest wavelength increment possible for the monochromator system, which is usually determined by the size of the smallest step produced by a stepper motor-driven wavelength drive. However, the greater data density requires greater digital storage and greater time to accumulate the spectrum.

4.4. OPTICAL ELEMENTS AND STRAY POLARIZATION EFFECTS

The optical elements, including the mirrors, gratings or prisms of the monochromator, sample cells, and even the sample under study, will introduce some measure of polarization or depolarization of the light as it traverses the optical path. It is, of course, desirable to minimize stray polarization effects from the optical elements, but most polarization effects cannot be eliminated entirely; fortunately, they are constant with time, although they will probably vary with wavelength. Therefore, optical elements should be carefully selected with their polarization characteristics in mind. The net result of stray polarization effects is to alter the measured signal in a more or less constant way. These effects can usually be assessed by determining the background spectrum of the signals in the absence of a sample (an experimental blank). They will be manifested by lack of baseline flatness and uniformity of signal strength. Such effects can be tolerated if they are not too large. By determining the spectrum of the sample and then subtracting a suitable blank, such effects can usually be eliminated because they are present in both the sample and blank. Furthermore, if a sample is optically active (such as a protein or other biomolecule or a chiral complex), then the CD spectrum can be determined with the field off and then subtracted from the MCD + CD spectrum with the field on (the CD and MCD signals are additive). It is always sound practice to determine a solvent (or other supporting matrix) blank and subtract it from the sample spectrum in order to remove any spectral features not due to the sample alone.

MCD measurements are best made for gases, liquid solutions, cubic crystals, or uniform glasses, where the chromophore environment is completely isotropic. Solid samples present severe problems if they are not isotropic. Birefringence (anisotropic index of refraction) in the sample can produce strong spurious signals that are sometimes orders of magnitude larger than the MCD signals and thus completely obliterate the desired measurement. These effects can result from light propagation through biaxial crystals or along a direction different from the unique axis in a uniaxial crystal, or from strains within either a crystal or a noncrystalline matrix. The birefringence in such solids will depolarize the light and alter the linear and circular polarization of the light as it passes through the sample. Depolarization effects can be assessed by placing an optically active sample with a known natural CD spectrum between the MCD sample of interest and the detector. With the field off, the measured CD can then be compared with the known CD of the optically active substance; depolarization by the MCD sample will reduce the expected CD signal. If the depolarization effects are small, corrections can be made; but often the effects will be large, making the MCD measurement impossible. The value of a blank is further emphasized here, because any small stray birefringence in the sample matrix can immediately be determined.

4.5. CALIBRATION

The calibration of CD and MCD spectrometers is usually performed by measuring ΔA for standard solutions. A typical and convenient standard for the UV–vis region is camphor sulfonic acid (CSA), which is a stable optically active organic compound that exhibits a strong positive CD at 290 nm ($\Delta\varepsilon = +3.37\,\mathrm{L\ mol^{-1}\ cm^{-1}}$) and a strong negative CD at 192 nm ($\Delta\varepsilon \approx -6.7\,\mathrm{L\ mol^{-1}\ cm^{-1}}$) (see ref. 5, which includes the literature references for calibration values). MCD measurements, especially those at variable temperature, can be calibrated by measuring the temperature-dependent positive (400 nm) and negative (285 nm) MCD features (C terms) of the $Fe(CN)_6^{3-}$ ion, for example (see Chapter 7, Section 7.1.1).

4.6. MAGNET SYSTEMS

Magnet systems for MCD measurements can be as simple as strong permanent magnets (fields up to ~1 T) with holes bored in tapered poles to allow the light passage, or they can involve electromagnets (fields to 1.5–2 T). The latter are somewhat cumbersome because of the size necessitated by the windings. In both cases the light aperture through the magnet poles should be as large as feasible for a good signal-to-noise ratio for the measurement, but not so large as to create an inhomogeneous field along the sample. Again, trade-off compromise is necessary for good results.

The best magnets are superconducting systems (fields up to 10 T), but these require liquid helium cryogenics for cooling the coils to below the critical temperatures of the windings. Figure 4.3 shows a sketch of a typical single coil superconducting magnet. Current is introduced into the magnet windings by opening the superconducting (SC) switch. The SC switch consists of a length of superconducting wire short-circuiting the magnet windings, which can be heated to slightly above the wire's critical temperature by passing a small current through a heater resistor located next to a portion of the wire, which will make it resistive, thereby

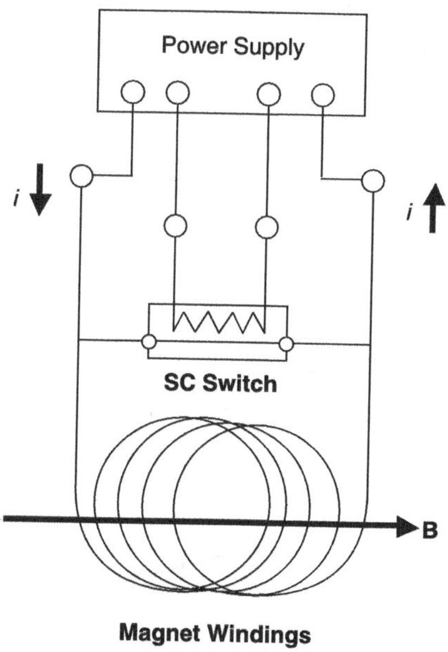

Figure 4.3. Single-coil superconducting magnet.

opening the switch. With the switch open, current from a suitable high-current power supply can then be impressed on the magnet coil. Currents of ~80–120 A (depending upon the magnet specifications) can be used to energize the windings by means of 1–2 V. Once the current is present in the windings, the SC switch is then closed by removing the small heater current, allowing the portion of wire to revert to superconducting and thus closing a current loop consisting of the magnet windings and the superconducting wire of the SC switch. The power supply current can now be removed by lowering the voltage to zero. The current in the windings loop now will persist, as long as the wires remain superconducting. This is the so-called "persistent mode" of operation: Once energized, power from the supply is removed and the field will persist (as long as liquid helium is present to maintain the temperature below the superconducting critical temperature). In order to deenergize the magnet, the power supply is again turned on, and the applied current is matched to that in the superconducting loop. Now the SC switch is opened again and the power supply can remove the current by application of a negative voltage. The single coil design is suitable for samples in cylindrical cells inserted along the coil axis. This design has the advantage that long cells (up to ~10 cm), suitable for gas samples, for example, can be easily accommodated. Figure 4.4 shows a sketch of a split coil design. The total field strength is the sum of the fields produced by two separate coils positioned on either side of the sample. This design has the advantage that a light beam can be oriented both parallel and perpendicularly to the magnet axis; the perpendicular orientation is useful for magnetic linear dichroism MLD measurements

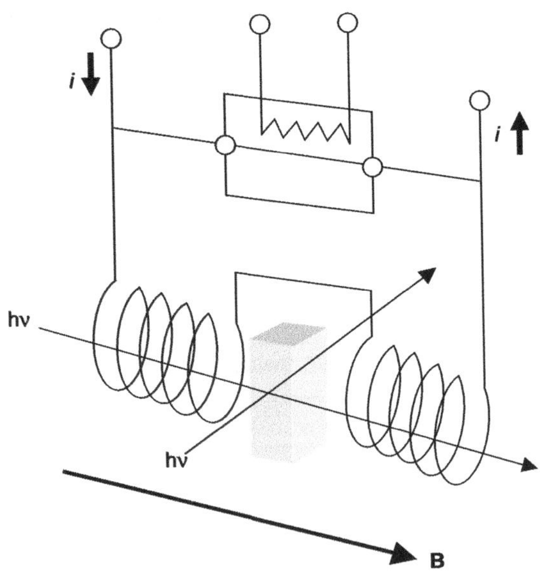

Figure 4.4. Split coil superconducting magnet that allows light propagation through the sample either parallel or perpendicular to the field direction **B**.

(see Chapter 9). This design usually accommodates rectangular cells of 1-cm path, or smaller, and the cells are introduced perpendicularly to the magnet axis.

Stray fields must also be considered when high field superconducting magnets are used. The stray fields can affect the electron trajectory within PMT detectors if they are placed too close to the magnet. Magnetic fields decrease as the cube of the distance from the magnet. Therefore, PMT detectors should be placed some distance (often ~1–2 m) away from high-field magnet systems in order to avoid such stray field problems. Similarly, arc sources are affected by strong fields. A xenon arc source can be extinguished by strong stray fields. Source lamps also must be placed 1–2 m from high-field systems.

Useful superconducting systems are often engineered with variable temperature sample holders where the liquid helium is also used as a sample refrigerant. The best systems are those that immerse the sample in a stream of cold helium gas or liquid helium below the normal boiling point (4.2 K). Temperature can then be controlled from below the lambda point of liquid helium (2.2 K, with pumping) to over 300 K by means of small heating coils attached to the sample holder to counterbalance the refrigeration of the cold helium. For low-temperature work, care must be taken to ensure that the matrix holding the sample does not produce spurious signals due to strain birefringence due to contraction or phase changes upon cooling to very low temperatures.

Finally, the field direction depends upon the direction of the current in the windings. The orientation of the field should be such that the N → S vector points toward the detector. If it points in the opposite direction, the signs of all the MCD signals will be reversed. The signal phase in the lock-in amplifier can be adjusted up to ±180 degrees to compensate for this orientation if necessary. If there is doubt about the orientation or phase setting, a standard spectrum (such as the $Fe(CN)_6^{3-}$ ion, for example, noted above in Section 4.5) can be determined and compared with known signs of the signals.

5 The Interpretation of MCD Spectra

The most important use of MCD spectra in the UV–vis–near-IR region is to assist in the interpretation of electronic spectra and provide experimentally based information about the electronic states involved in the observed transitions. This use requires the companion use of absorption spectra, ideally obtained simultaneously and under the same experimental conditions as the MCD spectra. MCD spectra alone without the companion absorption spectrum can provide nothing more than a "fingerprint" measurement for the system under investigation. Electronic state properties can often be determined by suitable ratios of ΔA MCD data to A absorption data in the form of term ratios described by A_1/D_0, B_0/D_0, and C_0/D_0. These ratios of MCD data to absorption data remove band-shape dependence and, in an approximate way, the medium effects that affect both absorption and MCD measurements in condensed media (solution for example). The relevant ratios are determined from a theoretical model for the transitions visualized for the system, and then compared with experiment.

5.1. TERM ASSIGNMENTS

Among the first tasks of interpreting MCD spectra is the assignment of the appropriate term type. This task can be straightforward if the transitions are separated from others by energies greater than their bandwidths, but can be quite difficult if the spectrum is congested with several overlapping transitions lying within their respective bandwidths. The nature of the chemical system also provides important information. For example, C terms will be absent for diamagnetic systems; paramagnetic systems will exhibit temperature-dependent C terms, which will likely dominate the MCD spectrum; high-symmetry molecules can give rise to electronic degeneracies and therefore to A terms; lower-symmetry molecules will not show true A terms, but may exhibit pseudo A terms (closely spaced B terms of opposite sign) if magnetically interacting states are close in energy (near degeneracy); virtually all transitions will exhibit B terms of some intensity, although they may be overshadowed by more intense A or C terms. The interpretation of MCD spectra requires intelligent consideration of the electronic energy levels of the molecular system, including the

A Practical Guide to Magnetic Circular Dichroism Spectroscopy, by W. Roy Mason

symmetries of the electronic states which are expected in the energy region under investigation. A transition hypotheses can then be advanced and tested within the framework of the molecule's energy level scheme. For example, the sign and the approximate magnitude of the MCD terms expected can be predicted for a given transition and then compared with experiment. Thus, transitions that predict the wrong sign for the observed MCD term can be ruled out, and thereby narrow the possibilities. There is no substitute for experience in the initial approach to interpreting MCD spectra, but in many respects the visualization of a self-consistent and theoretically sound interpretation of observed MCD and absorption spectra is an "art form."

5.2. PREDICTION OF SIGN AND APPROXIMATE MAGNITUDE OF MCD TERMS

In order to predict MCD term sign and magnitude, the matrix elements in the appropriate term equations must be evaluated. This task can be quite complicated if the symmetry is low or if the orbitals involved in the transition are heavily mixed. In cases where orbitals can be approximated as nearly pure atomic orbitals and the molecular symmetry is high, elegant symmetry techniques are available to assist in the evaluation of the relevant matrix elements. In some high-symmetry cases the signs of the MCD terms are completely symmetry-determined and can be evaluated relatively easily by using tables of symmetry coupling coefficients. The development of the irreducible tensor theory of symmetry coupling coefficients is beyond the scope of the present text; a very complete description is given by Piepho and Schatz [ref. 1, Chapters 8–20]. A more descriptive development of matrix element evaluation will be presented here with the goal of determining MCD term signs using simplified orbital approximations for non-complicated, high-symmetry cases (simply reducible ambivalent groups—those which do not have complex irreps or repeated representations).

5.3. THE WIGNER–ECKART THEOREM AND REDUCED MATRIX ELEMENTS (RMEs)

The matrix elements in the MCD term equations (Eqs. (3.15), (3.18)–(3.20) for isotropic molecules of O or T symmetry, and Eqs. (3.23)–(3.26) for space-averaged nonisotropic molecules) can be converted to reduced matrix elements (RMEs) by using the Wigner–Eckart theorem. This theorem relates a component-dependent matrix element to a component-independent RME together with a $3jm$ coupling coefficient and a $2jm$ phase factor:

$$\langle a\alpha|O^f_\phi|b\beta\rangle = \begin{pmatrix} a \\ \alpha \end{pmatrix}\begin{pmatrix} a^* & f & b \\ \alpha^* & \phi & \beta \end{pmatrix}\langle a\|O^f\|b\rangle \tag{5.1}$$

matrix element	$2jm$ phase	$3jm$	RME

where α, β, and ϕ are components (irrep partners) corresponding to the a, b, and f irreps for the functions a and b and for the operator O, respectively; the RME is noted by the double vertical lines surrounding the operator. The $2jm$ phase and the $3jm$ coefficients are symmetry-dependent quantities. Some quantities may be imaginary and the asterisk (*) indicates a conjugate value. The advantage of using RMEs and coupling coefficients is that not all component combinations need to be considered. One needs only a table of coefficients and a single value for the RME in order to write the matrix element for any components. The $3jm$ values that are not included in tables are assumed to be zero, which indicates that the components for that case give a zero matrix element. The value of the RME can be determined from a single, convenient set of components by means of Eq. (5.1). Tables of $3jm$ values and $2jm$ phases for single-valued irreps for O, T_d, D_4, and D_3 are found in Appendixes A, B, and C at the end of this book. More extensive tables, including values for double-group irreps that are needed for half-integral spin cases, are found in the appendixes of Piepho and Schatz [ref. 1], or in Butler's book [ref. 6]. Care must be taken, however, to use only tables based on the same phase assumptions when working problems—for example, the standard basis adopted in ref. 1—in order to be self-consistent.

5.4. MCD-TERM EQUATIONS INVOLVING RMEs

The properties of the $3jm$ coefficients allow some simplification to be made for the MCD-term equations. These involve the consolidation of several $3jm$ coefficients into $6j$ coefficients, which can also be tabulated, but greatly simplify the expressions. Furthermore, several phase factors associated with the basis chosen can be combined into a single numerical factor that is also symmetry-dependent. Again the use of these factors assumes that the underlying assumptions as to basis functions remain the same throughout. The MCD-term equations are given here for simply reducible groups (single-valued irreps) for the transition $A \rightarrow J$ for A and C terms and for $A \rightarrow J$ and $A \rightarrow K$ for B terms in Eqs. (5.2)–(5.4).

$$A_1 = [(\text{MCD factor})2\{A\}/(\mu_B|A|)\langle J\|\mu^{f'}\|J\rangle \begin{Bmatrix} f' & f & f \\ A^* & J^* & J^* \end{Bmatrix} \times \langle A\|m^f\|J\rangle\langle J\|m^f\|A\rangle] + C_0 \tag{5.2}$$

$$\begin{aligned} B_0 = \text{Re}\Bigg\{&(\text{MCD factor})4/(\mu_B|A|)\Bigg[\Bigg(\sum_{K\neq A}\{J\}/(W_K - W_A)\langle K\|\mu^{f'}\|A\rangle \\ &\times \begin{Bmatrix} f' & f & f \\ J^* & A^* & k^* \end{Bmatrix}\langle A\|m^f\|J\rangle\langle J\|m^f\|K\rangle\Bigg) \\ &+\Bigg(\sum_{K\neq J}\{A\}/(W_K - W_J)\langle J\|\mu^{f'}\|K\rangle \begin{Bmatrix} f' & f & f \\ A^* & K^* & J^* \end{Bmatrix} \\ &\times \langle A\|m^f\|J\rangle\langle K\|m^f\|A\rangle\Bigg)\Bigg]\Bigg\} \end{aligned} \tag{5.3}$$

$$C_0 = (\text{MCD factor})\ 2\{J\}/(\mu_B|A|)\langle A\|\mu^{f'}\|A\rangle \begin{Bmatrix} f' & f & f \\ J^* & A^* & A^* \end{Bmatrix}$$
$$\times \langle A\|m^f\|J\rangle\langle J\|m^f\|A\rangle \tag{5.4}$$

In these equations the MCD factor expresses the μ_0 and $m_{\pm 1}$ operators in terms of the symmetry group basis operators and is given by

$$\text{MCD factor} = \langle f'\phi'_0|\mu_0\rangle\langle f\phi_1|m_1\rangle\langle f\,\phi_{-1}|m_{-1}\rangle \begin{pmatrix} f' & f & f \\ \phi 0' & \phi - 1 & \phi 1 \end{pmatrix} \tag{5.5}$$

where the factors $\langle f'\phi'_0|\mu_0\rangle$, $\langle f_{\phi 1}|m_1\rangle$, and $\langle f\phi_{-1}|m_{-1}\rangle$ are the operator basis transformation coefficients and $\begin{pmatrix} f' & f & f \\ \phi 0' & \phi - 1 & \phi 1 \end{pmatrix}$ is a $3jm$ for the operator irreps and partners. The quantities within $\{\ldots\}$—for example, $\begin{Bmatrix} f' & f & f \\ A^* & J^* & J^* \end{Bmatrix}$—are $6j$ symbols (the elements within the $6j$ symbol are irreps for the operators f' and f, and for the states A and J), and symbols between vertical lines—$|A|$, for example—indicate the dimension of the irrep to which the state belongs. The $2j$ phase, $\{A\} = \{J\} = 1$ for single-valued irreps. Finally the quantity D_0 is given by

$$D_0 = 1/(|f||A|)\delta(A^*fJ)|\langle A\|m^f\|J\rangle|^2$$
$$= -\{A\}/(|f||A|)\delta(A^*fJ)\{A^*fJ\}\langle A\|m^f\|J\rangle\langle J\|m^f\|A\rangle \tag{5.6}$$

where the quantity $\delta(A^*fJ)$ is 1 if the direct product $f \times J$ contains A; otherwise it is 0. The quantity $\{A^*fJ\}$ is a symmetry-dependent $3j$ phase factor. By using Eqs. (5.2), (5.4), and (5.6), expressions can be written for A_1/D_0 and C_0/D_0 for simply reducible groups where the irreps of A and J are single-valued. These expressions are given in Eqs. (5.7) and (5.8):

$$A_1/D_0 = -(\text{MCD factor})\ (2|f|/\mu_B)\{AfJ\}\left(\begin{Bmatrix} f' & f & f \\ A & J & J \end{Bmatrix}\langle J\|\mu^{f'}\|J\rangle\right.$$
$$\left.+\begin{Bmatrix} f' & f & f \\ J & A & A \end{Bmatrix}\langle A\|\mu^{f'}\|A\rangle\right) \tag{5.7}$$

$$C_0/D_0 = -(\text{MCD factor})(2|f|/\mu_B)\{AfJ\}\begin{Bmatrix} f' & f & f \\ J & A & A \end{Bmatrix}\langle A\|\mu^{f'}\|A\rangle \tag{5.8}$$

Equations (5.2)–(5.8) are for the oriented or isotropic case. If the molecule of interest does not belong to a cubic group (O, O_h, or T_d) or is in an oriented matrix environment, then space-averaged term equations must be used [see Eqs. (3.23)–(3.26)]. The MCD equations for the space-averaged case involving simply reducible groups are given below in Eqs. (5.9)–(5.13).

$$\overline{A}_1 = \left[\sum_{ijk} \varepsilon_{ijk}(\text{MCD}_{ijk}\text{ factor})\{A\}/(3\mu_B|A|)\langle J\|\mu^{fi'}\|J\rangle \begin{Bmatrix} fi' & fj & fk \\ A^* & J^* & J^* \end{Bmatrix} \right.$$
$$\left. \times \langle A\|m^{fj}\|J\rangle\langle J\|m^{fk}\|A\rangle \right] + \overline{C}_0 \qquad (5.9)$$

$$\overline{B}_0 = \text{Re}\left\{ \sum_{ijk} \varepsilon_{ijk}(\text{MCD}_{ijk}\text{ factor})2/(3\mu_B|A|)\left[\left(\sum_{K\neq A} \{J\}/(W_K - W_A)\langle K\|\mu^{fi'}\|A\rangle \right.\right.\right.$$
$$\left. \times \begin{Bmatrix} fi' & fj & fk \\ J^* & A^* & K^* \end{Bmatrix} \langle A\|m^{fj}\|J\rangle\langle J\|m^{fk}\|K\rangle \right) + \left(\sum_{K\neq J} -\{A\}/(W_K - W_J) \right.$$
$$\left.\left.\left. \times \langle J\|\mu^{fi'}\|K\rangle \begin{Bmatrix} fi' & fj & fk \\ A^* & K^* & J^* \end{Bmatrix} \langle A\|m^{fj}\|J\rangle\langle K\|m^{fk}\|A\rangle \right) \right] \right\} \qquad (5.10)$$

$$\overline{C}_0 = \sum_{ijk} \varepsilon_{ijk}(\text{MCD}_{ijk}\text{ factor})\{J\}/(3\mu_B|A|)\langle A\|\mu^{fi'}\|A\rangle \begin{Bmatrix} fi' & fj & fk \\ J^* & A^* & A^* \end{Bmatrix}$$
$$\times \langle A\|m^{fj}\|J\rangle\langle J\|m^{fk}\|A\rangle \qquad (5.11)$$

$$\overline{D}_0 = -\{A\}/(3|A|)\sum_j 1/|f_j|\delta(A^* f_j J)\{A^* f_j J\}\langle A\|m^{fj}\|J\rangle\langle J\|m^{fj}\|A\rangle \qquad (5.12)$$

where the ijk sums are over i, j, and k, each equal to -1, 0, and 1, respectively, and ε_{ijk} is an alternating tensor as defined for Eq. (3.22). The MCD_{ijk} factor is given by

$$\text{MCD}_{ijk}\text{ factor} = \langle f_i'\phi_i'|\mu_i\rangle\langle f_j\phi_j|m_j\rangle\langle f_k\phi_k|m_k\rangle \begin{pmatrix} f_i' & f_k & f_j \\ \phi_i & \phi_k & \phi_j \end{pmatrix} \qquad (5.13)$$

If a real basis set is used, then the sums taken over i, j, and k are each equal to x, y, and z, respectively, and the MCD_{ijk} factor must be multiplied by $i = (-1)^{1/2}$. As before, the space-averaged MCD terms for the cubic O, O_h, and T_d groups are the same as for the oriented or isotropic case: $\overline{A}_1 = A_1$; $\overline{B}_0 = B_0$; $\overline{C}_0 = C_0$; and $\overline{D}_0 = D_0$. For lower-symmetry groups, the values for the oriented and space averaged cases are not the same.

In order to illustrate, and to give an example of the use of the MCD term equations, consider D_4 symmetry with $A = A_1$ and $J = E$. In this case, C_0 or $\overline{C}_0$ are zero because A is nondegenerate and therefore $\langle A_1|\mu|A_1\rangle = 0$. For the space-averaged case using Eqs. (5.9) and (5.12) (the MCD factor here has been expanded for clarity), we have

$$\overline{A}_1 = \langle A_2 a_2|\mu_0\rangle\langle E1|m_1\rangle\langle E-1|m_{-1}\rangle \begin{pmatrix} A_2 & E & E \\ \alpha_2 & -1 & 1 \end{pmatrix} 2/(3\mu_B)$$
$$\times \begin{Bmatrix} A_2 & E & E \\ A_1 & E & E \end{Bmatrix} \langle E\|\mu^{A_2}\|E\rangle \times \langle A_1\|m^E\|E\rangle\langle E\|m^E\|A_1\rangle$$
$$\overline{D}_0 = -(1/3)\langle A_1\|m^E\|E\rangle\langle E\|m^E\|A_1\rangle \qquad (5.14)$$

The symmetry-determined operator transformation coefficients (first three terms in the $\overline{A}_1$ equation above) are found in Appendix B, Table B.1. The $3j$, $2j$, and $2jm$ phase factors are given in Table B.2, while the $3jm$, $\begin{pmatrix} A_2 & E & E \\ \alpha_2 & -1 & 1 \end{pmatrix} = -(2)^{-1/2}$, can then be obtained from Table B.3. Finally the $6j$, $\begin{Bmatrix} A_2 & E & E \\ A_1 & E & E \end{Bmatrix} = -1/2$ can be found in Table B.4. When the appropriate substitutions are made, we have the following:

$$\begin{aligned}\overline{A}_1/\overline{D}_0 &= (+1)(-1)(-1)(-2^{-1/2})(2/3\mu_B)(-1/2) \\ &\quad \times \langle A_1\|m^E\|E\rangle\langle E\|m^E\|A_1\rangle/(-1/3)\langle A_1\|m^E\|E\rangle\langle E\|m^E\|A_1\rangle \\ &= -1/(2^{1/2}\mu_B)\langle E\|\mu^{A_2}\|E\rangle \end{aligned} \tag{5.15}$$

The determination of A_1/D_0 for the oriented case by means of Eq. (5.7) gives

$$\begin{aligned} A_1/D_0 &= -(\text{MCD factor})(2|E|/\mu_B)\{A_1EE\}\left(\begin{Bmatrix} A_2 & E & E \\ A_1 & E & E \end{Bmatrix}\langle E\|\mu^{A_2}\|E\rangle\right) \\ &= -(-2^{-1/2})(2(2)/\mu_B)(+1)(-1/2)\langle E\|\mu^{A_2}\|E\rangle \\ &= -2/(2^{1/2}\mu_B)\langle E\|\mu^{A_2}\|E\rangle \end{aligned} \tag{5.16}$$

Thus the A_1/D_0 ratio for the oriented case is twice that of the $\overline{A}_1/\overline{D}_0$ ratio for the space-averaged case for D_4.

5.5. EVALUATION OF RMEs FOR MCD TERMS

The determination of the sign and approximate magnitude of MCD terms, or the ratios A_1/D_0 or C_0/D_0, involves the evaluation of the RMEs for the excited state $\langle J\|\mu^f\|J\rangle$ and/or ground state $\langle A\|\mu^f\|A\rangle$. These are, in general, multielectronic RMEs and must first be reduced to one-electron RMEs, and then by approximating the wavefunctions in terms of atomic orbitals, the one-electron RMEs must then be evaluated. This process involves several steps, depending upon the complexity of the model and whether or not spin is a contributor to the states' total angular momentum. Thus, for example, if spin singlet ($S = 0$) states are under consideration, the operator μ^f can be written simply as $\mu^f = -\mu_B L^f$, but if the spin multiplicity is greater than one, the appropriate operator is $\mu^f = -\mu_B(L + 2S)^f$. Thus even nondegenerate states with zero orbital angular momentum, but with spin angular momentum, can exhibit degeneracies and therefore A or C terms. The symmetry properties of a given electronic state can be viewed as the coupling of the symmetry properties of all of the electrons, including their spin, that make up the configuration which gives rise to the state in question. Completely filled orbitals, of course, are totally symmetric and do not contribute to the overall symmetry of the state. Partly filled orbitals determine the symmetry properties of the electronic state. The simplification process can thus take advantage of the coupling properties of the electronic wavefunctions of the partly filled orbitals, together with the symmetry of the angular momentum operators. The process, which is developed and described in some detail in Piepho and Schatz [ref. 1, Chapters 18–20], can be expressed in terms of symmetry irreps for the electronic functions (from the model) and coupling coefficients and phase factors (from appropriate tables), much the same way as the MCD term equations were assembled.

The evaluation the RMEs in a spin–orbit coupled $|(\mathcal{S}S,h)t\rangle$ basis for simply reducible groups that do not have repeated irreps can be described by the following steps: (1) the separation of spin and orbital parts of the RME of the magnetic moment μ^f; (2) the reduction of multielectronic state functions to one-electron RMEs; and (3) the evaluation of the one-electron RMEs. The first step is given by Eqs. (5.17) and (5.18), where $\mathcal{S}$ refers to spherical SO_3 symmetry and S, h, and t refer to the spin, orbital, and spin–orbit irreps, respectively, of the symmetry group of interest ([ref. 1, Eqs. (18.1.8) and (18.1.9)], which are more general and cover non-simply reducible groups and those with repeated representations):

$$\langle(\mathcal{S}S,h)t\|O^f\|(\mathcal{S}'S',h')t'\rangle = \delta_{\mathcal{S}\mathcal{S}'}\delta_{SS'}|t|^{1/2}|t'|^{1/2}\{S'h't'\}\{hfh'\} \times \begin{Bmatrix} f & t' & t \\ S & h & h' \end{Bmatrix}\langle h\|O^f\|h'\rangle \tag{5.17}$$

$$\langle(\mathcal{S}S,h)t\|S^f\|(\mathcal{S}'S',h')t'\rangle = \delta_{hh'}|t|^{1/2}|t'|^{1/2}\{Sht\}\{S'\}\{SfS'\} \times \begin{Bmatrix} f & t' & t \\ h & S & S' \end{Bmatrix}\langle\mathcal{S}S\|S^f|\mathcal{S}'S'\rangle \tag{5.18}$$

The second step for the orbital RME $\langle h\|O^f\|h'\rangle$ involves the reduction to one-electron RMEs and is given by Eq. (5.19) [ref. 1, Eq. (20.1.7)]:

$$\begin{aligned}
&\langle\mathscr{A}(a^m(\mathcal{S}_1h_1), b^n(\mathcal{S}_2h_2))\mathcal{S}h\mathscr{M}\|O^f\|\mathscr{A}(a^m(\mathcal{S}_1'h_1'), b^n(\mathcal{S}_2'h_2'))\mathcal{S}'h'\mathscr{M}'\rangle \\
&= \delta(\text{spin})[\delta_{h_2h_2'}|h|^{1/2}|h'|^{1/2}\{h_1h_2h\}\{h_1fh_1'\}\begin{Bmatrix} f & h' & h \\ h_2 & h_1 & h_1' \end{Bmatrix} \\
&\times g(a^m,\mathcal{S}_1h_1fh_1')\langle a\|o^f\|a\rangle + \delta_{h_1h_1'}|h|^{1/2}|h'|^{1/2}\{h_1'h_2'h'\}\{h_2fh_2'\} \\
&\times \begin{Bmatrix} f & h' & h \\ h_1 & h_2 & h_2' \end{Bmatrix} g(b^n, \mathcal{S}_2h_2fh_2')\langle b\|o^f\|b\rangle]
\end{aligned} \tag{5.19}$$

In Eq. (5.19), $\mathscr{A}(\ldots)$ indicates antisymmetrization of the multielectronic configuration from which h is derived; $\langle a\|o^f\|a\rangle$ and $\langle b\|o^f\|b\rangle$ are one-electron RMEs; and $g(a^m,\mathcal{S}_1h_1fh_1')$ and $g(b^n,\mathcal{S}_2h_2fh_2')$ are coefficients of fractional parentage (cfp's) that are symmetry-determined factors that relate electron configurations to "hole" configurations. The next step is to unreduce the one-electron RMEs by using the Wigner–Eckhart theorem, Eq. (5.1). Orbital approximations must be made, and then a convenient case involving specific orbital components must be selected in order to use Eq. (5.1). However, once the RME is evaluated for one set of components, then the matrix element for any set of components may be evaluated with the use of a an appropriate $3jm$ table. The $\langle \mathcal{S}S\|S^f\|\mathcal{S}'S'\rangle$ spin RME of Eq. (5.18) can be conveniently evaluated by relating the spin in the group G of interest to the spherical SO_3 symmetry group of all rotations. This procedure is described in ref. 1, Chapter 15 and in ref. 6 and uses tables of $3jm$ factors that relate symmetries in a chain from the group SO_3 to G. For example, if a chain is described by $SO_3 \supset G_1 \supset G_2 \supset G$, three $3jm$ factors are used to relate the RME in the group G to the RME in SO_3, by using Eq. (5.20):

$$\langle \mathcal{S}S_1S_2S_G\|S^{ff_1f_2f_G}\|\mathcal{S}''S_1S_2S_G\rangle^G = \underset{3jm\text{ factor}}{\begin{pmatrix} \mathcal{S} & f & \mathcal{S} \\ a_1 & f_1 & b_1 \end{pmatrix}\begin{matrix} SO_3 \\ G_1 \end{matrix}} \underset{3jm\text{ factor}}{\begin{pmatrix} a_1 & f_1 & b_1 \\ a_2 & f_2 & b_2 \end{pmatrix}\begin{matrix} G_1 \\ G_2 \end{matrix}}$$

$$\times \underset{3jm\text{ factor}}{\begin{pmatrix} a_2 & f_2 & b_2 \\ a_G & f_G & b_G \end{pmatrix}\begin{matrix} G_2 \\ G \end{matrix}} \langle \mathcal{S}\|S^f\|\mathcal{S}'\rangle^{SO_3} \tag{5.20}$$

Once the spin RME in the group G, $\langle \mathcal{S}S_1S_2S_G\|S^{ff_1f_2f_G}\|\mathcal{S}''S_1S_2S_G\rangle^G$ is related to SO_3, $\langle \mathcal{S}\|S^f\|\mathcal{S}'\rangle^{SO_3}$ then the latter can be evaluated by Eq. (5.21):

$$\langle \mathcal{S}\|S^f\|\mathcal{S}''\rangle^{SO_3} = \delta_{\mathcal{S}\mathcal{S}'}[\mathcal{S}(\mathcal{S}+1)(2\mathcal{S}+1)]^{1/2} \tag{5.21}$$

In order to illustrate these steps, consider the evaluation of the $\overline{A}_1/\overline{D}_0$ ratio for the D_4 symmetry case introduced in Eq. (5.15). In a planar metal complex with nd^8 electronic configuration, an $^1A_1 \rightarrow {}^1E$ transition may be visualized from the excitation from the occupied nd_{xz}, nd_{yz} degenerate e orbitals to the non degenerate $a_2(n+1)p_z$ orbital: $(a_1)^2(e)^4(b_2)^2 \rightarrow (a_1)^2(e)^3(b_2)^2(a_2)$. This is shown in Figure 5.1. In this case the operator $\mu^{A_2} = -\mu_B L^{A_2}$ is required since only singlet states are involved, and the multielectronic $\langle {}^1E\|\mu^{A_2}\|{}^1E\rangle$ RME must be reduced to a one-electron RME $\langle e\|\mu^{A_2}\|e\rangle$ and finally evaluated. The first step is to uncouple the spin from the overall state function (application of Eq. (5.17), using the dimension of $E(=2)$ and Tables B.2 and B.4):

$$\begin{aligned}
\langle {}^1E\|\mu^{A_2}\|{}^1E\rangle &= \langle (0A_1,E)E\|\mu^{A_2}\|(0A_1,E)E\rangle = \delta_{00}\delta_{A_1A_1}|E|^{1/2}|E|^{1/2} \\
&\quad \times \{A_1EE\}\{EA_2E\}\begin{Bmatrix} A_2 & E & E \\ A_1 & E & E \end{Bmatrix}\langle E\|\mu^{A_2}\|E\rangle \\
&= (1)(1)(2^{1/2})(2^{1/2})(+1)(-1)(-1/2)\langle E\|\mu^{A_2}\|E\rangle \\
&= \langle E\|\mu^{A_2}\|E\rangle
\end{aligned}$$

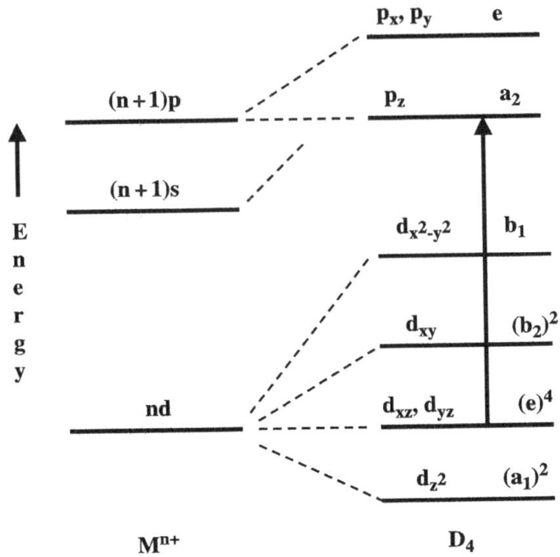

Figure 5.1. Energy levels for a $nd^8 D_4$ metal complex. The parentheses and superscripts around the irreps on the right side indicate the electron occupancy of the ground state.

where the singlet spin function $S = 0$ belongs to the A_1 irrep of D_4. The next step is to reduce the $\langle E\|\mu^{A_2}\|E\rangle$ RME to one-electron form (application of Eq. (5.19), again using Tables B.2 and B.4):

$$\langle (e^3(1/2e), a_2(1/2a_2)E\|\mu^{A_2}\|(e^3(1/2e), a_2(1/2a_2)E\rangle = \delta_{\text{spin}}\delta_{a_2a_2}|E|^{1/2}|E|^{1/2}$$

$$\times \{EA_2E\}\{EA_2E\}\begin{Bmatrix} A_2 & E & E \\ A_2 & E & E \end{Bmatrix} g(e^3,1/2ea_2e)\langle E\|\mu^{A_2}\|E\rangle + 0$$

$$= (1)(1)(2^{1/2})(2^{1/2})(-1)(-1)(+1/2)(+1)\langle e\|\mu^{A_2}\|e\rangle = \langle e\|\mu^{A_2}\|e\rangle$$

where

$$g(e^3, 1/2ea_2e) = -(-1)^{a_2} g(e,1/2ea_2e) = +1$$

The final step is to unreduce the one-electron $\langle e\|\mu^{A_2}\|e\rangle$ RME. This may be accomplished by using the Wigner–Eckart theorem, Eq. (5.1), together with Tables B.1, B.2, and B.3.

$$\langle e\|\mu^{A_2}\|e\rangle = \langle e_x| - \mu_B l_z^{A_2}|e_y\rangle \Big/ \left[\begin{pmatrix} e \\ x \end{pmatrix}\begin{pmatrix} e & A_2 & e \\ x & a_2 & y \end{pmatrix}\right] = \langle e_x| - \mu_B l_z^{A_2}|e_y\rangle(-1)(2^{1/2})$$

$$= (2^{1/2})\mu_B\langle e_x|l_z^{A_2}|e_y\rangle$$

If the orbitals of the ground configurations can be approximated by metal-centered atomic nd orbitals and the a_2 orbital of the excited configuration as an atomic $(n+1)p$ orbital for example, then the angular momentum matrix element can be evaluated. Using a standard basis for the atomic orbital functions (Table B.1), we have

$$\langle e_x | l_z^{A_2} | e_y \rangle = \langle i d_{yz} | l_z^{A_2} | d_{zx} \rangle = (i)(i) = -1$$

Therefore, $\bar{A}_1/\bar{D}_0$ will be given from Eq. (5.15) by

$$\bar{A}_1/\bar{D}_0 = -1/2(2^{1/2}\mu_B)\langle E \| \mu^{A_2} \| E \rangle = -1/(2^{1/2}\mu_B)(2^{1/2})\mu_B(-1) = +1$$

Thus the MCD for this transition should exhibit a positive A term, and the experimental $\bar{A}_1/\bar{D}_0$ ratio should be near +1.

5.6. EVALUATION OF MATRIX ELEMENTS FOR LCAO–MO FUNCTIONS

It should be noted in passing that the evaluation of the $\langle e_x | l_z^{A_2} | e_y \rangle$ matrix element in this case is quite easy because the e_x and e_y functions are atomic functions centered on the atom at the center of the coordinate system (the metal at the center of the square complex). In general, the wavefunctions for a molecule will be LCAO–MO functions and will contain contributions from several atoms in the molecule, some or all of which will not be located at the center of the molecular coordinate system. The LCAO functions are of the form

$$|a\alpha\rangle = \sum c_i \phi_i \quad \text{and} \quad |b\beta\rangle = \sum c_j \phi_j$$

where ϕ_i and ϕ_j are atomic orbitals on centers i and j, respectively. The angular momentum matrix elements will be given by

$$\langle a\alpha | l_\gamma | b_\beta \rangle = \sum_{ij} (c_i)^* c_j \langle \phi_i | l_\gamma | \phi_j \rangle$$

If ϕ_i and ϕ_j are on different atomic centers then to a first approximation we can assume that

$$\langle \phi_i | l_\gamma | \phi_j \rangle \approx 0 \qquad \text{for } i \neq j$$

These matrix elements are proportional to the overlap integral $\langle \phi_i | \phi_j \rangle$ and will undoubtedly be small compared to the nonzero elements on a single center $\langle \phi_k | l_\gamma | \phi_k \rangle$, for example. These single-centered elements must still be evaluated even though the center k and γ are different. Thus there is a dependence on the molecular geometry, but the evaluation usually involves expressing the operator l_γ in terms of the coordinate origin of ϕ_k. This may be accomplished through the definitions of angular momentum operators:

$$l_x = -i\hbar(y\partial/\partial z - z\partial/\partial y)$$
$$l_y = -i\hbar(z\partial/\partial x - x\partial/\partial z)$$
$$l_z = -i\hbar(x\partial/\partial y - y\partial/\partial x)$$

by substituting the coordinates at center k in terms of center γ. For example, if atom k is displaced from the origin (center γ) along the $+z$ axis by a constant distance R, it can be shown that $x = x_k$, $y = y_k$, and $z = z_k + R$, and therefore $\partial/\partial x = \partial/\partial x_k$, $\partial/\partial_y = \partial/\partial_{y_k}$, and $\partial/\partial z = \partial/\partial z_k$. Then we have the following operators in terms of center k:

$$l_x = l_{x_k} + i\hbar R\,\partial/\partial y_k$$
$$l_y = l_{y_k} - i\hbar R\,\partial/\partial x_k$$
$$l_z = l_{z_k}$$

Therefore if $\gamma = z$ we have simply

$$\langle\phi_k|l_z|\phi_k\rangle = \langle\phi_k|l_{z_k}|\phi_k\rangle$$

but if we require $\gamma = x$, for example, we would have

$$\langle\phi_k|l_x|\phi_k\rangle = \langle\phi_k|l_{x_k}|\phi_k\rangle + i\hbar R\,\langle\phi_k|\partial/\partial_{y_k}|\phi_k\rangle$$

and a specific form of the atomic orbital radial function would be required for evaluation. Often an intelligent choice of coordinate system will allow simplification.

Returning to symmetry aspects, it is worthwhile to note that if the ${}^1A_1 \rightarrow {}^1E$ transition discussed above were to result from the excitation $(a_1)^2(e)^4(b_2)^2 \rightarrow (a_1)^2(e)^3(b_2)^2(b_1)$, where the initially empty b_1 orbital has the symmetry of the $nd_{x^2-y^2}$ orbital, then the reduction of the $\langle {}^1E\|\mu^{A_2}\|{}^1E\rangle$ RME to one-electron RME would proceed in an identical way except there would be a sign change:

$$\langle(e^3(1/2e), b_1(1/2b_1)E\|\mu^{A_2}\|(e^3(1/2e), b_1(1/2b_1)E\rangle = \delta_{\text{spin}}\delta_{b_1b_1}|E|^{1/2}|E|^{1/2}$$
$$\times\ \{EB_1E\}\{EA_2E\}\begin{Bmatrix} A_2 & E & E \\ B_1 & E & E \end{Bmatrix} g(e^3, 1/2ea_2e)\langle E\|\mu^{A_2}\|E\rangle$$
$$= (1)(1)(2^{1/2})(2^{1/2})(+1)(-1)(+1/2)(+1)\langle e\|\mu^{A_2}\|e\rangle$$
$$= -\langle e\|\mu^{A_2}\|e\rangle$$

This change is due to the 3j phase factor (Table B.2) $\{EB_1E\} = +1$ instead of $\{EA_2E\} = -1$ when the empty orbital was a_2 symmetry in the above case. All of the other factors, including the 6j, have the same values as before. The evaluation of the one-electron RME is the same as before, and therefore the $\overline{A}_1/\overline{D}_0$ ratio will be -1 in this case. Thus the sign is affected by the symmetry of the empty orbital into which the electron is excited, but the magnitude is determined by the angular momentum of the degenerate occupied e orbitals. Therefore the sign of the A term in this case can be a useful probe of orbital symmetry and the excited-state configuration.

5.7. SPIN–ORBIT COUPLING CONSIDERATIONS

In those cases where spin–orbit coupling is strong and states can be represented by the coupling of spin functions with orbital functions, a careful consideration of the nature of the coupling can be helpful in determining whether spin angular momentum plays a role in determining MCD terms. For example, the triplet spin functions in D_4 symmetry belong to A_2 and E irreps. In the $(e)^4 \to (e)^3(a_2)$ case above, the ^3E state will give an E spin–orbit state from $A_2 \times E$ and will give A_1, A_2, B_1, and B_2 spin–orbit states from $E \times E$. The degenerate $E(^3E)$ spin–orbit state should give an MCD A term, but the degeneracy results from the orbital part of the function, and not the spin. In the first step where the spin and orbital parts of the RME are separated, we will have with the triplet spin function $S = 1$ which belongs to the A_2 irrep:

$$\begin{aligned}
\langle {}^3E \| \mu^{A_2} \| {}^3E \rangle &= \langle (1A_2, E)E \| \mu^{A_2} \| (1A_2, E)E \rangle = \delta_{11}\delta_{A_2A_2}|E|^{1/2}|E|^{1/2} \\
&\quad \times \{A_2EE\}\{EA_2E\} \begin{Bmatrix} A_2 & E & E \\ A_2 & E & E \end{Bmatrix} \langle E \| \mu^{A_2} \| E \rangle \\
&= (1)(1)(2^{1/2})(2^{1/2})(-1)(-1)(+1/2)\langle E \| \mu^{A_2} \| E \rangle \\
&= \langle E \| \mu^{A_2} \| E \rangle = \langle E \| -\mu_B L^{A_2} \| E \rangle
\end{aligned}$$

The remainder of the evaluation of the $\bar{A}_1/\bar{D}_0$ ratio will be the same as before.

If, on the other hand, we consider the spin–orbit states that arise, for example, from the excitation $(a_1)^2 \to (a_1)(a_2)$ in D_4, then we obtain $A_2(^1A_2)$, $A_1(^3A_2)$, and $E(^3A_2)$. The degeneracy of the $E(^3A_2)$ spin–orbit state results from the degenerate E spin function and the non degenerate A_2 orbital function. In this case the MCD term must result from spin angular momentum. The separation of the spin and orbital parts of the RME shows that the orbital RME is zero. Not only is $\langle A_2 \| \mu^{A_2} \| A_2 \rangle = 0$, but also the required $6j$ symbol is $\begin{Bmatrix} A_2 & E & E \\ E & E & E \end{Bmatrix} = 0$. The spin RME is determined [Eq. (5.18)] as

$$\begin{aligned}
\langle E(^3A_2) \| \mu^{A_2} \| E(^3A_2) \rangle &= \langle (1E, A_2)E \| \mu^{A_2} \| (1E, A_2)E \rangle = \delta_{EE}|E|^{1/2}|E|^{1/2} \\
&\quad \times \{EA_2E\}\{E\}\{EA_2E\} \begin{Bmatrix} A_2 & E & E \\ A_2 & E & E \end{Bmatrix} \langle 1E \| \mu^{A_2} \| 1E \rangle \\
&= (1)(2^{1/2})(2^{1/2})(-1)(1)(-1)(+1/2)\langle 1E \| \mu^{A_2} \| 1E \rangle \\
&= \langle 1E \| -\mu_B (2S)^{A_2} \| 1E \rangle
\end{aligned}$$

This spin-only RME can be evaluated directly by the Chain of Groups method [ref. 1, Chapter 15; ref. 6] by means of the $SO_3 \supset O \supset D_4$ chain [see Eq. (5.20)]. This requires two $3jm$ factors that relate the irreps of the higher symmetry group (SO_3, spherical group of all rotations) to lower ones (O and D_4 in this case) (Appendix D, Tables D.1 and D.2):

$$\langle 1E\| -\mu_B(2S)^{A_2}\|1E\rangle = -2\mu_B \underbrace{\begin{pmatrix} 1 & 1 & 1 \\ T_1 & T_1 & T_1 \end{pmatrix}^{SO_3}_{O}}_{3jm\text{ factor}} \underbrace{\begin{pmatrix} T_1 & T_1 & T_1 \\ E & A_2 & E \end{pmatrix}^{O}_{D_4}}_{3jm\text{ factor}} \langle 1\|S^1\|1\rangle^{SO_3}$$

By substituting $3jm$ factors $\begin{pmatrix} 1 & 1 & 1 \\ T_1 & T_1 & T_1 \end{pmatrix}^{SO_3}_{O} = -1$ (Table D.1) and $\begin{pmatrix} T_1 & T_1 & T_1 \\ E & A_2 & E \end{pmatrix}^{O}_{D_4} = -3^{-1/2}$ (Table D.2) and evaluating the $\langle 1\|S^1\|1\rangle^{SO_3}$ spin RME in spherical symmetry using Eq. (5.21), the spin only RME can be determined:

$$\begin{aligned}\langle 1E\| -\mu_B(2S)^{A_2}\|1E\rangle &= -2\mu_B(-1)(-3^{-1/2})\delta_{11}[1(1+1)(2(1)+1)]^{1/2} \\ &= -2(2^{1/2})\mu_B\end{aligned}$$

Therefore

$$\bar{A}_1/\bar{D}_0 = -(2\mu_B)^{-1/2}\langle E(^3A_2)\|\mu^{A_2}\|E(^3A_2)\rangle = -(2\mu_B)^{-1/2}(-2(2^{1/2})\mu_B) = +2$$

The orbital symmetry still affects the MCD term sign even if the RMEs are spin-only. For example, the excitation $(b_2)^2 \rightarrow (b_2)(a_2)$ in D_4 gives $B_1(^1B_1)$, $B_2(^3B_1)$, and $E(^3B_1)$ spin–orbit states. As above, the MCD A term for the $E(^3B_1)$ state will be due to a spin-only RME with a zero orbital RME. Proceeding as before, we have

$$\begin{aligned}\langle E(^3B_1)\|\mu^{A_2}\|E(^3B_1)\rangle &= \langle (1E,B_1)E\|\mu^{A_2}\|(1E,B_1)E\rangle = \delta_{EE}|E|^{1/2}|E|^{1/2} \\ &\quad \times \{EB_1E\}\{E\}\{EA_2E\}\begin{Bmatrix} A_2 & E & E \\ B_1 & E & E \end{Bmatrix}\langle 1E\|\mu^{A_2}\|1E\rangle \\ &= (1)(2^{1/2})(2^{1/2})(+1)(1)(-1)(+1/2)\langle 1E\|\mu^{A_2}\|1E\rangle \\ &= -\langle 1E\| -\mu_B(2S)^{A_2}\|1E\rangle\end{aligned}$$

The negative sign preceding the RME results from the $3j$ factor $\{EB_1E\} = +1$, whereas $\{EA_2E\} = -1$ for the $E(^3A_2)$ case; the $6j$ symbols for both cases have the same value of $+1/2$. The remainder of the evaluation is independent of the orbital symmetry for both cases, and therefore for $E(^3B_1)$ the value of $\bar{A}_1/\bar{D}_0$ is -2.

5.8. HERZBERG–TELLER APPROXIMATION FOR VIBRONIC TRANSITIONS

Finally, transitions that are formally forbidden in the Franck–Condon (FC) approximation are considered. This category includes parity-forbidden ligand field (LF) $d \rightarrow d$ or $f \rightarrow f$ transitions and certain symmetry-forbidden charge transfer (CT) transitions. The problem can be complicated and is dependent upon the type of intensity-borrowing mechanism. Here we consider only the simplest case of vibronic coupling in a Herzberg–Teller approximation where only excited-state mixing and a single symmetry for the allowed excited states is involved in the coupling (a more general treatment of the Herzberg–Teller approximation can be found in ref. 1, Chapter 22). In such cases the symmetry of the enabling vibration and the allowed excited state can affect the A_1/D_0 ratio, but C_0/D_0 is independent of the vibrational symmetry. In the case of simply reducible groups with no repeated irreps for the oriented or isotropic case, the ratios are given by Eqs. (5.22) and (5.23) [ref. 1, Eqs. (22.4.3) and (22.4.4)]

$$A_1/D_0 = [-(\text{MCD factor})\,(2|f|/\mu_B)|N|\{J\}\{AFN\}\{NhJ\}\{Jf'J\} \times \begin{Bmatrix} f' & f & f \\ A & N & N \end{Bmatrix} \begin{Bmatrix} N & f' & N \\ J & h & J \end{Bmatrix} \langle J\|\mu^{f'}\|J\rangle] + C_0/D_0 \qquad (5.22)$$

$$C_0/D_0 = -(\text{MCD factor})\,(2|f|/\mu_B)\{AfN\} \begin{Bmatrix} f' & f & f \\ N & A & A \end{Bmatrix} \langle A\|\mu^{f'}\|A\rangle \qquad (5.23)$$

where h is the vibrational symmetry of the enabling vibration and N is the symmetry of the allowed mixing state(s). In this case the mixing-state symmetry is restricted to a single symmetry. The nonisotropic space-averaged case is simply 1/2 of Eqs. (5.22) or (5.23).

6 Case Studies I. Diamagnetic Systems: *A* and *B* Terms

In this chapter and the next, a number of case studies will be described in order to illustrate some of the information that can be obtained from experimental MCD spectra. This information can assist in the interpretation of electronic spectra and provide some properties of the electronic states and energy levels of the systems under investigation. This chapter will focus on examples with diamagnetic ground states and therefore will offer examples that exhibit *A* and *B* terms in MCD spectra. Chapter 7 will describe examples of paramagnetic samples which feature temperature-dependent *C* terms. The examples in these two chapters are intended to be descriptive with an emphasis on experimental spectra. The reader is referred to the original literature for detailed arguments focused on subtle aspects of theory or spectral interpretation in individual cases. The selection of examples is intended to display a breadth of applications for MCD applied to interpretation of electronic spectra, electronic structure, and properties of electronic states for a variety of atomic and molecular systems.

6.1. *A* AND *B* TERMS FOR DIAMAGNETIC ATOMS AND MOLECULES

For atoms and molecules that have diamagnetic, and therefore nondegenerate, ground states, *C* terms will be absent. If the molecule possesses a proper or improper rotation axis of order greater than three, excited-state degeneracies are possible, and *A* terms are expected for transitions to those degenerate states. There will also be *B* terms for such transitions with varying intensity proportional to differences in energy between states with which coupling can occur in the presence of the magnetic field. For transitions to non degenerate excited states, even for high-symmetry molecules, only *B* terms are expected. The lowest level of interpretation can therefore be to simply detect transitions to degenerate states by the observation of an *A* term in the MCD spectrum. However, in order to provide a convincing case for an assignment, the sign and approximate magnitude of the *A* term must be determined from the expectations of an electronic structural model, and this must be in accord with the observation. The sign and magnitude of B terms are not always easily determined because they are dependent upon the relative placement in energy of nearby states that can interact with the state in question. If only a few states are close in energy, an analysis can prove useful to make a reasoned case for assignment, but there will always be some uncertainty about the interpretation of the sign and magnitude of B terms. In the case of low-symmetry molecules with rotation axes of order two or one, there will be only nondegenerate excited states, and only B terms are expected in the MCD spectra. However, if two such states are close in energy and have a nonzero Zeeman interaction in the presence of the magnetic field, the MCD spectrum can exhibit a pseudo A term due to the two overlapping (within their bandwidths) B terms of opposite sign, even though a true degenerate excited state does not exist. Such "near-degeneracy" cases in low-symmetry molecules are often complicated and difficult to interpret. Again, if only a few states are involved, the sign of the pseudo A term can be predicted, which can provide support for the interpretation.

6.2. ATOMIC MERCURY VAPOR

An interesting example that illustrates the relationship between the atomic Zeeman effect and MCD A and B terms is given by mercury vapor [ref. 7]. Figure 6.1 shows the MCD and absorption spectra for Hg vapor in the region of the ${}^1S_0(6s^2) \rightarrow {}^3P_1(6s6p)$ atomic resonance transition (253.65 nm) at different field strength. The heavy Hg atom provides strong spin–orbit coupling ($\zeta_{6p} = 4260\,\text{cm}^{-1}$) that will intermix the ${}^3P_1(6s6p)$ state with the spin-allowed ${}^1P_1(6s6p)$ state of the ${}^1S_0(6s^2) \rightarrow {}^1P_1(6s6p)$ resonance transition that occurs at shorter

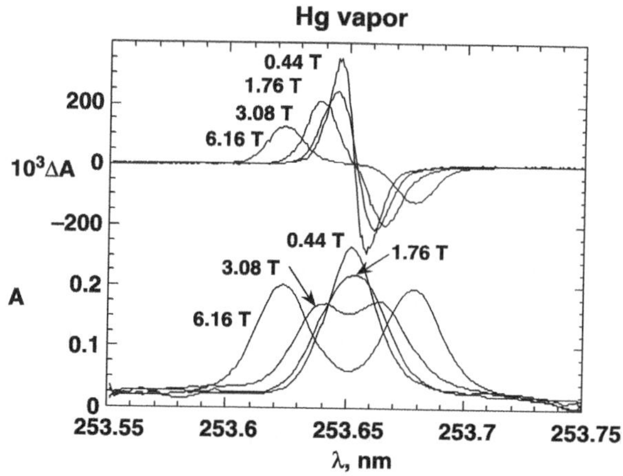

Figure 6.1. The ${}^1S_0(6s^2) \rightarrow {}^3P_1(6s6p)$ atomic resonance transition (253.65 nm) for mercury vapor at different magnetic fields. Adapted from ref. 7.

wavelength (184.957 nm)s. The intermixed "triplet" and "singlet" spin–orbit states can be described as

$$\text{"triplet"}\,|P_{0,\pm 1}\rangle = -a|{}^1P_{0,\pm 1}\rangle + b|{}^3P_{0,\pm 1}\rangle$$

$$\text{"singlet"}\,|P_{0,\pm 1}\rangle = b|{}^1P_{0,\pm 1}\rangle + a|{}^3P_{0,\pm 1}\rangle$$

where the subscripts on $P_{0,\pm 1}$ refer to m_J values of 0, ± 1 and the mixing coefficients are constrained by $b \geq a > 0$ and $a^2 + b^2 = 1$. The spin–orbit coupling serves to give the 253.65-nm band substantial intensity even though the excited state is of triplet origin. The energy levels for the $|P_{0,\pm 1}\rangle$ Zeeman sublevels are pictured in Figure 3.2 (Chapter 3, Section 3.4), apart from their spin multiplicity labels (the singlet states indicated in Figure 3.2 must be replaced by the mixed "triplet" or "singlet" $|P_{0,\pm 1}\rangle$ states, with no spin multiplicity labels, described here, but the Zeeman splitting for $B \neq 0$ will still apply). The MCD spectrum at low field exhibits a very strong positive *A* term while the absorption band broadens slightly. The atomic resonance band in the vapor state is very narrow, which enhances the *A* term signal as explained in Chapter 3 (Section 3.12). At higher field the band separates into two resolved Zeeman components, and the positive *A* term devolves into two oppositely signed *B* terms for the two allowed Zeeman transitions. This behavior illustrates clearly the relationship between (a) the *A* term at low field where the Zeeman splitting is smaller than the bandwidth and (b) the B terms of opposite sign at high field where degeneracy is lifted and the splitting is greater than the bandwidth. It must be admitted that the narrow line of the atomic resonance transition for Hg vapor is unique; most molecular samples accessible to MCD study have much broader bands, and the separation into resolved Zeeman components is not easily observed.

At low field (<0.5 T) where the Zeeman splitting is less than the linewidth, a moment analysis of the *A* term allowed an estimate of the spin–orbit mixing coefficients and thus an assessment of the singlet character of the 253.65-nm transition. It can be shown that the $\overline{A}_1/\overline{D}_0$ value for the spin–orbit transitions is given by [ref. 7]

$$\overline{A}_1/\overline{D}_0 = 2|a|^2 + (1+2)|b|^2$$

if the 6*s* and 6*p* orbitals are approximated by pure atomic orbitals on Hg. A moment analysis of the MCD and absorption spectra at 0.44 T gave a value $\overline{A}_1/\overline{D}_0 = +2.88 \pm 0.19$, which, in turn, implies values of $a = 0.35$ and $b = 0.94$, or about 12% singlet character in the "triplet" transition. At higher field the separation of the two Zeeman absorptions or MCD *B* terms can give an estimate of the Zeeman splitting which is given by

$$\Delta E = 2g\mu_B B$$

where g is the excited-state g factor, μ_B = Bohr magneton, and B = magnetic field. The observed energy splittings are 4.2 ± 0.2 and $8.6 \pm 0.2\,\text{cm}^{-1}$ at 3.08

and 6.16 T, respectively. These values of ΔE compare favorably with the splittings calculated from the g factor of 1.479 derived from earlier atomic spectral data (4.25 and 8.51cm^{-1}, respectively) [ref. 7]. Also since $\overline{A}_1/\overline{D}_0 = 2g$ [Eq. (3.34)], so that even at low field the g value of $+2.88/2 = +1.44 \pm 0.1$ agrees favorably with the g factor from atomic spectral data.

6.3. THE SODIDE ION Na^- IN A SOLID NH_3 MATRIX

When an alkali metal dissolves in liquid ammonia, alkali metal cations M^+ and solvated electrons e^-_{solv} are produced. As the alkali metal concentration is increased, the formation of various ionic species have been suggested, including M^- ions by an electron transfer: $2M \rightarrow M^+ + M^-$. Such ions have been characterized in amine and ether solvents, but they don't seem to be present in liquid NH_3. The sodide Na^- ion, however, was identified by an MCD spectrum for a solid NH_3 matrix formed by codeposition of Na and NH_3 on a sapphire window. The matrix isolated ion was characterized by an absorption band at 16,950 cm^{-1} which was accompanied by a positive MCD A term [ref. 8]. The absorption and MCD spectra are shown in Figure 6.2. The MCD spectrum also revealed a

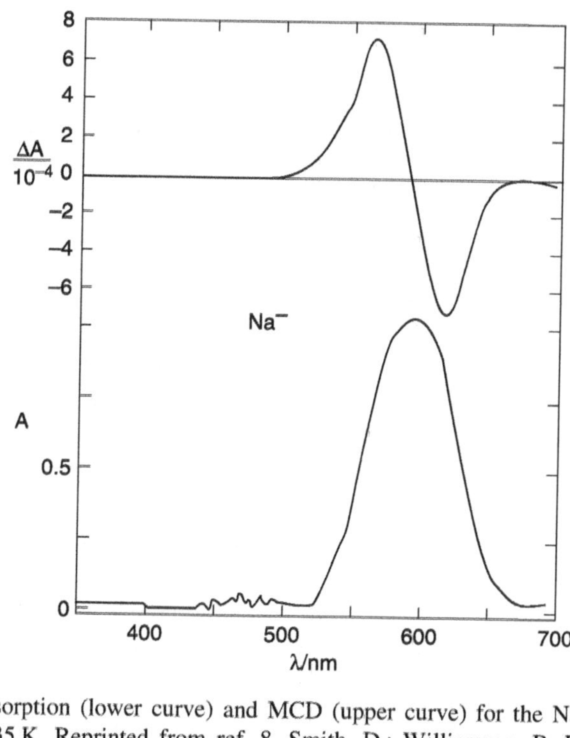

Figure 6.2. Absorption (lower curve) and MCD (upper curve) for the Na^- ion in a solid NH_3 matrix at 35 K. Reprinted from ref. 8. Smith, D.; Williamson, B. E.; Schatz, P. N. *Chem. Phys. Letters* **1986**, *131*, 457. Copyright 1986 with permission from Elsevier.

positive *A* term for the solvated electron in the near-IR region at 8000 cm^{-1} (not shown in Figure 6.2) which is comparable to that observed for other alkali metals in liquid NH_3. The absorption and MCD spectra shown in Figure 6.2 were found to be temperature-independent in the range 3.3–35 K, which eliminates the possibility of the band being due to the presence of paramagnetic neutral Na atoms and the expected *C* term. The observed spectra are consistent with the $^1S(3s^2) \rightarrow {}^1P(3s3p)$ transition for the Na^- ion. A moment analysis of the spectra gave $A_1/D_0 = 1.3 \pm 0.25$; the predicted value for an isolated Na^- ion, neglecting spin–orbit coupling and assuming the excited state $g = +1$ for ^1P, is $A_1/D_0 = 2.0$ (see Chapter 3, Section 3.4 and Figure 3.2). It was argued that the solid NH_3 matrix causes quenching of the orbital angular momentum of the isolated ion due to the interaction of the excited 3*p* orbitals with the environment.

6.4. SQUARE COMPLEXES OF D_{4h} SYMMETRY

Transition metal complexes of nd^8 electron configuration $(nd_z^2)^2(nd_{xz}nd_{yz})^4(nd_{xy})^2$, or in terms of D_{4h} orbital symmetries $(a_{1g})^2(e_g)^4(b_{2g})^2$, typically form square complexes with diamagnetic $^1A_{1g}$ ground states. Examples include $PtX_4{}^{2-}$ and AuX_4^-, $X = Cl^-$, Br^-, and I^-; $M(NH_3)_4^{2+}$, M = Pt(II) or Pd(II); and $M(CN)_4^{2-}$, M = Pt(II), Pd(II), and Ni(II). Figure 6.3 shows the relative energies of the metal orbitals for a square complex, assuming the complex lies in the xy plane and the fourfold axis is along the *z* axis. The transitions shown are ligand field (LF) transitions which are only vibronically allowed and are normally weak. Dipole allowed electronic transitions are possible only to A_{2u} (*z*-polarized) or E_u (xy-polarized)

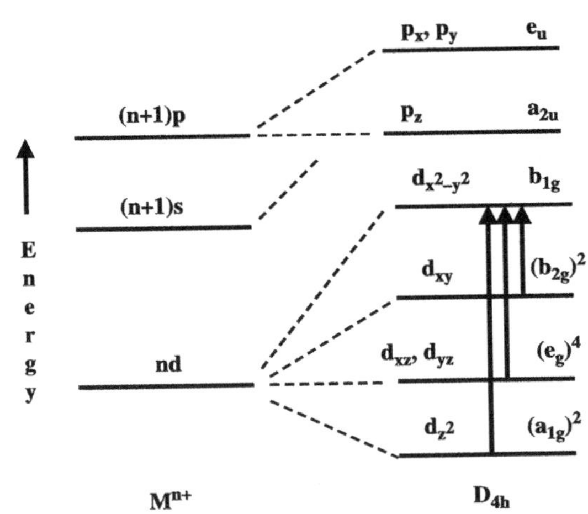

Figure 6.3. Metal orbitals for a square D_{4h} complex in the *xy* plane with the fourfold axis along *z*. The HOMO is the $b_{2g}d_{xy}$ orbital; three ligand field (LF) transitions to the empty $b_{1g}d_{x^2-y^2}$ orbital are shown.

excited states. The electronic states in D_{4h} symmetry for square complexes can be easily correlated with D_4 and irreducible tensor methods [ref. 1] can be used to evaluate MCD terms or term ratios as illustrated in Chapter 5. In the examples discussed below, several types of transitions for square metal complexes will be considered, including LF, charge transfer, and metal-centered $nd \rightarrow (n+1)p$-type transitions.

6.4.1. Ligand Field Spectra for $PtCl_4^{2-}$

An early use of *A* terms can be found in the study of the lower-energy and lower-intensity LF transitions for the $PtCl_4^{2-}$ ion [refs. 9–11]. Figure 6.4 shows the absorption and MCD spectrum for this ion in aqueous HCl solution. The MCD spectrum in this case allowed an interpretation from among several possibilities. As shown in Figure 6.3, three vibronic, parity-forbidden ($g \rightarrow g$), spin-allowed

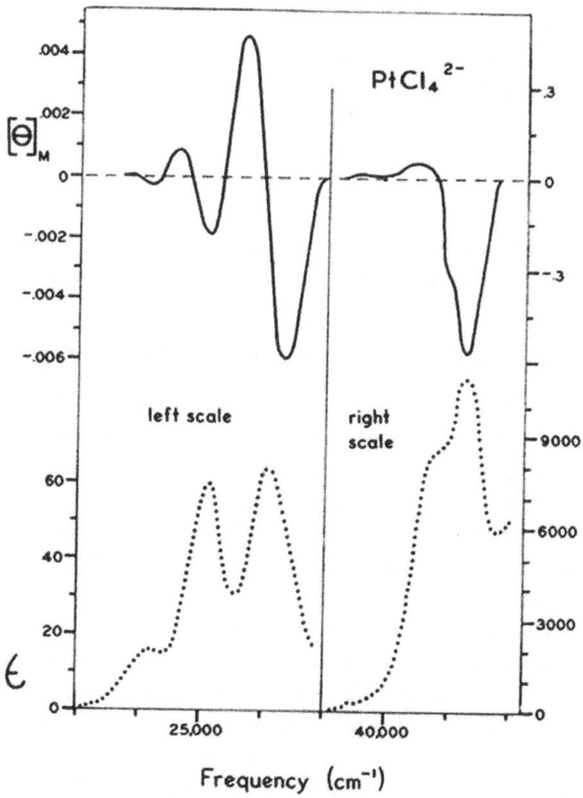

Figure 6.4. Absorption and MCD spectra for $PtCl_4^{2-}$ in 0.5 M HCl (aq). $[\theta]_M$ is the molar ellipticity in degrees deciliter mol^{-1} decimeter^{-1} per gauss. Reprinted with permission from ref. 10. McCaffery, A. J.; Schatz, P. N.; Stephens, P. J. *J. Amer. Chem. Soc.* **1968**, *90*, 5730. Copyright 1968 American Chemical Society.

LF transitions are expected from the excitations into the empty $b_{1g}(d_{x^2-y^2})$ orbital (LUMO): ${}^1A_{1g} \to {}^1B_{1g}[a_{1g} \to b_{1g}]$; ${}^1A_{1g} \to {}^1E_g[e_g \to b_{1g}]$; and ${}^1A_{1g} \to {}^1A_{2g}[b_{2g} \to b_{1g}]$. The $e_g \to b_{1g}$ transition thus will have a degenerate excited state and therefore should present an *A* term in the MCD spectrum. In the case of the $PtCl_4^{2-}$ ion, the three spin-allowed LF transitions have been observed at 2.55–3.03 μm^{-1}, and a shoulder band at ~3.76 μm^{-1} (1 μm^{-1} = 10^4 cm^{-1}). The MCD spectrum (Figure 6.4) shows a negative *B* term for the 2.55 μm^{-1} band, a clear negative *A* term for the band at 3.03 μm^{-1}, and a small positive *B* term for the 3.76-μm^{-1} shoulder. The logical assignment for the 3.03-μm^{-1} band is therefore ${}^1A_{1g} \to {}^1E_g[e_g \to b_{1g}]$. The assignment of the 2.55-μm^{-1} band to ${}^1A_{1g} \to {}^1A_{2g}[b_{2g} \to b_{1g}]$ was based on crystal polarization and by acquired optical activity in an optically active solvent; the *B* term observed is not conclusive, but it is consistent with an assignment to a transition to a nondegenerate state [refs. 12, 13]. The higher-energy LF assignment to the shoulder at ≈3.76 μm^{-1} was made, more or less, by default because the MCD spectrum in this region was rather uninformative. The *A* term for the 3.03-μm^{-1} band, however, is evidence for a degenerate excited state. Furthermore, the negative *A* term for the ${}^1A_{1g} \to {}^1E_g[e_g \to b_{1g}]$ transition is consistent with a Herzberg–Teller model (Eq. 5.22) if the enabling vibration is the out-of-plane ν_{a2u} normal bending mode that effectively removes the center of symmetry in the $PtCl_4^{2-}$ ion. This interpretation of the LF spectra for $PtCl_4^{2-}$ has provided an experimental basis for concluding that the energy ordering of the occupied 5*d* orbitals in the chloro complex is $d_{xy} > d_{xz,yz} > d_{z^2}$, as illustrated in Figure 6.3. The energy ordering of the occupied orbitals appears to be ligand-dependent in square complexes with the relative energy of the d_{z^2} orbital exhibiting variability (see the examples discussed below).

6.4.2. Intense Bands and Allowed Transitions for Square Complexes

A terms have also been used in several cases to aid in the interpretation of allowed transitions observed for square complexes. In particular for platinum(II) or gold(III) complexes, three types of allowed transitions have been identified in the UV region, depending upon the type of ligand present: (1) Ligand-to-metal charge transfer (LMCT) is observed for halo complexes PtX_4^{2-} or AuX_4^-, X = Cl^-, Br^-, I^-, (2) Metal-to-ligand charge transfer (MLCT) is exhibited by $Pt(CN)_4^{2-}$ or $Pt(PR_3)_4^{2+}$ complexes, and (3) $5d \to 6p_z$ metal localized transitions have been assigned for sigma-only N donors such as in the $Pt(NH_3)_4^{2+}$ or $Pt(en)_2^{2+}$ (en = ethylenediamine) cations. The symmetry aspects of the $d \to p$ and MLCT transitions are essentially the same; both of these processes involve excitation from the occupied metal orbitals to a higher-energy empty a_{2u} orbital. The empty orbital may be primarily a metal-centered $6p_z$ orbital or a π-type ligand combination with a_{2u} symmetry, which overlaps with and therefore contains some contribution from the metal $6p_z$. The procedure for the evaluation of the MCD terms, however, is insensitive to the nature of the a_{2u} orbital composition and would be identical to the D_4 example discussed above (Chapter 5, Sections 5.4 and 5.5); the final l_z RME depends upon the composition of only the occupied orbitals. The LMCT transitions, on the other hand, involve excitation from occupied nonbonding (n), π, or σ bonding ligand orbitals to the empty σ^* $d_{x^2-y^2}$ LUMO. Both LMCT and MLCT transitions are sensitive to the charge on the metal ion (metal oxidation state). This sensitivity is due to the stability of the metal orbitals which increases as the metal ion charge increases. The orbitals that are primarily ligand-based are less sensitive to metal ion charge. Therefore, LMCT transitions will red shift as the metal ion charge increases (oxidation state increases) and the metal orbitals become more stable, while MLCT transitions will blue shift. The shifts are often $\sim 1\ \mu m^{-1}$ or more for a change in metal ion oxidation state of one unit.

6.4.3. LMCT Transitions for PtX_4^{2-} and AuX_4^-, X = Cl^-, Br^-, I^-

The LMCT transitions for square complexes may be visualized with the help of Figure 6.5, which shows the filled nonbonding (n), π bonding (π), and σ bonding (σ) orbitals for tetrahalo complexes. In order to have non-zero A terms (nonzero l_z matrix elements in a one-center approximation), an intermixing of σ- and π-type orbitals is required [ref. 14]. Ignoring any ligand spin–orbit coupling and considering only spin-allowed transitions to the $\sigma^* b_{1g} d_{x^2-y^2}$ LUMO, the LMCT transitions that give rise to A terms are of the type $e_u \rightarrow b_{1g}$ and give ${}^1E_u$ states. There are two of these transitions which are designated σ-LMCT and π-LMCT,

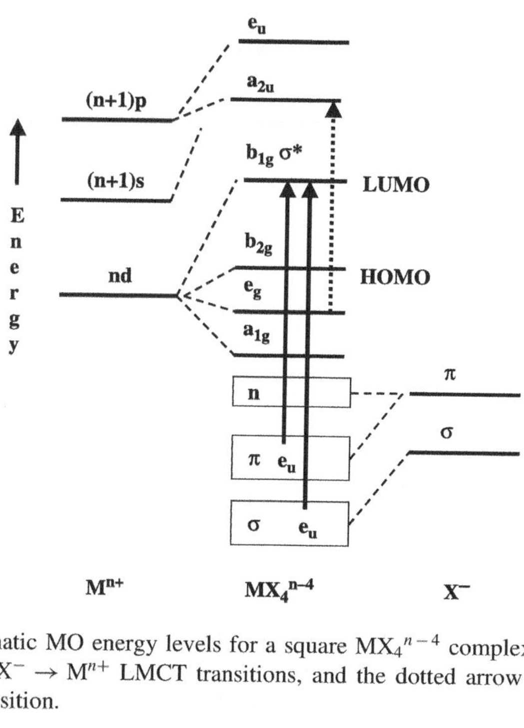

Figure 6.5. Schematic MO energy levels for a square $MX_4{}^{n-4}$ complex. The solid arrows show the π and $\sigma X^- \rightarrow M^{n+}$ LMCT transitions, and the dotted arrow indicates a $d \rightarrow p$ metal-centered transition.

depending upon whether the major component of the occupied MO is a σ or a π ligand orbital. Transitions to these two states are, of course, dependent upon the ligand type and are usually separated by about ~1 μm^{-1}, with the π-LMCT (which is of $\pi \rightarrow \sigma^*$ type) lower in energy and weaker than the σ-LMCT ($\sigma \rightarrow \sigma^*$ type). Since the occupied $e_u\sigma$ and $e_u\pi$ MOs have the same e_u symmetry, respectively, the intermixing (configuration interaction, CI) can be described as

$$e'_u\sigma = \lambda_1 e_u\sigma + \lambda_2 e_u\pi$$
$$e'_u\pi = -\lambda_2 e_u\sigma + \lambda_1 e_u\pi$$

where the primed MOs are of mixed σ–π composition with λ_1 and λ_2 mixing coefficients such that $\lambda_1, \lambda_2 > 0$ and $\lambda_1^2 + \lambda_2^2 = 1$. This leads to $\bar{A}_1/\bar{D}_0 = +2\lambda_1\lambda_2$ for the σ-LMCT ($e'_u\sigma \rightarrow b_{1g}$) and $-2\lambda_1\lambda_2$ for the π-LMCT ($e'_u\pi \rightarrow b_{1g}$) if ligand spin–orbit coupling is ignored (a fair approximation for $X = Cl^-$). Figure 6.6 shows the absorption and MCD spectra for AuX_4^-, $X = Cl^-$ and Br^- in acetonitrile solution [ref. 14]. A moment analysis of experimental MCD and absorption

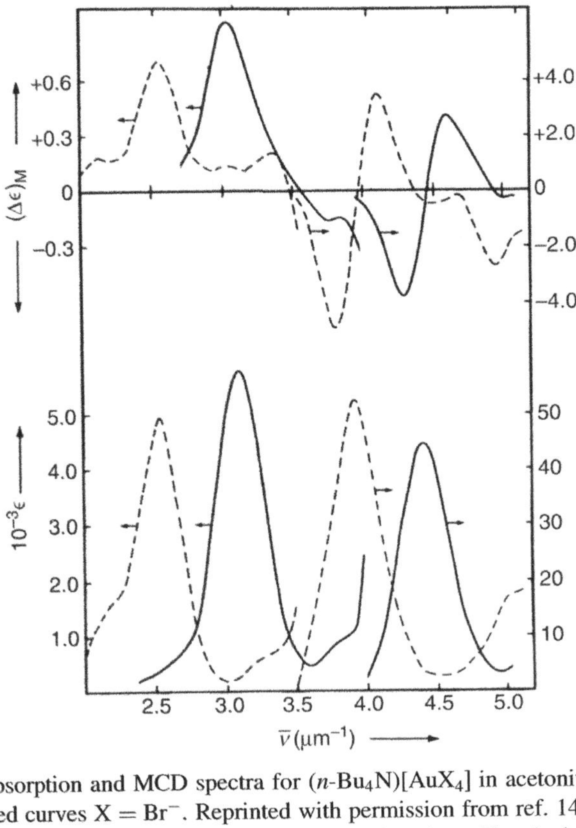

Figure 6.6. Absorption and MCD spectra for $(n\text{-}Bu_4N)[AuX_4]$ in acetonitrile. Solid curves $X = Cl^-$, dashed curves $X = Br^-$. Reprinted with permission from ref. 14. Isci, H.; Mason, W. R. *Inorg. Chem.* **1983**, 22, 2266. Copyright 1983 American Chemical Society.

spectra for $AuCl_4^-$ gave a value of +0.44 for the σ-LMCT band assigned at 4.4 μm^{-1} which, in turn, led to an estimate of 0.98 and 0.23 for λ_1 and λ_2, respectively [ref. 15]. A small negative value was found for the band assigned as the π-LMCT transition at 3.1 μm^{-1}, but the band also contained a large *B*-term contribution. Figure 6.7 shows the absorption and MCD spectra for the intense, higher energy bands for $PtCl_4^{2-}$ ion in acetonitrile solution. For $PtCl_4^{2-}$, a shoulder at 4.2 μm^{-1} on a more intense maximum at 4.4 μm^{-1} exhibited a negative *A* term, overlapping with a stronger positive *A* term [refs. 11, 15]. The lower-energy shoulder was interpreted as the π-LMCT transition, while the stronger maximum was assigned as a $d \rightarrow p$ transition (dotted arrow in Figure 6.5), analogous to that observed for $Pt(NH_3)_4^{2+}$ or $Pt(en)_2^{2+}$ (see below). The $d \rightarrow p$ assignment was justified because the band was only slightly higher in energy than the π-LMCT and therefore much too low in energy for a σ-LMCT. The negative *A* term is consistent with and provided experimental support for the π-LMCT assignment of the 4.2-μm^{-1} shoulder band. Similar assignments have been advanced from MCD

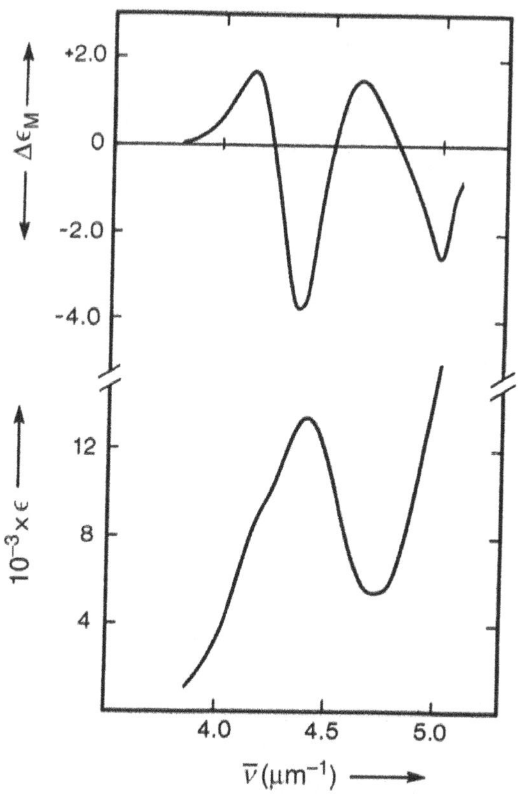

Figure 6.7. Absorption and MCD spectra for $(n\text{-}Bu_4N)_2[PtCl_4]$ in acetonitrile. Reprinted with permission from ref. 15. Isci, H.; Mason, W. R. *Inorg. Chem.* **1984**, *23*, 1565. Copyright 1984 American Chemical Society.

studies of $PtBr_4^{2-}$ and PtI_4^{2-} [refs. 15, 16]; in these cases, however, the absorption and MCD spectra are more complex due to the increased importance of ligand spin–orbit coupling, which gives intensity to transitions to spin–orbit states of triplet origin.

6.4.4. MLCT Transitions for $Pt(CN)_4^{2-}$

The absorption and MCD spectra for the $Pt(CN)_4^{2-}$ ion are displayed in Figure 6.8 and show three positive *A* terms at 3.54 μm^{-1} (band I, weaker and unsymmetrical), 3.84 μm^{-1} (II, strong and broad), and 4.55 μm^{-1} (IV, stronger) [refs. 17, 18]. The absorption bands associated with these MCD A terms are all intense with absorptivities of 1230, 12,900, and 29,300 $M^{-1}cm^{-1}$, respectively. They have been interpreted as MLCT from the occupied a_{1g} $5d_{z^2}$ and e_g $5d_{xz}$, $5d_{yz}$ metal orbitals to the empty $a_{2u}\pi^*$ cyanide orbital, though as noted the metal 6_{pz} orbital also has a_{2u} symmetry and therefore the empty orbital is almost certainly of mixed character. A schematic MO energy-level diagram is shown in Figure 6.9; the

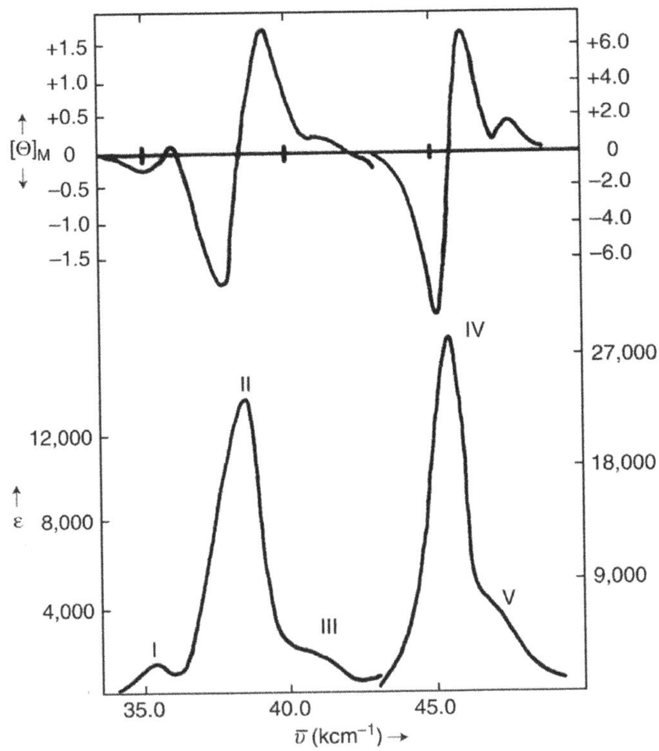

Figure 6.8. Absorption (lower curves) and MCD (upper curves) spectra for $(n\text{-}Bu_4N)_2[Pt(CN)_4]$ in acetonitrile solution. Reprinted with permission from ref. 18. Isci, H.; Mason, W. R. *Inorg. Chem.* **1975**, *14*, 905. Copyright 1975 American Chemical Society.

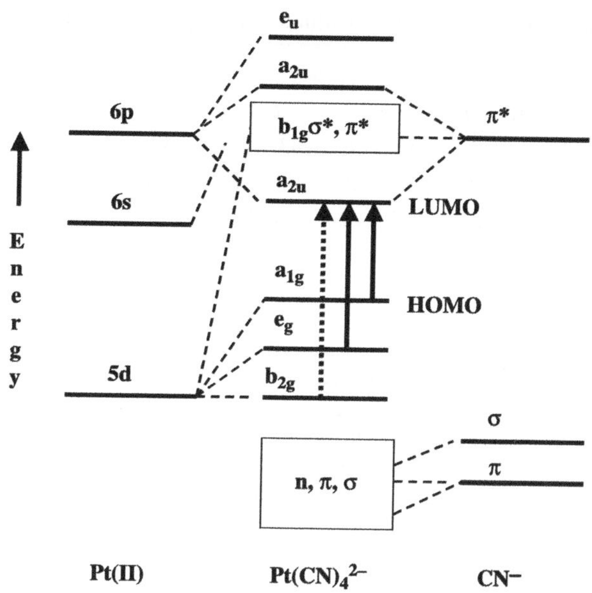

Figure 6.9. Schematic MO energy-level diagram for the $Pt(CN)_4^{2-}$ ion showing MLCT from occupied 5*d* Pt(II) orbitals to the a_{2u} LUMO, which is of mixed CN^- π^* and Pt(II) $6p_z$ character.

LUMO is now a MO with CN^- π^* and Pt $6p_z$ character. The ground state is totally symmetric $^1A_{1g}$ with electric-dipole-allowed transitions possible to A_{2u} (*z*-polarized) and E_u (*xy*-polarized) excited states. The two transitions indicated by the solid arrows in the diagram correspond to the a_{1g} $(d_{z^2}) \rightarrow a_{2u}$ and $e_g(d_{xz,yz}) \rightarrow a_{2u}$ MLCT transitions, which give allowed $^{1,3}A_{2u}$ and $^{1,3}E_u$ states, respectively. A third transition indicated by the dotted arrow is the $b_{2g}(d_{xy}) \rightarrow a_{2u}$ MLCT transition, which gives orbitally forbidden $^{1,3}B_{1u}$ states, transitions to which should be quite weak. The interpretation of the MCD spectrum included platinum spin–orbit coupling (ζ_{5d} ~3500 cm^{-1}), which serves to strongly intermix the singlet and triplet MLCT states. There are two A_{2u} spin–orbit states from $^1A_{2u}$ and ^3E_u which are intermixed to give $A_{2u}(1)$ and $A_{2u}(2)$ and four E_u spin–orbit states from ^1E_u, ^3E_u, $^3A_{2u}$, and $^3B_{1u}$ which give $E_u(1)$, $E_u(2)$, $E_u(3)$, and $E_u(4)$. Diagonalization of the spin–orbit secular determinants (Table 6.1) using energies for the zero-order singlet and triplet states in the absence of spin–orbit coupling (see ref. 18 for details) is illustrated in Figure 6.10, where the energies of the calculated spin–orbit states are compared with the energies of the experimental spectrum. The diagonalization places the $E_u(1)$ spin–orbit state (predominantly $E_u(^3A_{2u})$ at lowest energy near 3.54 μm^{-1} (band I in Figure 6.8) and predicts a positive *A* term. The $E_u(2)$ and $A_{2u}(1)$ spin–orbit states ($E_u(^3E_u)$ and $A_{2u}(^3E_u)$ states, respectively) are predicted to be close in energy near 3.84 μm^{-1} (band II) with the prediction of a positive *A* term for $E_u(2)$ and a positive pseudo *A* term from the

TABLE 6.1. Spin–Orbit Secular Determinants for D_{4h} MLCT States in $Pt(CN)_4^{2-}$ [a]

E_u

$$\begin{vmatrix} {}^3B_{1u} - E & 0 & i\zeta/2 & -\zeta/2 \\ 0 & {}^3A_{2u} - E & -i3^{1/2}\zeta/2 & -3^{1/2}\zeta/2 \\ -i\zeta/2 & i3^{1/2}\zeta/2 & {}^3E_u - E & -i\zeta/2 \\ -\zeta/2 & -3^{1/2}\zeta/2 & i\zeta/2 & {}^1E_u - E \end{vmatrix} = 0$$

A_{2u}

$$\begin{vmatrix} {}^1A_{2u} - E & -(3/2)^{1/2}\zeta \\ -(3/2)^{1/2}\zeta & {}^3E_u - \zeta/4 - E \end{vmatrix} = 0$$

B_{1u}

$$\begin{vmatrix} {}^1B_{1u} - E & 2^{1/2}\zeta/2 \\ 2^{1/2}\zeta/2 & {}^3E_u - E \end{vmatrix} = 0$$

[a] ζ is the spin–orbit coupling constant for the Pt 5*d* orbitals ($\sim$3500 cm^{-1}); singlet and triplet state symbols refer to their respective zero-order energies; E is the spin–orbit energy. See ref. 18 for the input zero-order energy values used for the diagonalization for the determinants and the calculation of the spin–orbit eigenvectors.

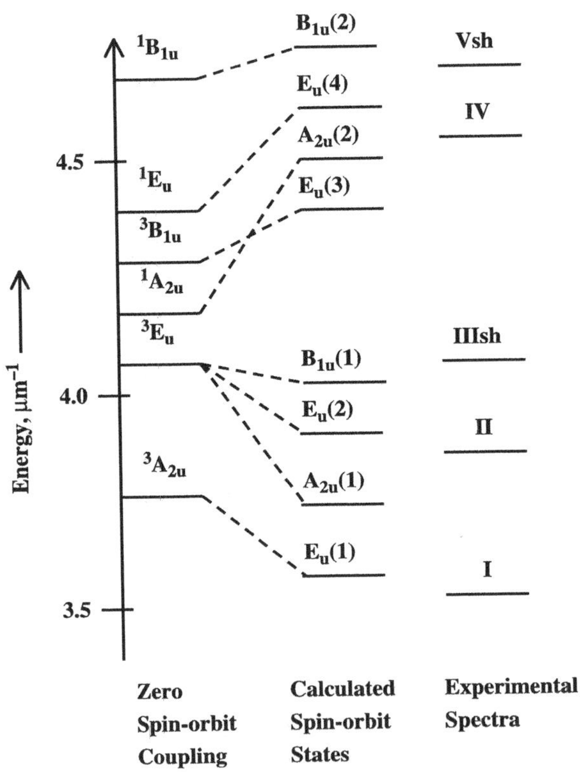

Figure 6.10. Calculated spin–orbit states and a comparison with the experimental spectra. (Adapted from ref. 18.)

overlapping B terms from $E_u(2)$ and $A_{2u}(1)$, assuming that the B terms for these two states have the greatest interaction because they are closest in energy. Finally, the $E_u(4)$ and $A_{2u}(2)$ spin–orbit states ($E_u(^1E_u)$ and the $A_{2u}(^1A_{2u})$ states) are also found to be very close in energy near 4.55 μm-1 (band IV) and are predicted to give a positive A term for $E_u(4)$ and another positive pseudo A term from $E_u(4)$ and $A_{2u}(2)$. Spin–orbit $B_{1u}(1)$ and $B_{1u}(2)$ were also placed near the enrgies of the shoulder bands III and V, respectively, but it is not clear that these bands are due to the forbidden MLCT transitions. An alternative assignment is to ν_{CN} vibrations (~2000 cm^{-1}) built upon the stronger bands II and IV; bands III and V were thus assigned to $E_u(2) + \nu_{CN}$ and $E_u(4) + \nu_{CN}$ [ref. 18]. The interpretation for the $Pt(CN)_4^{2-}$ spectra represents an interesting case where platinum spin–orbit coupling mixes the MLCT excited states to a significant extent, and B terms of opposite sign for allowed spin–orbit A_{2u} and E_u states give rise to positive pseudo A terms that overlap positive A terms for the E_u states and therefore enhance their intensity. The metal spin–orbit model thus appears to be in accord with the intense features in the experimental spectrum. It might be emphasized again that the A terms and pseudo A terms in this approximation do not depend upon the character of the empty a_{2u} orbital, only on its symmetry. The interpretation cannot assess the metal $6p_z$ versus CN^- π^* composition of empty the a_{2u} orbital. Finally, it may be remarked that the interpretation indicates that the relative energy ordering of the occupied $5d$ orbitals is $a_{1g}(d_{z^2}) > e_g(d_{xz,yz}) > b_{2g}(d_{xy})$, as shown in Figure 6.9, with the σ symmetry d_{z^2} orbital for $Pt(CN)_4^{2-}$ as the highest-energy occupied $5d$ orbital. In contrast, the d_{z^2} orgital is placed lowest of the occupied $5d$ orbitals for $PtCl_4^{2-}$ (see above). The higher energy of the d_{z^2} orbital in the cyano complex is likely due to the greater σ donor strength of the CN^- ligand compared to Cl^-.

6.4.5. *d* → *p* Transitions for $Pt(NH_3)_4^{2+}$ and $Pt(en)_2^{2+}$

The $Pt(NH_3)_4^{2+}$ and $Pt(en)_2^{2+}$ complexes have Pt(II) coordinated by four sigma-only N donors. The absence of π bonding by the ligands dictates that the $5d_{xz,yz}$ and $5d_{xy}$ orbitals will be nonbonding and be at nearly the same energy. The $5d_{z^2}$ orbital has σ symmetry and therefore will be involved to some extent in σ bonding to the N donor ligands and will differ in energy from the $5d_{xz,yz}$ and $5d_{xy}$. The observed absorption and MCD spectra [ref. 19] for $Pt(NH_3)_4^{2+}$ and $Pt(en)_2^{2+}$ show an intense band at 5.1 μm^{-1} and 5.0 μ^{m-1}, respectively, which is accompanied by a strong positive A term in both cases. As an example, Figure 6.11 presents the absorption and MCD spectra for $Pt(en)_2^{2+}$ in acetonitrile solution; similar results were found for aqueous solutions and for $Pt(NH_3)_4^{2+}$ in aqueous or acetonitrile solutions. The observed values of $\overline{A}_1/\overline{D}_0$ were +0.8 and +1.0, for $Pt(NH_3)_4^{2+}$ and $Pt(en)_2^{2+}$, respectively. A weaker band is also observed for both complexes, as a shoulder at 4.6 μm^{-1} for $Pt(NH_3)_4^{2+}$ in acetonitrile solution and at 4.47 μm^{-1} for $Pt(en)_2^{2+}$ in acetonitrile (band III in Figure 6.11) or in 0.10 M aqueous $HClO_4$. A moment analysis of this weaker band revealed a small positive A term together with a larger negative B term. The A term $\overline{A}_1/\overline{D}_0$

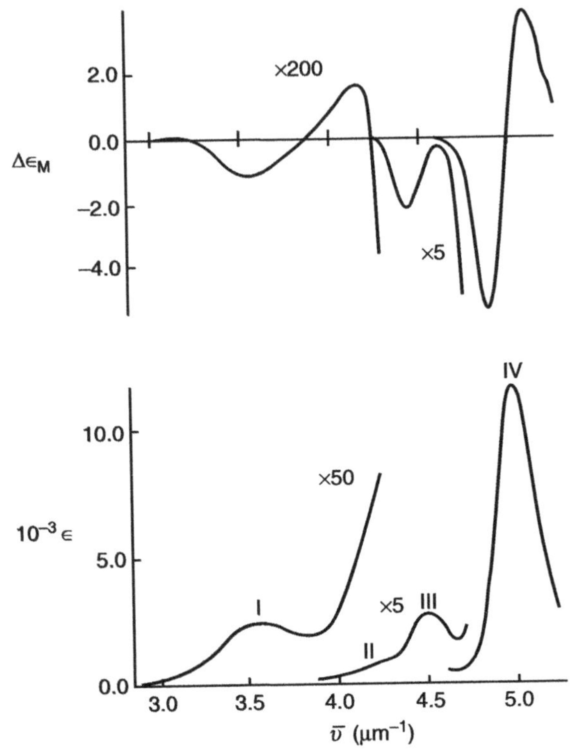

Figure 6.11. Absorption and MCD spectra for $[Pt(en)_2](ClO_4)_2$ in acetonitrile solution. Reprinted with permission from ref. 19. Mason, W. R. *Inorg. Chem.* **1986**, *25*, 2925. Copyright 1986 American Chemical Society.

values were +0.8 and +0.5 for $Pt(NH_3)_4^{2+}$ and $Pt(en)_2^{2+}$, respectively. These intense bands have been interpreted as metal localized $d \rightarrow p$ transitions. Since the occupied orbitals are predominantly metal in character, platinum spin–orbit coupling is expected to be significant. Therefore the model used to interpret the $d \rightarrow p$ spectra for these N donor Pt(II) complexes was similar to that used for $Pt(CN)_4^{2-}$ as discussed above, except that the empty a_{2u} orbital in this case was assumed to be entirely metal $6p_z$. Based on some spin–orbit calculations similar to those for $Pt(CN)_4^{2-}$, the weaker low-energy band was assigned as a transition to the $E_u(^3A_{2u})$ spin–orbit state resulting from the $a_{1g} \rightarrow a_{2u}$ $5d_{z^2} \rightarrow 6p_z$ excitation, while the more intense band was interpreted as nearly overlapping transitions to the $A_{2u}(^1A_{2u})$ $[a_{1g} \rightarrow a_{2u}]$ $5d_{z^2} \rightarrow 6p_z$ and $E_u(^3E_u)$ $[e_g \rightarrow a_{2u}]$ $5d_{xz,yz} \rightarrow 6p_z$ spin–orbit states. The intensity of the band was interpreted as being carried by the singlet character; the positive *A* term was interpreted as resulting from the combination of the positive *A* term from the $E_u(^3Eu)$ state; and a positive pseudo *A* term was interpreted as resulting from an $A_{2u}(^1A_{2u})$ – $Eu(^3E_u)$ interaction. The conclusion reached in this study was that energetically the σ bonded

$5d_{z^2}$ > nonbonding $5d_{xz,yz}$ and $5d_{xy}$ and that the spectra for the N-donor Pt(II) complexes and $Pt(CN)_4^{2-}$ were remarkably similar, but that the bands for $Pt(CN)_4^{2-}$ were shifted by about 1 μm^{-1} to lower energy. This red shift is in accord with a stabilizing interaction of the a_{2u} π^* CN^- MO with the metal $6p_z$ orbital, thereby giving a lower-energy a_{2u} LUMO for $Pt(CN)_4^{2-}$ than for $Pt(NH_3)_4^{2+}$and $Pt(en)_2^{2+}$ [ref. 19].

6.5. LINEAR TWO-COORDINATE $D_{\infty h}$ COMPLEXES

Two-coordinate $D_{\infty h}$ complexes are found for heavy metal ions of closed-shell nd^{10} electron configuration—for example, Hg(II), Au(I), and Pt(0). These complexes have diamagnetic, totally symmetric ground states, ${}^1\Sigma_g^+$. Studies have shown that a number of different types of complexes of these metal ions exhibit MCD *A* terms, indicating the presence of degenerate excited states. The nature of these states, of course, depends upon the type of ligand coordinated to the metal ion. As for square complexes, halide ligands may be expected to exhibit LMCT, whereas for CN^- or PR_3 ligands, MLCT might be anticipated. In addition, metal-centered $d \rightarrow s$ and $d \rightarrow p$ transitions have also been observed. The HOMOs for

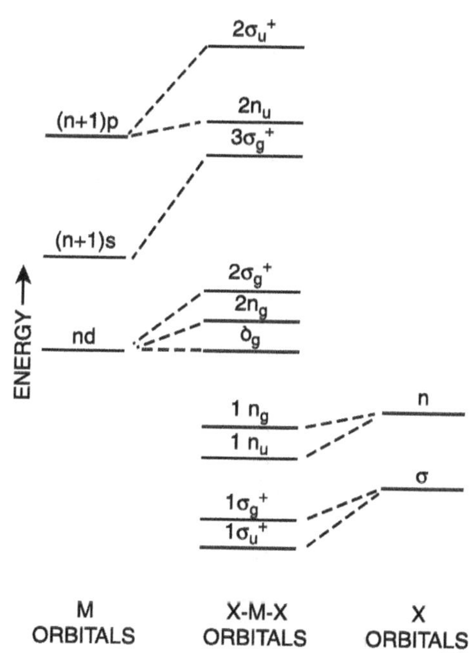

Figure 6.12. Energy levels for linear ($D_{\infty h}$) MX_2 halide complexes. Reprinted with permission from ref. 21. Koutek, M. E.; Mason, W. R. *Inorg. Chem.* **1980**, *19*, 648. Copyright 1980 American Chemical Society.

these complexes are metal-localized *nd* orbitals, while the LUMOs are visualized as $(n+1)s$ and $(n+1)p$ predominantly metal orbitals for halide complexes, or MOs involving ligand π^* orbitals for π acceptor ligands. As an example, Figure 6.12 shows the energy levels for MX_2; X = halide. For the linear geometry, transitions from a totally symmetric Σ_g^+ ground state to Σ_u^+ (*z*-polarized) or Π_u (*xy*-polarized) excited states are electric dipole allowed. The latter may exhibit *A* terms, while both the former and latter both will have *B* terms of some magnitude in MCD spectra. Two studies that illustrate the use of *A* terms from MCD spectra are briefly discussed below.

6.5.1. HgX_2 and AuX_2^-, X = Cl^-, Br^-, and I^-

The HgX_2 molecules and AuX_2^- ions represent linear two-coordinate halo complexes of Hg(II) and Au(I), respectively. In the case of the AuX_2^- ions, the electronic spectra exhibit both weak and intense bands in the UV region. For example, the absorption and MCD spectra for the $AuCl_2^-$ in Figure 6.13 show a weak band at

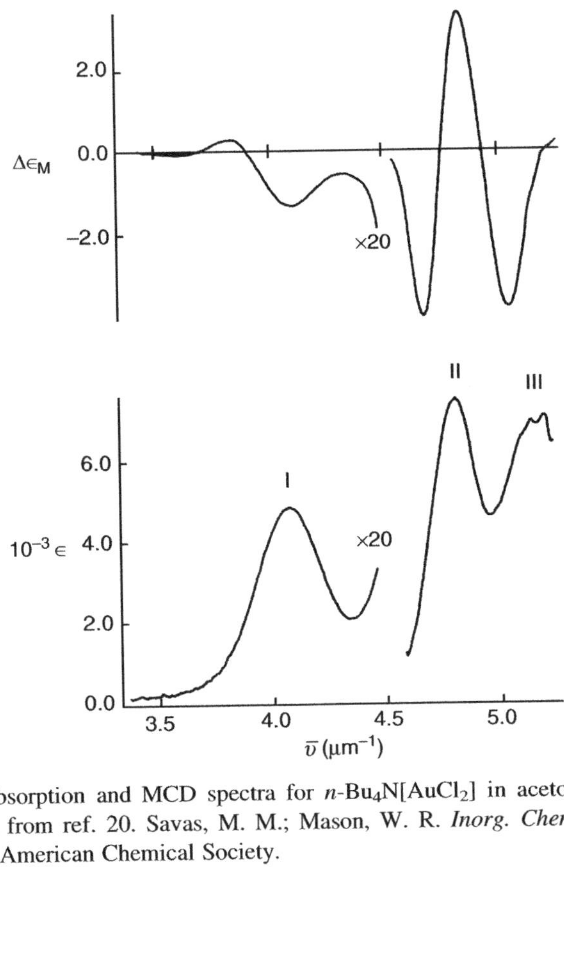

Figure 6.13. Absorption and MCD spectra for *n*-$Bu_4N[AuCl_2]$ in acetonitrile. Reprinted with permission from ref. 20. Savas, M. M.; Mason, W. R. *Inorg. Chem.* **1987**, *26*, 301. Copyright 1987 American Chemical Society.

4.05 μm^{-1}, which corresponds to a negative *A* term in the MCD spectrum, and two intense bands at 4.78 and 5.14 μm^{-1}; the former exhibits a clear positive *A* term and the latter appears to have a positive *A* term, but the spectral cutoff obscures the positive portion of the term [ref. 20]. The weak band was assigned as a metal-centered $d \rightarrow s$ transition, ${}^1\Sigma_g^+ \rightarrow {}^1\Pi_g$ $[2\pi_g(5d_{xz,yz}) \rightarrow 3\sigma_g^+\ (6s)]$. The negative *A* term is in accord with a Herzberg–Teller model involving the $\nu_3(\sigma_u^+)$ normal mode. The two intense bands were assigned as $d \rightarrow p$-type transitions because analogous bands in $AuBr_2^-$ and AuI_2^- are found at nearly the same energy; the $d \rightarrow p$ transition is metal localized and is not expected to be strongly ligand-dependent. The detailed interpretation takes into account extensive metal spin–orbit interactions, which make the assignment to a single configuration state difficult. However, it is likely that the band at 4.78 μm^{-1} is predominantly a transition to the Π_u state from the $2\sigma_g^+\ (5d_{z^2}) \rightarrow 2\pi_u(6p_x 6p_y)$ one-electron excitation. The calculated $\bar{A}_1/\bar{D}_0$ value for this transition (+1) is smaller than observed (+1.8), which argues for extensive mixing with other states of spin-forbidden (triplet) origin. An assignment to another state of mixed composition was given for the 5.14-μm^{-1} band. In addition to the $d \rightarrow p$ transitions, the spectra for $AuBr_2^-$ and AuI_2^- become

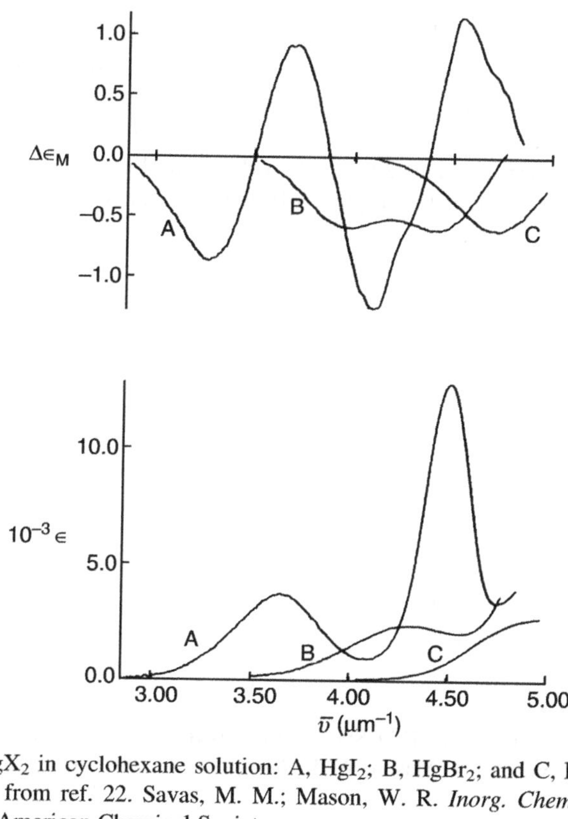

Figure 6.14. HgX_2 in cyclohexane solution: A, HgI_2; B, $HgBr_2$; and C, $HgCl_2$. Reprinted with permission from ref. 22. Savas, M. M.; Mason, W. R. *Inorg. Chem.* **1988**, *27*, 658. Copyright 1988 American Chemical Society.

progressively more complicated, and the presence of LMCT transitions was inferred. The LMCT states are also intermixed due to the increased halide spin–orbit coupling for Br^- and I^- [ref. 21].

In contrast to the $d \rightarrow s$ and $d \rightarrow p$ transitions for AuX_2^- ions, the UV spectra for the HgX_2 complexes are sensitive to the halide and show the energy ordering $Cl^- > Br^- > I^-$, which is typical for LMCT spectra (Figure 6.14 [ref. 22]). The red shift of the spectra of the Hg(II) molecules compared to the Au(I) ions is also consistent with LMCT. Unfortunately, the band energies for $HgCl_2$ and to some extent $HgBr_2$ are near the limit of measurement, so that it was not possible to obtain more details from absorption and MCD spectra. The spectra for HgI_2 were lower in energy and were obtained satisfactorily; two clear positive A terms in the MCD spectra (Figure 6.14, curves A) were observed. The spectral interpretation, however, is complicated by strong I^- spin–orbit coupling. A study of the LMCT spin–orbit states for the I^- complex revealed extensive mixing of Σ_u^+ and Π_u states of singlet and triplet origin which result from the $1\pi_u \rightarrow 3\sigma_g^+$ and $1\pi_g \rightarrow 2\pi_u$ excitations. The MCD spectrum was interpreted as positive *A* terms for Π_u states, together with positive pseudo *A* terms from close proximity of Π_u states and Σ_u^+ states [ref. 22].

6.5.2. MLCT Transitions for $Pt(PBu^t_3)_2$

The absorption and MCD spectra for $Pt(PBu^t_3)_2$ in acetonitrile or hydrocarbon solutions show several strong bands that are accompanied by prominent *A* terms (Figure 6.15). These bands have been interpreted [refs. 23, 24] in terms of transitions from the occupied $1\pi_g$, δ_g, and $2\sigma_{g+}$ metal orbitals to $1\pi_u$, an orbital that has both ligand π and metal *p* character (Figure 6.16); the model is similar to that used for

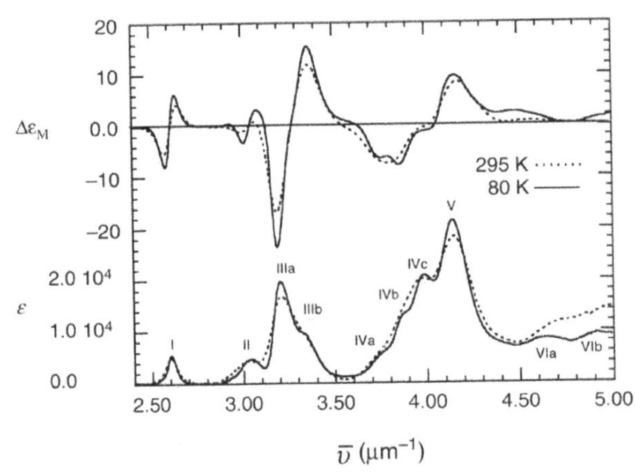

Figure 6.15. Absorption and MCD Spectra for $Pt(PBu^t_3)_2$ in 2-methylpentane. Reprinted with permission from ref. 24. Mason, W. R. *Inorg. Chem.* **2001**, *40*, 6316. Copyright 2001 American Chemical Society.

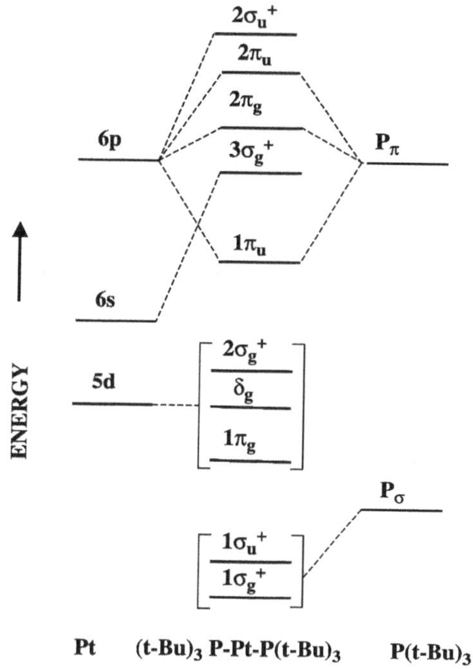

Figure 6.16. Energy levels for $Pt(PBu^t_3)_2$. Reprinted with permission from ref. 24. Mason, W. R. *Inorg. Chem.* **2001**, *40*, 6316. Copyright 2001 American Chemical Society.

$Pt(CN)_4^{2-}$ discussed above for square complexes. The MLCT character of the spectra is clearly demonstrated by the large (1.5 μm^{-1}) red shift of the spectrum for the Pt(0) complex compared with the spectrum for the analogous Au(I) complex $Au(PBu^t_3)_{2+}$ [ref. 23]. The detailed interpretation requires that the strong metal spin–orbit coupling be taken into account. The bands are ascribed to transitions to allowed Σ_u^+ and Π_u spin–orbit states of mixed singlet–triplet origin. The interpretation is complicated because several states are expected to be close in energy. For example, band IIIa and IIIb are assigned to transitions to the two $\Pi_u(^1\Pi_u)$ states from the $2\sigma_g^+ \rightarrow 1\pi_u$ and $\delta_g \rightarrow 1\pi_u$ excitations; the positive A terms predicted for the two states are assumed to overlap because the two states are expected to have zero magnetic interaction between them (no B terms). This interpretation is supported by the observed $\bar{A}_1/\bar{D}_0$ value for bands IIIa and IIIb together, which is found to be +2.5, whereas the $\bar{A}_1/\bar{D}_0$ value for each positive A term for the two $\Pi_u(^1\Pi_u)$ states is separately calculated to be +1. The conclusions drawn from the detailed interpretation of the spectra suggest that the range of energies covered by the excited states is likely due to Pt(0) spin–orbit coupling and that the energy separation of the 5d orbitals (ligand field splitting) is small, indicating only a minimal involvement of these occupied orbitals in metal–ligand bonding.

6.6. OCTAHEDRAL COMPLEXES

Diamagnetic octahedral complexes are found for several kinds of metal ions. These include transition metal complexes of low-spin nd^6 electron configuration and main group metal complexes of $nd^{10}(n+1)s^2$ or nd^{10} configurations. In all of these cases, allowed transitions consist of transitions from the totally symmetric ${}^1A_{1g}$ ground state to ${}^1T_{1u}$ excited states, but strong spin–orbit coupling can intermix T_{1u} states of singlet and triplet origin, as discussed above. The types of transitions depend upon the type of complex, with LMCT observed for halide complexes, MLCT for cyano complexes, and metal-localized $s \rightarrow p$ for main group metal ions. Vibronically allowed ligand field (LF) transitions have also been studied for transition metal complexes, but the MCD spectra often consist of *B* terms and are difficult to characterize.

6.6.1. LMCT Transitions for nd^6 Halide Complexes

A number of halide complexes of Rh(III), Ir(III), Pd(IV), and Pt(IV) have been investigated; also, *A* terms have been observed for the intense bands for all except Ir(III), where only weaker LF band were observed [ref. 25]. The intense bands have been assigned as LMCT on the basis of energy shifts with oxidation state, and they can be divided into two systems: π-LMCT and σ-LMCT. The former are to lower energy by about 1 μm^{-1} and are less intense than the latter. The LMCT pattern is quite analogous to that observed for square complexes discussed above. As an example, Figure 6.17 shows the MCD for $PtCl_6^{2-}$, and Figure 6.18 presents a schematic energy-level diagram for octahedral halo complexes [ref. 25], where the HOMO is the metal $2t_{2g}$. The band at 3.8 μm^{-1} was assigned as the π-LMCT ${}^1A_{1g} \rightarrow {}^1T_{1u}$ $[2t_{1u}(\pi)^5 e_g]$, and the band near 4.9 μm^{-1} was assigned as the σ-LMCT ${}^1A_{1g} \rightarrow {}^1T_{1u}$ $[1t_{1u}(\sigma)^5 e_g]$. The interpretation of the negative *A* term observed in the MCD for the 3.8-μm^{-1} band does, however, require mixing of the chloride $t_{1u}(\sigma)$ and $t_{1u}(\pi)$ orbitals, just as described above for the square chloro complexes. This was considered necessary because, in the absence of strong ligand spin–orbit coupling (a reasonable assumption for the chloro ligand), the calculated $\bar{A}_1/\bar{D}_0$ value for the π-LMCT without σ–π mixing (−0.25) was smaller than observed (−0.58); the value with σ–π mixing (−0.49) was found to be closer to the observed value. It is interesting that LMCT from the chloride $t_{2u}(\pi)$ nonbonding orbitals can be ruled out by symmetry considerations because the transition to the ${}^1T_{1u}$ $[(t_{2u})^5 e_g]$ state is expected to have a positive *A* term, which is not in accord with experiment. The LMCT absorption and MCD spectra for $PtBr_6^{2-}$ and PtI_6^{2-} are more complicated, and inclusion of ligand spin–orbit coupling in the analysis was necessary; extensive mixing of states of singlet and triplet parentage, including those from the $[(t_{2u})^5 e_g]$ configuration, was postulated.

6.6.2. LMCT Transitions for $4d^{10}$ Complexes $SbCl_6^-$ and $SnCl_6^{2-}$

Another example of octahedral chloro complexes involves the main group Sb(V) and Sn(IV) $4d^{10}$ metal ions. The $SbCl_6^-$ and $SnCl_6^{2-}$ complexes of these metal

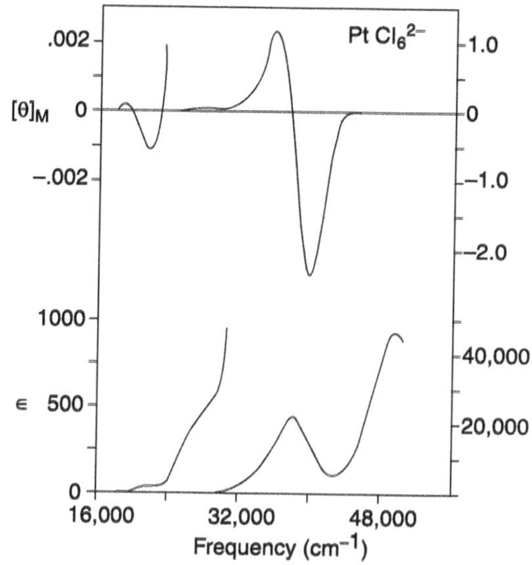

Figure 6.17. Absorption and MCD for $PtCl_6^{2-}$ in 2M HCl. Reprinted with permission from ref. 25. Henning, G. N.; Dobosh, P. A.; McCaffery, A. J.; Schatz, P. N. *J. Amer. Chem. Soc.* **1970**, *92*, 5377. Copyright 1970 American Chemical Society.

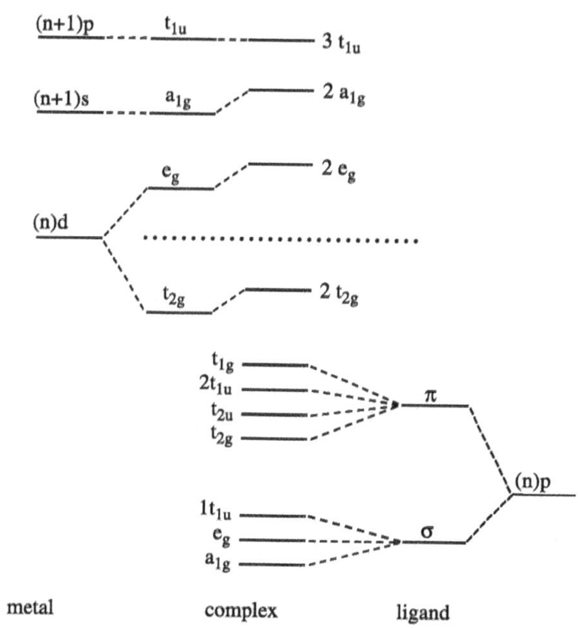

Figure 6.18. Energy levels for octahedral halo complexes. Reprinted with permission from ref. 25. Henning, G. N.; Dobosh, P. A.; McCaffery, A. J.; Schatz, P. N. *J. Amer. Chem. Soc.* **1970**, *92*, 5377. Copyright 1970 American Chemical Society.

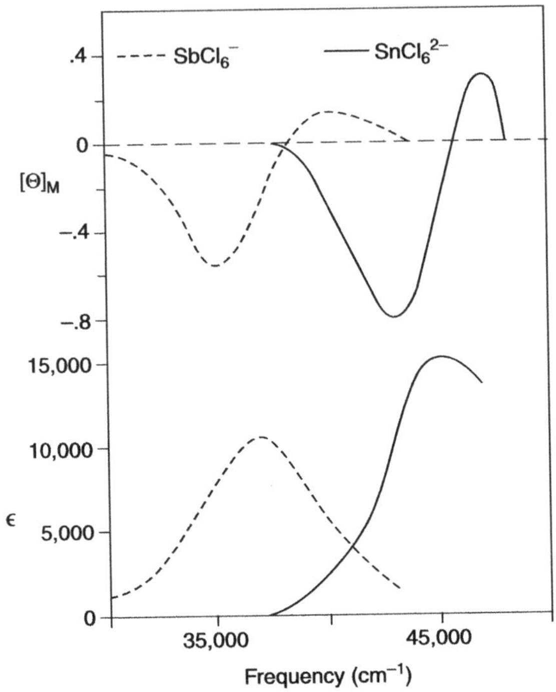

Figure 6.19. Absorption and MCD spectra for $SbCl_6^-$ (dashed lines) and $SnCl_6^{2-}$ (solid lines) in 12 M HCl. Reprinted with permission from ref. 26. Schatz, P. N.; Stephens, P. J.; Henning, G. N.; McCaffery, A. J. *Inorg. Chem.* **1968**, *7*, 1246. Copyright 1968 American Chemical Society.

ions both exhibit positive *A* terms in the MCD spectra for intense bands at 3.7 and 4.5 μm^{-1}, respectively, as shown in Figure 6.19 [ref. 26]. The observed $\bar{A}_1/\bar{D}_0$ values were +0.2 to +0.7 (the precision was limited by solvent cutoff). Two possible assignments have been considered for these bands: LMCT from $2t_{1u}(\pi) \rightarrow 2a_{1g}$ metal 5*s* and $d \rightarrow p$ from the HOMO $2e_g \rightarrow 3t_{1u}$ 5*p* (see Figure 6.18). A simple, single-configuration approximation, without spin–orbit coupling, predicts a positive *A* term with $\bar{A}_1/\bar{D}_0 = +0.5$ for the former, while for the latter a negative *A* term is predicted with $\bar{A}_1/\bar{D}_0 = -0.5$. In this case the sign and approximate magnitude of the observed *A* term allows experimental support for one assignment (LMCT) while eliminating the other plausible interpretation ($d \rightarrow$ p) [ref. 26].

6.6.3. Metal-Centered $s \rightarrow p$ and LMCT Transitions in $BiCl_6^{3-}$

The MCD spectra for octahedral halo Bi(III) complexes show a prominent low-energy positive *A* term in their UV-Vis spectra. This term is also accompanied by higher-energy intense bands ascribed to LMCT, which have negative *A* terms.

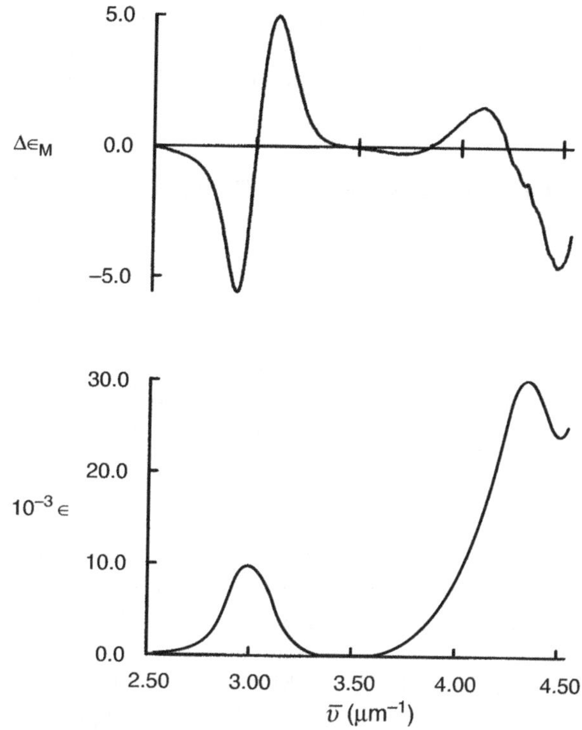

Figure 6.20. Absorption and MCD spectra for $(n\text{-}Bu_4N)_3[BiCl_6]$ in 50-fold excess Et_4NCl in acetonitrile solution. Reprinted with permission from ref. 27. Mason, W. R. *Inorg. Chem.* **1999**, *38*, 2742. Copyright 1999 American Chemical Society.

For example, the MCD spectrum for $BiCl_6^{3-}$ in Figure 6.20 shows a strong positive A term at 3.0 μm^{-1} and at higher energy a negative A term near 4.3 μm^{-1} [ref. 27]. The low-energy A term has been assigned to a metal-centered $6s \rightarrow 6p$ transition $^1A_{1g} \rightarrow {}^3T_{1u}$ $[2a_{1g}3t_{1u}]$ (see Figure 6.18; the HOMO for BiX_6^{3-} is $2a_{1g}$). Strong Bi(III) spin–orbit coupling allows mixing of this triplet state with the higher-energy (unobserved) singlet state $^1T_{1u}$ $[2a_{1g}3t_{1u}]$ of the same configuration. The $\bar{A}_1/\bar{D}_0$ value was found to be +1.6 for the 3.0-μm^{-1} band, compared to +3.0 for the $T_{1u}(^3T_{1u})$ and +2.0 for the $T_{1u}(^1T_{1u})$. It was argued that there must be some quenching of the orbital magnetic moment due to covalent bonding of the ligands to the $3t_{1u}$ metal p orbitals; a complete quenching would be expected to give a value of +2.0. However, the lower value must also include the extensive triplet–singlet spin–orbit intermixing of the $^3T_{1u}$ $[2a_{1g}3t_{1u}]$ and $^1T_{1u}$ $[2a_{1g}3t_{1u}]$ states which gives the transition its considerable intensity. The higher-energy negative A term is in accord with the LMCT $^1A_{1g} \rightarrow {}^1T_{1u}$ $[(t_{2g})^5 3t_{1u}]$; an alternative transition involving the nonbonding t_{1g} ligand orbitals, $^1A_{1g} \rightarrow {}^1T_{1u}$ $[(t_{1g})^5 3t_{1u}]$, can be excluded because the A term is predicted to be positive and therefore not in agreement with experiment. This is another example of where the sign of the MCD *A* term can assist between alternative assignments. The LMCT spectra for $BiBr_6^{3-}$ and BiI_6^{3-} show increased complexity compared to $BiCl_6^{3-}$ in a fashion that indicates the increasing importance of halide spin–orbit interaction as discussed above.

6.6.4. MLCT and $d \rightarrow p$ Transitions for nd^6 $M(CO)_6$ and $M(CN)_6^{4-}$ Complexes

The hexacarbonyl $M(CO)_6$ (M = Cr(0), Mo(0), and W(0)) and hexacyano $M(CN)_6^{4-}$(M = Fe(II), Ru(II), and Os(II)) complexes of nd^6 electron configuration exhibit intense bands in the UV which have been assigned as MLCT from the occupied ndt_{2g} orbitals (HOMO) to empty ligand $t_{1u}\pi^*$ orbitals (LUMO). Figure 6.21 shows schematically the MOs involved. However, when the MCD spectra were measured [ref. 28], the $M(CO)_6$ complexes showed a rather different spectral pattern than that for the $M(CN)_6^{4-}$ ions. For example, the absorption and MCD spectra for $W(CO)_6$ are shown in Figure 6.22, which are nearly identical to that found for both the Cr(0) and Mo(0) hexacarbonyl complexes. The MCD spectrum is dominated by two prominent *B* terms. In contrast, the spectra for the hexacyano $M(CN)_6^{4-}$ complexes (Figure 6.23) show a positive *A* term for the intense band at high energy presented by each complex, indicating a transition to a degenerate excited state. The explanation for the difference was interpreted in terms of the composition of the t_{1u} LUMO, which contains both ligand π^* and metal $(n+1)p$ character. In the case of the $M(CO)_6$ complexes, the d–p energy separation was estimated from atomic spectral data to be 2.5–3 μm^{-1} and thus lower than the energies of the observed spectral bands. Therefore the t_{1u} LUMO must be of mixed π^* and metal p character. In contrast, for the M(II) hexacyano complexes the separation is about 7–8 μm^{-1} and thus considerably higher than the energies of the bands in the observed spectra. In this case the t_{1u} LUMO was

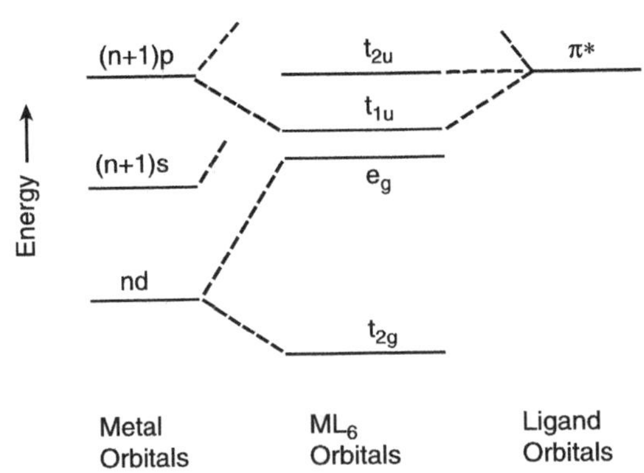

Figure 6.21. Schematic MOs for the nd^6 $M(CO)_6$, and $M(CN)_6^{4-}$ complexes. Reprinted with permission from ref. 28. Chastain, S. K.; Mason, W. R. *Inorg. Chem.* **1981**, *20*, 1395. Copyright 1981 American Chemical Society.

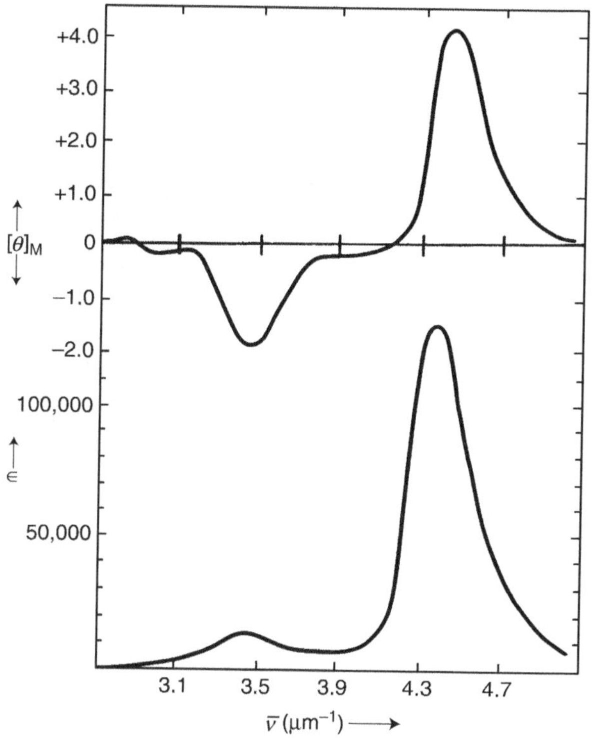

Figure 6.22. Absorption and MCD spectra for $W(CO)_6$ in acetonitrile solution. Reprinted with permission from ref. 28. Chastain, S. K.; Mason, W. R. *Inorg. Chem.* **1981**, *20*, 1395. Copyright 1981 American Chemical Society.

considered to be mostly ligand π^*. Therefore, the $t_{1u}(i)$ LUMO orbitals are visualized as mixed π^* and metal p character given by

$$|t_{1u}(i)\rangle = a_i|\pi^*\rangle + b_i|(n+1)p\rangle$$

where a_i and b_i are mixing coefficients. The $\bar{A}_1/\bar{D}_0$ value for the ${}^1A_{1g} \rightarrow {}^1T_{1u}$ $[(t_{2g})^5 t_{1u}]$ is then proportional to the magnitude of the a_i and b_i coefficients and is given by ref. 28

$$\overline{A}_1/\overline{D}_0 = (3/4)|a_i|^2 - (1/2)|b_i|^2$$

The MLCT and $d \rightarrow p$ character of the transition will affect the A-term sign because a positive A term is predicted for MLCT and a negative A term is predicted for $d \rightarrow p$. In fact, an a_i/b_i ratio of approximately 0.82 will give a zero $\bar{A}_1/\bar{D}_0$ value, and therefore only B terms will be seen in the MCD spectrum as in the case of the hexacarbonyl complexes. For the hexacyano complexes, if $b_i \approx 0$, the transition will be predominantly MLCT and a positive A term will result.

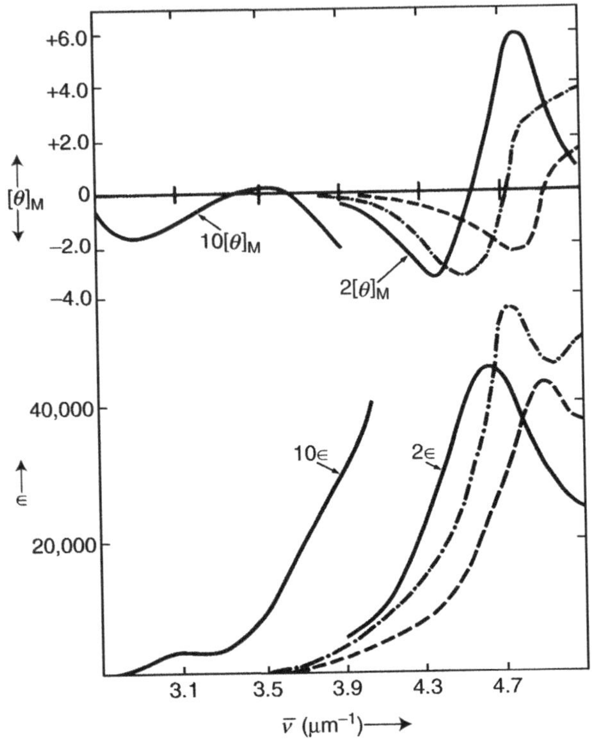

Figure 6.23. Absorption and MCD spectra for $M(CN)_6^{4-}$ in water: $Fe(CN)_6^{4-}$, solid lines; $Ru(CN)_6^{4-}$, dashed lines; and $Os(CN)_6^{4-}$, dashed–dotted lines. Reprinted with permission from ref. 28. Chastain, S. K.; Mason, W. R. *Inorg. Chem.* **1981**, *20*, 1395. Copyright 1981 American Chemical Society.

Thus the interpretation of the spectra cannot be given as simply MLCT (or $d \rightarrow p$) without the consideration of the composition of the t_{iu} LUMO.

6.7. METAL CLUSTER COMPLEXES

Cluster complexes consist of three or more metal centers that are bonded together through metal–metal bonding; the metal centers also are coordinated by one or more ligands of various types. The simplest trimetallic clusters often exhibit an equilateral triangular array of metal centers with nominally D_{3h} symmetry, and therefore two-dimensional degeneracy is possible for E'-type electronic states. The first two examples to be considered will consist of triangular arrays of metal ions: One involves Pt(0) having $5d$ valence electrons, and the other involves an Hg_3^{4+} core where the valence orbitals are $6s$ and $6p$. Following the discussion of these simple cluster complexes, some larger and more complicated gold cluster ions will be considered.

6.7.1. MLCT Transitions for Triangular $Pt_3(CO)_3(P(t\text{-}Bu)_3)_3$

The $Pt_3(CO)_3(P(t\text{-}Bu)_3)_3$ complex consists of an equilateral triangle of Pt atoms, with CO ligands bridging each side, and tri-*t*-butylphosphine ligands, $P(t\text{-}Bu)_3$, in-plane terminally bonded at each vertex of the triangle [ref. 29]. The complex is a member of an interesting class of 42 electron trinuclear Pt complexes with similar structures where typically the Pt–Pt distances fall in the range of 2.62–2.71 Å, which is shorter than the Pt–Pt distance in Pt metal (2.78 Å), indicative of Pt–Pt bonding. The electronic structures of these complexes have been of interest for their relevance as catalysts and to the structure of Pt metal surfaces. There have been several Hückel MO (HMO) calculations utilizing fragment molecular orbitals (FMOs) performed in order to model the electronic structures of these cluster complexes; however, spectroscopic results for corroboration are not extensive. The absorption and MCD spectra for $Pt_3(CO)_3(P(t\text{-}Bu)_3)_3$ in acetonitrile solution are shown in Figure 6.24. The spectra show that the MCD spectrum is better resolved and thus reveals more features than the absorption

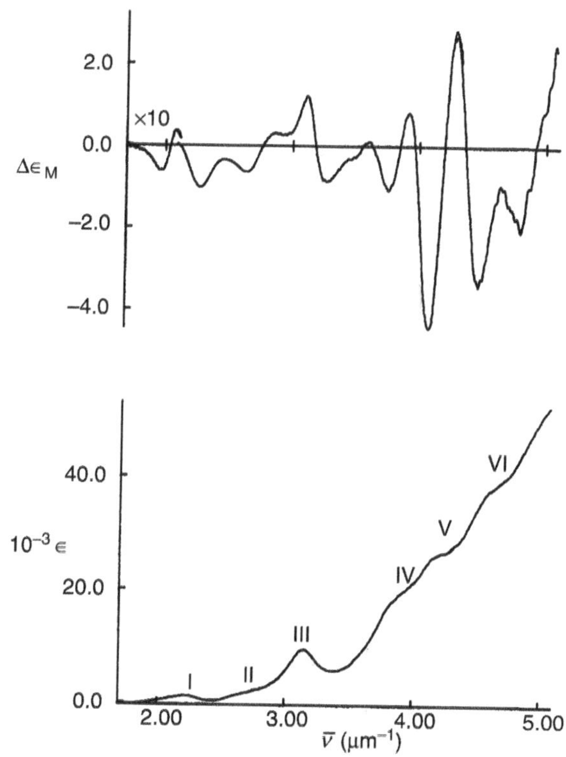

Figure 6.24. Absorption (lower curve) and MCD (upper curve) for $Pt_3(CO)_3(P(t\text{-}Bu)_3)_3$ in acetonitrile solution. The MCD data for 1.8–2.2 μm^{-1} were multiplied by 10 before plotting. Reprinted with permission from ref. 29. Jaw, H.-R. C.; Mason, W. R. *Inorg. Chem.* **1990**, *29*, 3452. Copyright 1990 American Chemical Society.

spectrum, which consists of a number of partly resolved shoulders. It is clear from the spectra that there are a number of transitions, and many of them are characterized by *A* terms signaling degeneracy. The HMO calculations provided a framework in which to interpret the MCD spectra. Figure 6.25 shows a schematic MO energy-level diagram with the highest filled levels (predominantly 5*d* platinum orbital combinations) and the lowest empty levels (predominantly ligand π^* P and π^*CO orbital combinations). The complex is diamagnetic and assumed to have D_{3h} core symmetry with an $^1A_1'$ ground state. Therefore, transitions to out-of-plane *z*-polarized A_2'' and in-plane xy-polarized E' excited states will be dipole allowed. Numerous allowed excited states can be visualized, especially when Pt spin–orbit coupling is included ($\zeta_{5d} \sim 3000$–$4000\ \mathrm{cm}^{-1}$). The interpretation, therefore is based on relative intensity and energies as judged by the HMO calculations and the MCD *A* term sign for transitions to E' states. For example, the lowest energy band I is relatively weak for a dipole allowed transition ($\varepsilon = 1500\ \mathrm{M}^{-1}\mathrm{cm}^{-1}$) and is accompanied by a negative *B* term in the MCD spectrum. The transition is therefore assigned to $^1A_1' \rightarrow A_2''(^1A_2')\ [(2a_1')(3a_2'')]$. The low

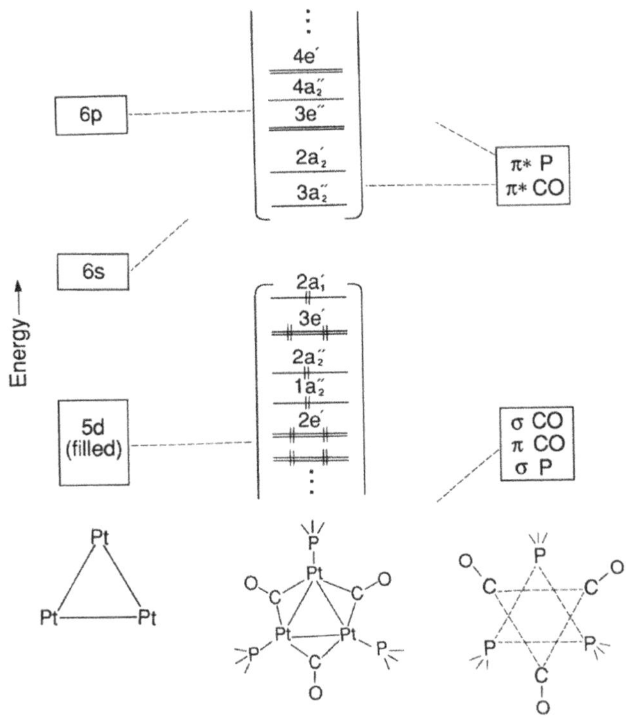

Figure 6.25. Molecular orbital energy-level diagram for $Pt_3(CO)_3(P(t\text{-}Bu)_3)_3$ assuming D_{3h} symmetry. Reprinted with permission from ref. 29. Jaw, H.-R. C.; Mason, W. R. *Inorg. Chem.* **1990**, *29*, 3452. Copyright 1990 American Chemical Society.

intensity was rationalized by the inplane σ character of the $2a_1'$ orbital and the out-of-plane π character of the $3a_2''$ orbital. The transition is of the metal-to-ligand charge transfer type (MLCT) because the filled orbital is predominantly metal 5d orbitals localized in the Pt triangle, while the empty orbital is ligand-based. Another fairly clear assignment is for band III, which is the only transition to exhibit a negative A term. The sign of the A terms for transitions to excited E' states can be determined from the $\bar{A}_1/\bar{D}_0$ parameter ratio, which, in turn, can be determined from

$$\overline{A}_1/\overline{D}_0 = -(2)^{-1/2}\mu_B\langle E'\|\mu\|E'\rangle$$

where $\mu = -\mu B(L+2S)$ and $\bar{D}_0 = (1/3|\langle A_1'\|m\|E'\rangle|^2$. The only negative A term expected among the low-energy transitions visualized in Figure 6.25 is for that to the $E'(^1E')$ state of the $(3e')^3(2a_2')$ excited configuration. This transition, like that for band I, is primarily MLCT. The assignment of band II was to the states corresponding to the spin–orbit states of triplet origin from the $(3e')^3(2a_2')$ excited configuration. The MCD was interpreted as due to a positive pseudo A term from the close energies of the $E'(^3E')$ and $A_2''(^3E')$ spin–orbit states. It was argued that the magnitude of the positive pseudo A term must be substantially larger than the A term expected for the transition to the $E'(^3E')$ state alone, because the A-term sign for this transition, like the predominantly singlet, is expected to be negative (see ref. 29. for the details of the B-term sign determinations). Finally, bands IV and V were also assigned as MLCT to $E'(^3E')$ together with $A_2''(^3E')$ (band IV) and $E'(^1E')$ (band V) spin–orbit states from the $(2e')^3(2a_2')$ excited configuration (see Figure 6.25). Reasoning similar to that for band II also suggested that there may be a component of a pseudo A term present from $E'(^3E')$ and $A_2''(^3E')$, but one that is predicted to be positive and the same sign as expected for the $E'(^3E')$ state. The MCD spectral observations, unfortunately, sometimes consist of contributions from several states, the interpretation of which make assignments challenging. However, because of the changes in sign of $\Delta\varepsilon_M$, the MCD spectrum in this case offers better resolved data than the absorption spectrum alone.

6.7.2. Metal-Centered Transitions in the $Hg_3(dppm)_3^{4+}$ Cluster Complex

An Hg_3^{4+} triangle is found in the center of solid $[Hg_3(dppm)_3](SO_4)_2$, dppm = bis(diphenylphosphino)methane (Figure 6.26 [ref. 30]). The Hg–Hg distance was found to be 2.76–2.80 Å. In contrast to the triangular $Pt_3(CO)_3(P(t\text{-}Bu)_3)_3$ complex considered above, the complex exhibits only two features in ethanol solution at room temperature: an intense broad absorption band at 3.01 μm^{-1} and a poorly resolved shoulder at ~3.3 μm^{-1}. An assignment based on D_{3h} symmetry for the broad band in solution was proposed earlier as $^1A_1' \rightarrow {}^1E'$ $[(a_1')^2 \rightarrow (a_1')(e')]$, a metal-centered $\sigma \rightarrow \sigma^*$-type excitation. Figure 6.27 shows the metal-centered MO energy levels formulated for the planar Hg_3^{4+} core in D_{3h} symmetry. The HOMO consists of the σ-bonding a_1' orbital, which is a symmetric

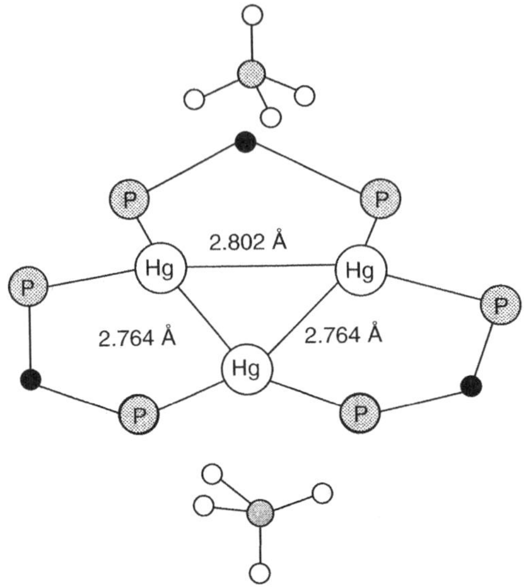

Figure 6.26. Structure of $[Hg_3(dppm)_3](SO_4)_2$ in the solid (phenyl substituents on the P donors and hydrogens on the C atoms have been omitted) with the SO_4^{2-} ions above and below the triangular plane of Hg atoms. Reprinted with permission from ref. 30. Mason, W. R. *Inorg. Chem.* **1997**, *36*, 1164. Copyright 1997 American Chemical Society.

combination primarily of Hg 6*s* orbitals. The lowest-energy-allowed transition visualized from this diagram is the $\sigma \rightarrow \sigma^*$ transition to the E' state (labeled I in the diagram), so the assignment of the lowest energy band seemed reasonable. The next lowest energy transition was visualized as a $\sigma \rightarrow \pi$ transition to an A_2'' state (labeled II in the diagram) and could conceivably be responsible for the poorly resolved shoulder absorption. If this interpretation were correct, then MCD spectra should show an *A* term for the observed lowest energy band at 3.01 μm^{-1} in the absorption spectrum and possibly a *B* term for the shoulder at $\sim$3.3 μm^{-1}. In order to sharpen the broad band, and possibly resolve the shoulder, in the absorption spectrum, along with the MCD spectrum, the cluster complex salt was dissolved in a solution that forms a glass when cooled to low temperature. The glass preserves the isotropic solution environment for the complex. Figure 6.28 shows the absorption and MCD spectra for $Hg_3(dppm)_3(SO_4)_2$ in 5 : 3 : 2 v/v ethanol–methanol–ether solution at 295 K (solid lines) and glass at 80 K (dashed lines). Indeed the spectra sharpen at low temperature, revealing two absorption bands labeled I and II, and is shifted slightly to higher energy. Also, the MCD is resolved into a positive *A* term for band I and a negative *B* term for band II. It is important to note here that the temperature changes displayed by the MCD are due to band narrowing and shift of the absorption at low temperature and are not due to temperature-dependent *C* terms and ground-state degeneracy. The changes on cooling are typical for allowed transitions.

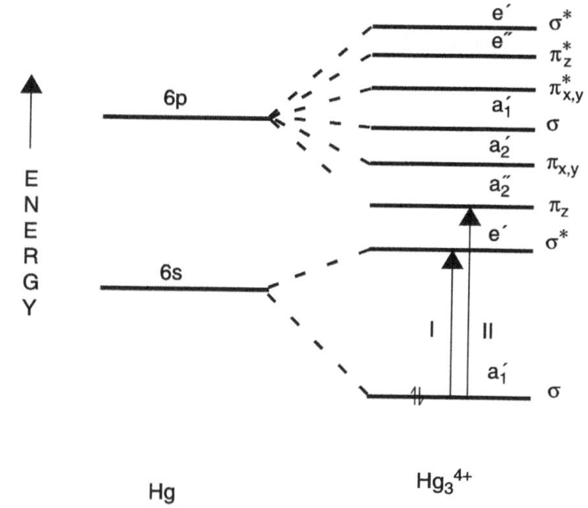

Figure 6.27. Metal-centered MO energy levels for Hg_3^{4+}, assuming D_{3h} symmetry. Reprinted with permission from ref. 30. Mason, W. R. *Inorg. Chem.* **1997**, *36*, 1164. Copyright 1997 American Chemical Society.

Though it may appear that the observed MCD is consistent with the two low-energy transitions indicated in Figure 6.27 and therefore consistent with the proposed $\sigma \rightarrow \sigma^*$ transition for band I and $\sigma \rightarrow \pi$ transition for band II, a careful consideration of the *A*-term sign for band I shows that the sign is *not* consistent with that expected for the $^1E'[(a_1')^2 \rightarrow (a_1')(e')]$ excited state. The *A* term for this

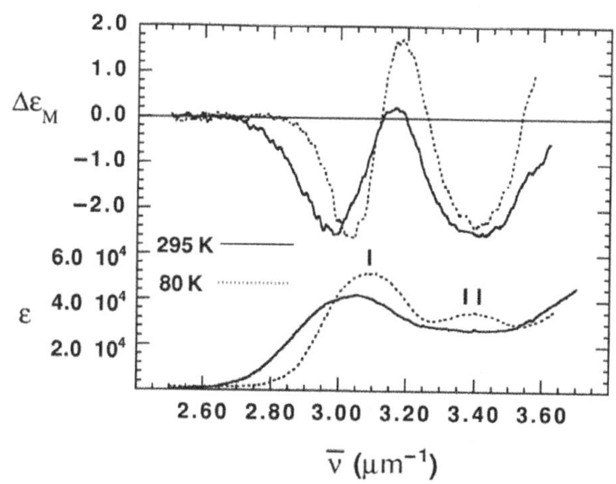

Figure 6.28. Absorption (lower curves) and MCD (upper curves) for $Hg_3(dppm)_3(SO_4)_2$ in 5:3:2 v/v ethanol–methanol–ether solution at 295 K (solid lines) and 80 K (dashed lines). Reprinted with permission from ref. 30. Mason, W. R. *Inorg. Chem.* **1997**, *36*, 1164. Copyright 1997 American Chemical Society.

$\sigma \rightarrow \sigma^*$ transition is predicted to be of small magnitude and, most importantly, *negative*. The *A*-term sign was determined by

$$\bar{A}_1/\bar{D}_0 = \frac{-1}{\sqrt{2}\mu_B}\langle E'\|\mu\|E'\rangle = \pm\langle e' \pm 1\|l_0\|e' \pm 1\rangle$$

where the one-electron $e' \pm 1$ MOs are in standard basis [ref. 1]. Since the e' orbitals are composed of Hg 6*s* atomic orbitals, the origin of the *A* term must arise from two-centered integrals because one-centered terms have zero angular momentum. The two-centered terms were approximated (in units of $h/2\pi$) by

$$\langle\phi_i|l_z|\phi_j\rangle = iR_{0j}\cos 60\langle\phi_i|\partial/\partial x|\phi_j\rangle$$

where ϕ_j (i, $j = 1$, 2, and 3) represent the 6*s* atomic orbitals on the three Hg atoms, $i = (-1)^{1/2}$, and R_{0j} is the distance from the origin to atom *j*. If the 6*s* orbitals are approximated by Slater-type functions, the right side of the equation above can be reduced to overlap integrals that can be evaluated for the average Hg–Hg distance in the Hg_3^{4+} triangle. The $\bar{A}_1/D_0$ ratio is found to be negative (~ -0.5 or smaller), determined by the magnitude of the overlap integrals (see ref. 30 for details for the approximations used). This observation illustrates, once again, that for MCD spectra to be consistent with and support a proposed assignment, the spectra must be consistent, not only in terms of energy and magnitude, but also in terms of the sign expected for the transition to a given excited state. In this case, in order to explain the MCD results, the assignment of band II was first considered in detail. On the basis of its higher energy and lower intensity than band I, band II was assigned as the $\sigma \rightarrow \pi$ transition to the nondegenerate A_2'' ($^1A_2'$) state of the ((a_1')(a_2'') excited configuration (II in Figure 6.27). This predominantly singlet excited state would then be expected to exhibit a *B* term in the MCD, the sign of which would depend upon higher energy states, transitions to which are obscured by absorptions from the phenyl substituents on the dppm ligands; the magnetic mixing with lower energy states was assumed to be very small based on approximation of matrix elements from orbital overlap arguments (see ref. 30 for details). It was argued that if this assignment is accepted, then an explanation for the positive *A* term for band I can involve the predominantly triplet states $E'(^3A_2'')$ and $A_1''(^3A_2'')$, which are expected to be to lower energy than the A_2'' ($^1A_2''$) state, just in the same region as the $^1E'$ state. The *A* term for the $E'(^3A_2')$ state is predicted to be positive, predominantly from spin angular momentum, and any contribution from the transition to the $A_1''(^3A_2'')$ state is expected to be very weak because it is dipole forbidden. Therefore, the interpretation of band I was that its intensity is carried by the $^1A_1 \rightarrow {^1E'}$ [$(a_1')^2 \rightarrow (a_1')(e')$] dipole allowed transition, but that the negative *A* term for this transition is weak and obscured by a stronger positive *A* term from the transition to the $E'(^3A_2'')$ state associated with the higher energy A_2'' ($^1A_2''$) state (band II). It was argued that the small magnitude of the *A* term for the $\sigma \rightarrow \sigma^*$ transition to the $E'(^1E')$ state provides indirect support for a 6*s* atomic orbital description of the Hg_3^{4+} σ and σ^* orbitals because if 6*p* orbitals were involved to any significant extent, then the negative *A* term for the $E'(^1E')$ state would have been much stronger.

6.7.3. Pt → Au Framework and Intraframework Transitions in the Gold Cluster Complexes $Pt(AuPPh_3)_8^{2+}$ and $Au(AuPPh_3)_8^{3+}$

As examples of larger cluster complexes, the gold cluster complexes are particularly interesting, featuring high-symmetry, heavy atoms, and complicated electronic absorption spectra. MCD has proved to be a useful tool in interpreting the complicated spectra and thereby establishing an experimental spectroscopic basis for a bonding model for these complexes [refs. 31–33]. The structures of the centered Pt (and also a related Pd) complex and the centered gold complex are shown in **1** and **2** below [ref. 33]. The bonding framework for these clustercomplexes

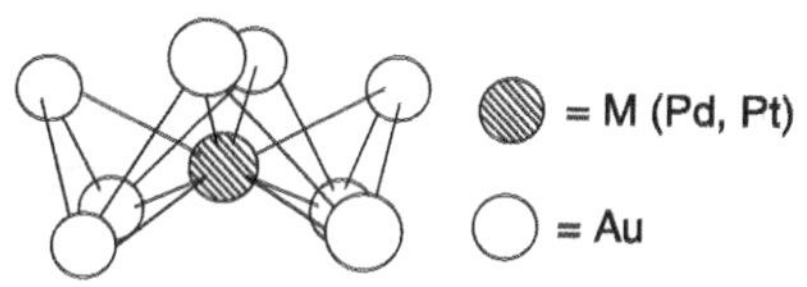

1

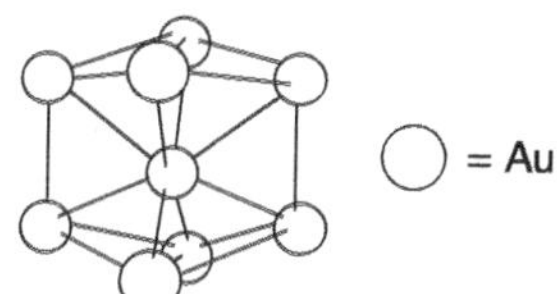

2

The symmetry of the centered crown structure **1** is D_{4d}, while that for the centered icosahedral fragment **2** is D_{2h}. Reprinted with permission from ref. 33. Mason, W. R. *Inorg. Chem.* **2000**, *39*, 370. Copyright 2000 American Chemical Society.

was formulated in terms of Au–Au bonding among the peripheral Au atoms of $(AuPPh_3)_8^{2+}$ primarily through the interaction of 6*s* atomic orbitals, which are referred to as framework orbitals. The nd^{10}-centered atom in both **1** and **2** was presumed to be bonded to the outer framework Au atoms by means of interaction with the central 6*s* orbital. Figure 6.29 shows a schematic MO energy-level diagram based on the D_{4d} symmetry for the $Pt(AuPPh_3)_8^{2+}$ cluster ion. In D_{4d} symmetry, transitions to E_1 (*xy*-polarized) and B_2 (*z*-polarized) excited states are dipole-allowed, the former can exhibit *A* terms and *B* terms in MCD spectra, while the latter can exhibit only *B* terms. Clearly, spin–orbit coupling is important because of the presence of the heavy metal atoms, and zero-order singlet and triplet states of the same symmetry will be intermixed. The low-energy absorption and MCD spectra for $Pt(AuPPh_3)_8^{2+}$ was found to be markedly different from that for $Au(AuPPh_3)_8^{3+}$ or $Pd(AuPPh_3)_8^{2+}$ [refs. 31, 32]. The differences observed for the Pt-centered cluster were attributed to the presence of higher-energy (less stable) filled 5*d* orbitals on the Pt(0) than expected for either the center 5*d* orbitals of Au(I) or 4*d* orbitals of Pd(0). Furthermore, the $Pt(AuPPh_3)_8^{2+}$ complex reacts with CO forming $Pt(CO)(AuPPh_3)_8^{2+}$, and this complex presents low-energy absorption and MCD more similar to the $Au(AuPPh_3)_8^{3+}$ and $Pd(AuPPh_3)_8^{2+}$ [ref. 32]. The change in the low-energy bands was attributed to the stabilizing effect of Pt–CO bonding on the central 5*d* Pt orbitals. The absorption and MCD spectra of the nitrate salts of $Pt(AuPPh_3)_8^{2+}$ and $Au(AuPPh_3)_8^{3+}$ embedded

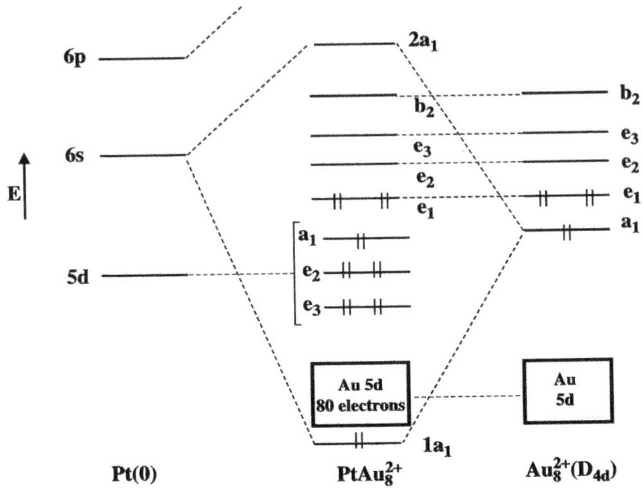

Figure 6.29. Molecular orbital energy levels in D_{4d} symmetry for the $Pt(AuPPh_3)_8^{2+}$ ion. Reprinted with permission from ref. 33. Mason, W. R. *Inorg. Chem.* **2000**, *39*, 370. Copyright 2000 American Chemical Society.

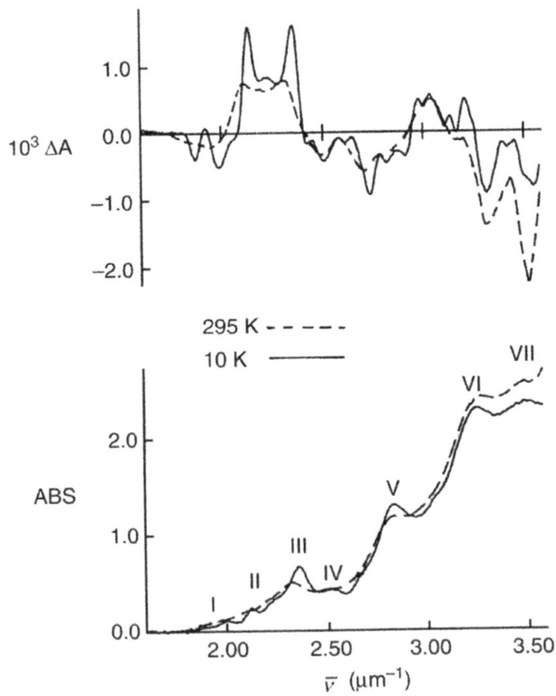

Figure 6.30. Absorption (lower curves) and MCD (upper curves) for $Pt(AuPPh_3)_8(NO_3)_2$ embedded in PMMA at 295 K (dashed lines) and 10 K (solid lines). Reprinted with permission from ref. 33. Mason, W. R. *Inorg. Chem.* **2000**, *39*, 370. Copyright 2000 American Chemical Society.

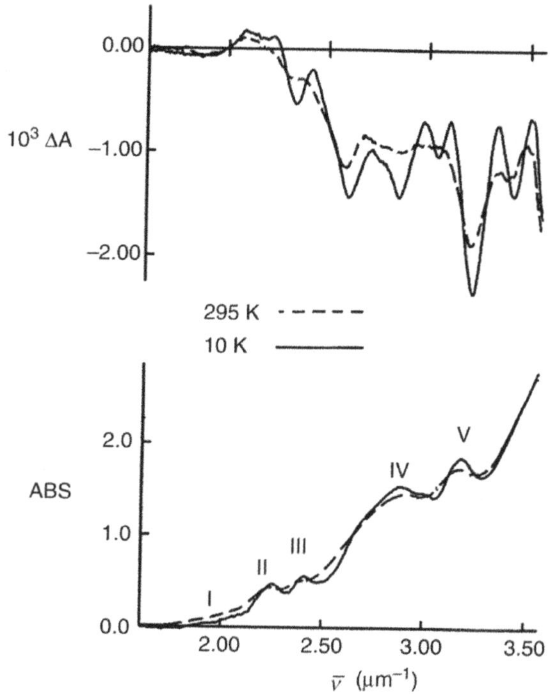

Figure 6.31. Absorption (lower curves) and MCD (upper curves) for $Au(AuPPh_3)_8(NO_3)_3$ embedded in PMMA at 295 K (dashed lines) and 10 K (solid lines). Reprinted with permission from ref. 33. Mason, W. R. *Inorg. Chem.*, **2000**, *39*, 370. Copyright 2000 American Chemical Society.

in poly(methyl methacrolate) (PMMA) thin films at 295 and 10 K [ref. 33] are displayed in Figures 6.30 and 6.31. Notice that the absorption spectra for the Pt- and Au-centered complexes are somewhat similar, and each consist of a number of poorly resolved bands. In contrast, the MCD spectra are better resolved and show a rich array of features. In the energy region $>2.6\ \mu m^{-1}$ the absorption and MCD for the Pt- and Au-centered complexes are somewhat similar, even though the MCD for the Au complex shows generally more negative signals. The bands in this region were assigned as intraframework (IF) transitions involving excitations from the HOMO e_1 orbital of the $Au(PPh_3)_8^{2+}$ framework to the empty framework-based Au orbitals e_2, e_3, b_2, and a_1 (see Figure 6.29). The complexity of the spectra prevent a high level of confidence for assignment detail, but a reasonable consistency was achieved [ref. 32]. A comparison of the MCD spectra between the Pt- and Au-centered complexes reveals the greatest differences at low-energy. This region $<2.6\ \mu m^{-1}$ was interpreted in terms of both IF transitions, which were generally weak, and Pt $5d \rightarrow$ Au $6s$ framework orbitals. Figure 6.32 shows an expanded view of the low-energy region for $Pt(AuPPh_3)_8^{2+}$ at 10 K and illustrates the complexity of the spectra. Bands IIb and IV, for

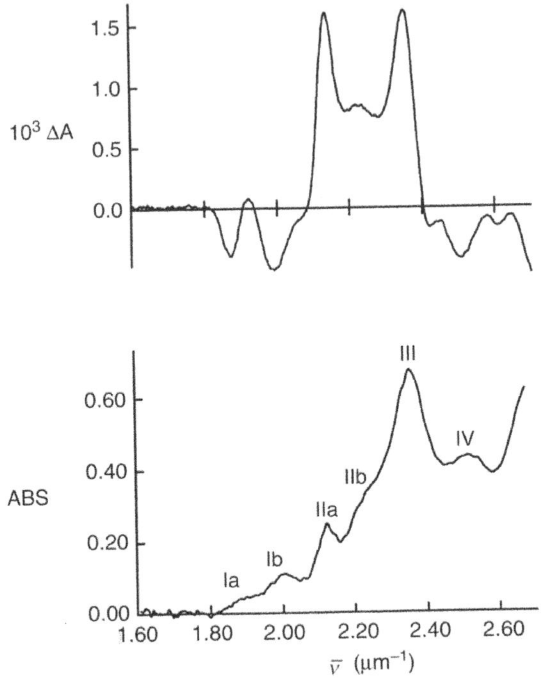

Figure 6.32. Expanded low-energy region of 10 K absorption (lower curve) and MCD (upper curve) for $Pt(AuPPh_3)_8(NO_3)_2$ embedded in PMMA. Reprinted with permission from ref. 33. Mason, W. R. *Inorg. Chem.*, **2000**, *39*, 370. Copyright 2000 American Chemical Society.

example were assigned as IF transitions analogous to those displayed by $Au(AuPPh_3)_8^{3+}$ at low energy. Bands IIa and III, however were interpreted as transitions to $E_1(^1E_1)$ and $B_2(^1B_2)$ of the e_3(Pt)e_2 and e_2(Pt)e_2 excited configurations. Bands Ia and Ib were ascribed to transitions to states of these configurations of predominantly triplet origin, consistent with their lower intensity (Table 3 of ref. 33 should be consulted for a detailed listing of all to the possible states and their MCD *A*-term signs; there are 19 dipole-allowed E_1 or B_2 states out of 55 from all of the IF and Pt → 6*s* framework configurations!). It was argued [ref. 33] that since bands IIb and III representing different occupied 5*d* orbitals on the center Pt atom are close in energy, there must be only a very weak "ligand field" exerted on the Pt from the framework Au atoms. This is consistent with the 5*d* electrons on the Pt as being essentially nonbonding, and exposed along the fourfold axis of the complex, which may explain the higher reactivity of the $Pt(AuPPh_3)_8^{2+}$ complex with electrophiles such as CO, H_2, CN–R, Hg(0), –HgCl, and so on [refs. 32, 33]. Such electrophiles could more readily interact with the 5*d* electron density of Pt(0) in the center of the D_{4d} structure than more stable and contracted d orbitals of the Au(I)- or Pd(II)-centered complexes.

6.8. TRIIODIDE ION: AN EXAMPLE OF A PSEUDO *A* TERM

The red, linear triiodide ion, I_3^-, which is formed from an equilibrium between I_2 and I^- in aqueous solution, exhibits a weak shoulder band at 531 nm and two intense bands in the UV region at 353 nm and 287 nm. The assignment of these bands had been controversial prior to an MCD study [ref. 34], which was unfortunate because the I_3^- ion was considered to be a good simple example of a three-centered bonded system. The MOs for the three-centered bonding were visualized as composed of 5*p* iodine valence orbitals with negligible involvement from 6*s* or 5*d* orbitals. The description leads to an I–I bond order of 0.5 and provides a prototype for a three-centered *p*-orbital system believed to be important in molecules of the *np* block elements. Figure 6.33 shows the relative energies for I_3^- MOs constructed from 5*p* atomic orbitals. The HOMO is the antibonding $2\pi_u$ and is fully occupied. The lowest unoccupied molecular orbital (LUMO) is the antibonding $2\sigma_u^+$. Since the ground state is ${}^1\Sigma_u^+$, the states of lowest-energy excited configuration $(2\pi_u)^3(2\sigma_u^+)$ are parity-forbidden ${}^{1,3}\Pi_g$, and transitions to them are expected to be weak. However, the states from the $(\pi_g)^3(2\sigma_u^+)$ and $(\sigma_g^+)(2\sigma_u^+)$ configurations, which involve the nonbonding π_g and σ_g^+, include Π_u and Σ_u^+ states, transitions to which are allowed and *xy*- or *z*-polarized, respectively. Because spin–orbit coupling is strong for I ($\zeta_{5p} = 5069\ \text{cm}^{-1}$), states of the same symmetry with singlet and triplet origin are expected to be strongly intermixed.

Figure 6.34 shows the absorption and MCD spectra for the I_3^- ion in acetonitrile solution, which, apart from a slight red shift, is substantially the same as for that observed in aqueous solution in the presence of excess I^- ion to prevent dissociation. The features at 4.07 and 4.85 μm^{-1} are ascribed to the presence of a low

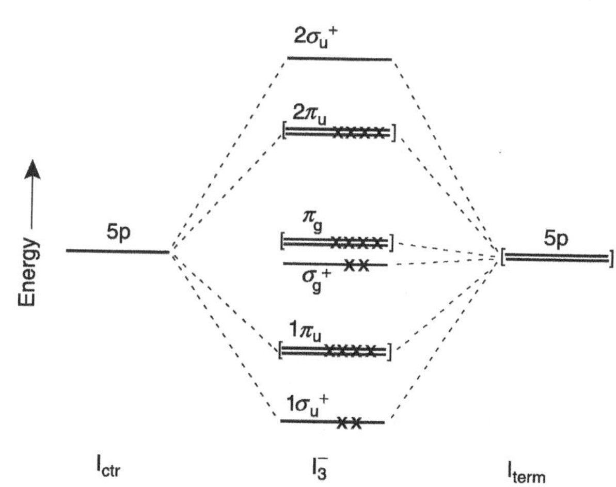

Figure 6.33. Molecular orbital energies for the I_3^- ion constructed from 5*p* atomic orbitals. Reprinted with permission from ref. 34. Isci, H.; Mason, W. R. *Inorg. Chem.* **1985**, *24*, 271. Copyright 2000 American Chemical Society.

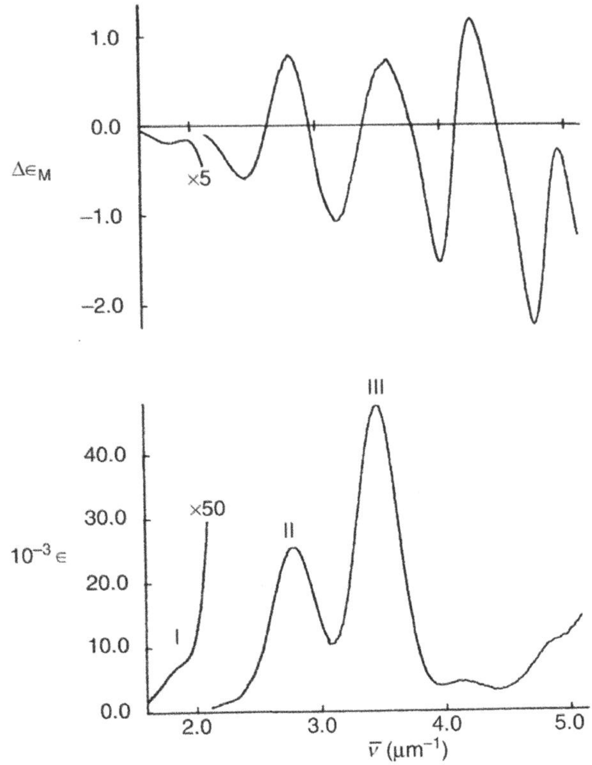

Figure 6.34. Absorption (lower curves) and MCD (upper curves) spectra for I_3- in acetonitrile solution. Reprinted with permission from ref. 34. Isci, H.; Mason, W. R. *Inorg. Chem.* **1985**, *24*, 271. Copyright 2000 American Chemical Society.

concentration of I^- ion which results from a slight dissociation: $I_3^- \rightleftarrows I_2 + I^-$. Both absorption and MCD in this region agree favorably with that measured for I^- alone. Bands I, II, and III in Figure 6.34 are attributed to the I_3^- ion. Consistent with its low intensity, band I was assigned to the transition to the parity-forbidden $\Pi(^1\Pi_g)$ state of the $(2\pi_u)^3(2\sigma_u^+)$ configuration, while the more intense bands II and III were assigned to transitions to the Π_u and Σ_u^+ states of the $(\pi_g)^3(2\sigma_u^+)$ and $(\sigma_g^+)(2\sigma_u^+)$ configurations, suitably intermixed by spin–orbit coupling. The detailed assignment of bands II and III and interpretation of the observed MCD spectra, however, require the inclusion of spin–orbit mixing of the three $|\Pi_u(j)\rangle$ and two $|\Sigma_u^+(j)\rangle$ spin–orbit states of these two configurations. The MCD for I_3^- in Figure 6.34 shows a negative *B* term with no corresponding absorption at 2.42 μm^{-1}, a positive *B* term at 2.77 μm^{-1}, corresponding to band II at 2.76 μm^{-1}, and a positive *A* term (or pseudo *A* term) centered at 3.36 μm^{-1}, corresponding to band III at 3.43 μm^{-1}. An earlier assignment of band II and III to transitions to $\Sigma_u^+(^3\Pi_u)$ and $\Sigma_u^+(^1\Sigma_u^+)$ was based on reasonable intensity arguments which predict that the *z*-polarized

$\sigma \rightarrow \sigma^*$-type transitions should be considerably more intense than the xy-polarized $\pi \rightarrow \sigma^*$-type transitions. This assignment, on the face of it, is not consistent with the observed *A* term in the MCD spectra because the Σ_u^+ states lack degeneracy. In order to develop a model for the interpretation of the MCD, some simple spin–orbit calculations were performed which provided calculated energies and coefficients for the spin–orbit eigenvectors (see ref. 34 for details). The magnitude of the MCD *A* and *B* terms could then be estimated from these parameters. The conclusion reached was that the *A* terms must be small due to the low intensity for transitions to the $|\Pi_u(j)\rangle$ states. However, the *B* terms for both $|\Pi_u(j)\rangle$ and $|\Sigma_u^+(j)\rangle$ spin–orbit states could have the magnitudes necessary to explain the observed spectra. Thus, the negative *B* term at 2.42 μm^{-1} was ascribed to $\Pi_u(1)$, which is largely $^3\Pi_u$; the positive *B* term at 2.77 μm^{-1} was interpreted as $\Sigma_u^+(1)$, which is a mixture of $^3\Pi_u$ and $^1\Sigma_u^+$; and the *A* term at 3.43 μm^{-1} was described as a pseudo *A* term consisting of two oppositely signed *B* terms resulting from the close-lying $\Sigma_u^+(2)$ and $\Pi_u(3)$ states, which were predominantly $^1\Sigma_u^+$ (lower energy) and $^1\Pi_u$, (higher energy), respectively. The *A* term for the $^1\Pi_u$ state is also expected to be positive and could combine with the pseudo *A* term, which may account for the observation that the absorption maximum for band III is slightly higher than the MCD zero crossing. The absorption intensity was assumed to be carried by the Σ_u^+ states, while the MCD features are due to the magnetic mixing of Π_u and Σ_u^+ states and the resulting *B* terms. The overall conclusion from the study is that the absorption and MCD spectra can be satisfactorily interpreted within the simple MO model as long as spin–orbit coupling and relative intensity considerations are included. The model assumes that the I_3^- ion remains linear in the excited states and that the π_g and σ_g^+ orbitals are energetically close, based on the terminal I atoms, and nearly nonbonding, as shown in Figure 6.33. An important lesson from this example is that it illustrates a case where observation of an *A* term in the MCD spectrum may actually be a pseudo *A* term and that considerations of relative intensity, together with MCD-term signs, are important in developing a logical interpretation.

6.9. METHYL IODIDE: $n \rightarrow \sigma^*$ AND IODINE-BASED $5p$ RYDBERG TRANSITIONS

Methyl iodide, CH_3I, a simple organic alkyl halide, exhibits two types of electronic transitions in the vapor phase which present spectra that appear distinctly different. At about 260 nm there is a broad, unstructured band, the so-called *A* band, that has been assigned to transitions from the iodine $5p$ nonbonding electrons to the molecular σ^* orbital [refs. 35, 36]. To higher energy, beginning near 210 nm and extending into the vacuum UV, a complicated series of structured bands appear which have been interpreted as iodine $5p \rightarrow 6s$, $6p$, and even $5d$ Rydberg transitions to empty, iodine-based atomic orbitals [refs. 37, 38]. These two types of transitions are illustrated in Figure 6.35. In both cases, MCD spectra have been found to be an extremely useful tool for developing a detailed spectral interpretation.

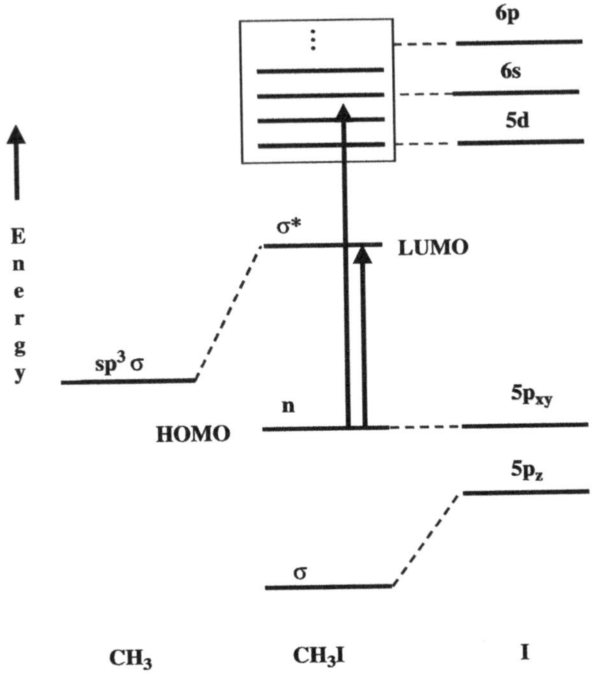

Figure 6.35. Schematic energy levels for CH_3I.

6.9.1. $n \rightarrow \sigma^*$ Transitions

In contrast to the single broad absorption band for the 260-nm band, the MCD spectrum reveals at least three overlapping terms as shown [ref. 36] in Figure 6.36 (dotted lines show digitized MCD spectra obtained for CH_3I vapor at 4 T from ref. 35). When iodine spin–orbit coupling is included, assignments were made from the $\Sigma(^1\Sigma^+)$ ground state to the unresolved transitions ($C_{\infty v}$ symbolism) to the three spin–orbit states $\Pi(^3\Pi)$, $\Sigma^+(^3\Pi)$, and $\Pi(^1\Pi)$, at 34,900, 38,900, and 40,900 cm^{-1}, respectively [refs. 35, 36]; a fourth state $\Delta(^3\Pi)$ is forbidden and therefore was not included. These transitions have sometimes been described in the literature as members of the $N \rightarrow Q$ complex with states symbolized as $^3Q_1$, $^3Q_0$, and $^1Q_1$. By means of positive A terms (using the standard sign convention) for the transitions to $^3Q_1$ and $^1Q_1$, a positive B term for $^3Q_0$, and a negative B term for $^1Q_1$, a satisfactory fit of the MCD spectrum was achieved. This is illustrated in Figures 6.36b and 6.36d (solid lines) [ref. 36].

6.9.2. Rydberg Transitions for CH_3I and CD_3I in the Vacuum UV

MCD spectra for methyl iodide [ref. 37] and perdeuteromethyl iodide [ref. 38] in the vapor phase have been particularly useful in the interpretation of the

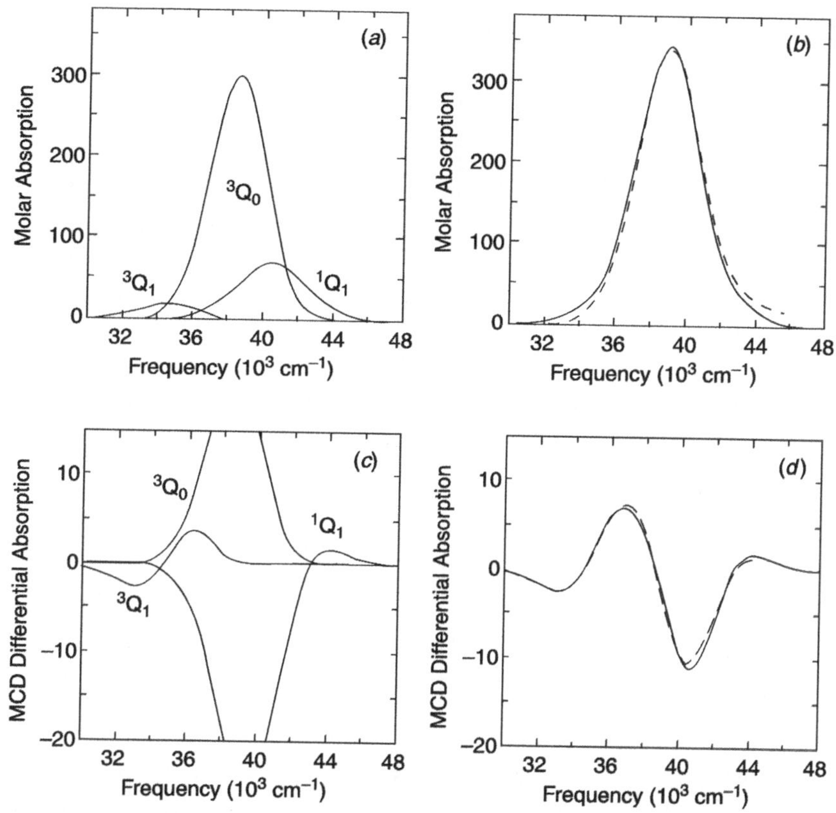

Figure 6.36. Absorption and MCD spectra for CH_3I: Calculated absorption (*a*) and MCD (*c*) components; total calculated (solid lines) and experimental (dashed lines, digitized from ref. 35) absorption (*b*) and MCD (*d*) spectra. Reprinted with permission from ref. 36. Johnson, B. R.; Kinsey, J. L. *J. Phys. Chem.* **1996**, *100*, 18937. Copyright 1996 American Chemical Society.

iodine localized Rydberg-like transitions found between 48,000 cm^{-1} and 60,000 cm^{-1}. This higher-energy vacuum UV (VUV) range required the use of a VUV monochromator and Hinteregger hydrogen discharge source, together with CaF_2 or LiF optical windows and a PEM with a CaF_2 modulator element. However, the spectrometer setup and the spectral acquisition otherwise was similar to that used for UV–vis measurements. Figure 6.37 shows the more recent absorption, negative derivative absorption, and the MCD spectra for CD_3I at ~7 T with a resolution of 20–30 cm^{-1}. Sample pressures were varied from 2 μm to 365 mm in order to give suitable signal strengths (ref. 38 should be consulted for details). No less than 76 origin and vibronic bands were resolved and identified. The earlier data for CH_3I [ref. 37] were similar and were augmented by some unpublished measurements and are also discussed in ref. 38. The numerous bands observed for both CH_3I and CD_3I were

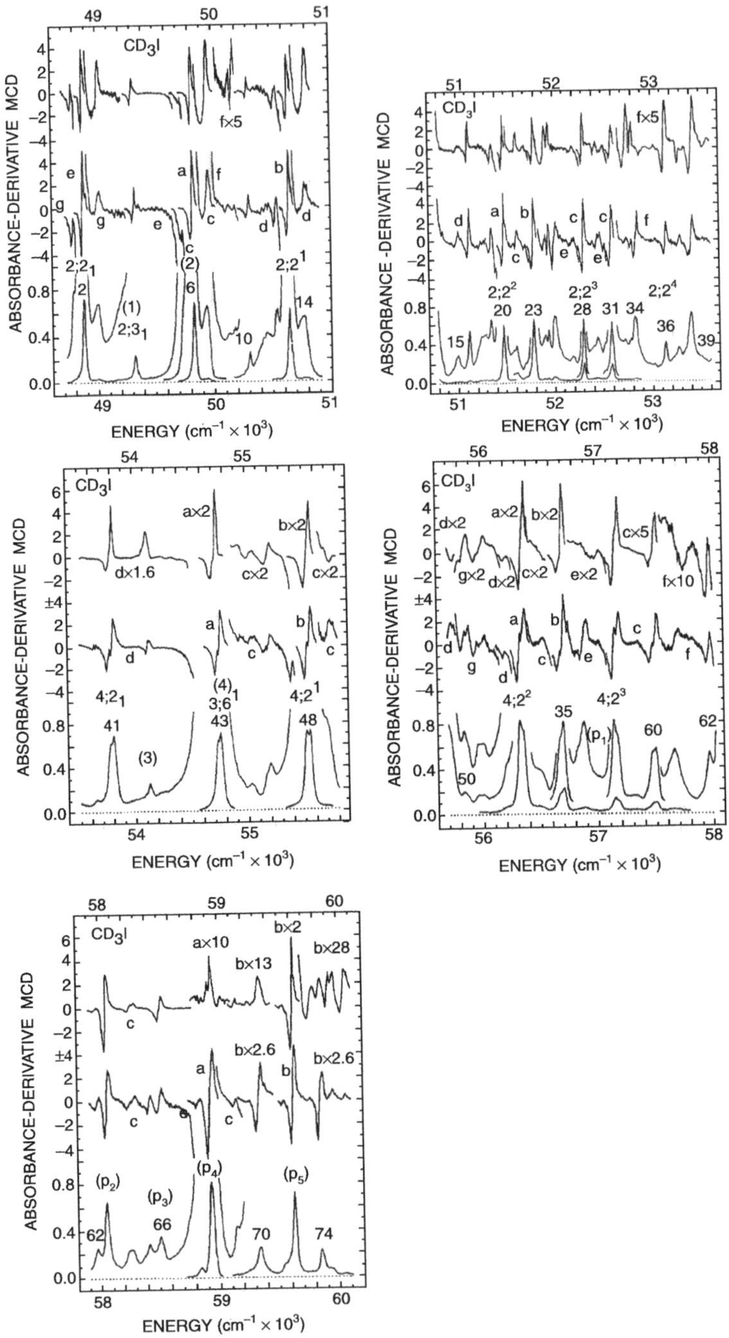

Figure 6.37. Absorption (lower curves), negative derivative (middle curves), and MCD at 7 T (upper curves) for CD_3I in the gas phase. The letters on the derivative curves indicate the sample pressures used (ref. 38 should be consulted for experimental details). Reprinted with permission from ref. 38. Felps, W. S.; Scott, J. D.; McGlynn, S. P. *J. Chem. Phys.* **1996**, *104*, 419. Copyright American Institute of Physics.

interpreted in terms of $5p \rightarrow 6s$ and $5p \rightarrow 6p$ manifolds: Rydberg transitions localized on the iodine atom. There are numerous states represented by these excitations. Origins for all four of the $6s$ states were identified, and as many as six of the $6p$ states were also assigned. There was also evidence for one or more states believed to be "interlopers" from the $5p \rightarrow 5d$ manifold. The specific assignments of the origins and intensity patterns were guided by calculated and observed Rydberg states for hydrogen iodide, HI. Degenerate excited states all exhibited MCD A terms; also, magnetic moments, determined by the method of moments, were presented for a number of the observed states. For example, three of the four origins for the $5p \rightarrow 6s$ states exhibit positive A terms, and the a_1 totally symmetric vibrationally excited states built upon these origins also exhibit positive A terms. However, an anomalous magnetic moment was observed for the origin of the first $5p \rightarrow 6s$ transition at 49,310 cm^{-1} for CD_3I (1.25 μ_B) compared with that for CH_3I at 49,220 cm^{-1} (3.04 μ_B). This apparent anomaly was traced to an accidental near-degeneracy between a vibronic band from the second transition in the former case. A moment of $\sim 3\mu_B$ is expected for the degenerate triplet state, and higher-energy vibronic states for the first band system for CD_3I do indeed exhibit moments near this value. Detailed assignments are given in tables in ref. 38.

6.10. BENZENE: $\pi \rightarrow \pi^*$ AND RYDBERG TRANSITIONS

Benzene with its planar D_{6h} C_6H_6 ring has long been of interest as the archprototype of a cyclic aromatic π-conjugated molecule with $4n + 2 = 6$ electrons. The electronic spectra of benzene in the UV and vacuum UV regions exhibit intense electronic transitions, which have been studied extensively in order to characterize the excited states. The lower-energy UV transitions have been assigned as molecular $\pi \rightarrow \pi^*$ type, while those at higher energy into the vacuum UV have been attributed to Rydberg-type transitions involving $\pi \rightarrow np$ and even $\pi \rightarrow nf$ excitations. The two types tend to overlap in the region of 150–185 nm.

6.10.1. $\pi \rightarrow \pi^*$ Transitions

The lower energy $\pi \rightarrow \pi^*$ transitions are based on the HOMO $\rightarrow$ LUMO excitation $(e_{1g}\pi)^4 \rightarrow (e_{1g}\pi)^3(e_{2u}\pi^*)$ (Figure 6.38), which gives rise to ${}^1A_{1g} \rightarrow {}^1B_{2u}$ (270–230 nm), ${}^1B_{1u}$ (210–190 nm), and ${}^1E_{1u}$ (170–185 nm) transitions, in order of increasing energy. Only the transition to ${}^1E_{1u}$ is electric dipole-allowed; the transitions to ${}^1B_{2u}$ and ${}^1B_{1u}$ require vibronic coupling to gain intensity. The lowest energy transition ${}^1A_{1g} \rightarrow {}^1B_{2u}$ transition in the 270- to 230-nm region provides a nice, if not complicated, example of vibronic structure. The vapor pressure of benzene at room temperature is sufficient to permit the observation of extensive vibronic bands for vapor samples. This structure is greatly reduced by band broadening for solution samples. The structure has been the subject of numerous studies since the late 1940s. Early attempts to use MCD to expand the understanding of the benzene spectra involved solution samples, which were unable to show

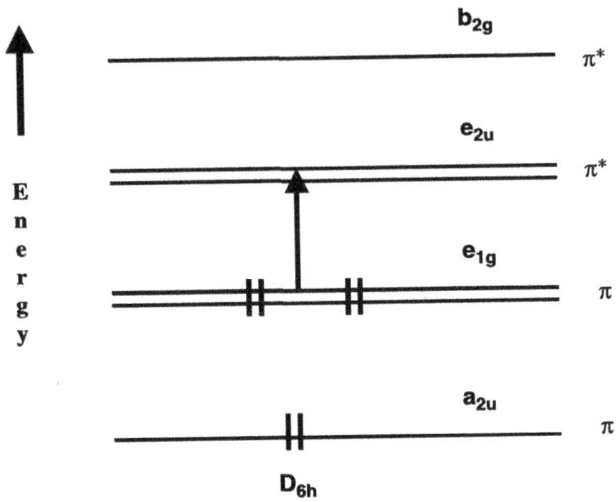

Figure 6.38. Energy levels for the π orbitals of benzene in D_{6h} symmetry.

the vibronic detail. It was not until the early 1970s that a vapor-phase MCD spectrum was reported for the lowest-energy ${}^1A_{1g} \rightarrow {}^1B_{2u}$ transition (Figure 6.39 [ref. 39]). The vapor spectra are quite complex showing a number of progressions in both absorption and MCD. Positive *A* terms are evident for the most intense

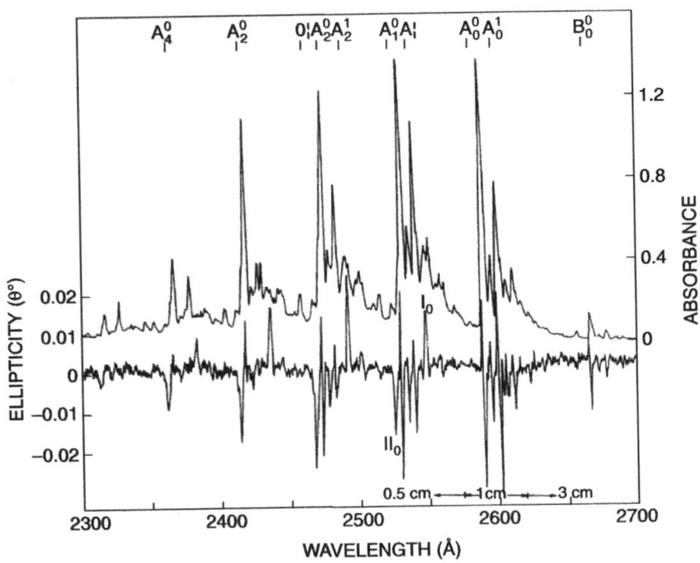

Figure 6.39. Absorption (upper curve) and MCD at 5.3 T (lower Curves) for benzene in the vapor phase with a spectral bandwidth of 0.4–0.6 Å. The path length for the MCD measurements are indicated just above the horizontal axis; the path length for the absorption was 1.0 cm throughout. Reprinted with permission from ref. 39, Douglas, I. N.; Grinter, R.; Thomson, A. J. *Molec. Phys.* **1973**, *26*, 1257. Copyright Taylor and Francis Ltd. (http://www.tandf.co.uk/journals).

features labeled A_n^0; also, a positive A term is exhibited by the low-energy band labeled B_0^0. These A terms imply degeneracy as a result of vibronic coupling since neither the ground-state $^1A_{1g}$ nor the excited-electronic-state $^1B_{2u}$ are degenerate. The A_0^0 and B_0^0 bands observed were assumed to represent false origins. The interpretation of the A_n^0 bands was based on vibronic coupling between the B_{2u} excited state and an excited-state e_{2g} normal mode (ring angle bending, $\nu_6 = 608\ \text{cm}^{-1}$ in the ground state) for the A_0 progression giving an $|E_{1u}(B_{2u}e_{2g})\rangle$ vibronic state, which is allowed and undergoes mixing with the higher-energy $^1E_{1u}$ excited state, thereby providing intensity to the formally forbidden $^1B_{2u}$ transition, and also accounting for the observed A term. The low-energy B_0^0 band was attributed to a "hot" band in which a one quantum of ν_6 is excited in the ground state. The ground-state e_{2g} mode, when coupled with the A_{1g} ground electronic state, gives an $|E_{2g}(A_{1g}e_{2g})\rangle$ vibronic ground state and therefore the hot band to lower energy of the true origin. The totally symmetric a_{1g} mode ($\nu_1 = 923\ \text{cm}^{-1}$) provides a clear progression to higher energy for the A_0^0 false origin (the strongest features in each of the prominent band systems in Figure 6.39). There is also a parallel progression $161\ \text{cm}^{-1}$ to the low-energy side of each of the A_n^0 bands, which is labeled A_1^n and is ascribed to an excitation of equal numbers of quanta of the e_{2u} vibration (ν_{16}) in ground and excited states. The $161\ \text{cm}^{-1}$ represents the difference between the frequency for this mode in the ground and excited electronic states. In addition, there are other progressions shown in Figure 6.39 which are not so easily interpreted. For example, the curious features labeled I_0 and II_0 near 255 and 252 nm, respectively, in the MCD spectrum appear to be B terms of opposite sign, but have very weak absorption bands associated with them. Perdeuterobenzene C_6D_6 was also studied in the vapor, but the absorption and MCD signals were much weaker and signal to noise considerations prevented any significant analysis [ref. 39].

An MCD study of benzene vapor in the $^1E_{1u}$ region revealed a complicated band shown in Figure 6.40 [ref. 40]; this band was later interpreted as the overlap of the $^1A_{1g} \rightarrow {}^1E_{1u}$ transition and the $^1A_{1g} \rightarrow 2R$ Rydberg transition from e_{1g} to the $3p$ carbon orbitals [ref. 41]. The absorption spectra in Figure 6.40 reveal a mixture of sharp and broad bands, in contrast to the vapor spectra in the $^1B_{2u}$ region, where the features were uniformlly sharp. A detailed assignment was not attempted for the vapor in this region. The later study of benzene isolated in low-temperature matrices showed a considerably different pattern in this region [ref. 41] (see below).

An absorption and MCD study of benzene isolated in nitrogen or argon matrices at low temperature revealed a markedly different pattern in the $^1B_{2u}$ region [ref. 41] compared to the vapor. Figure 6.41 shows absorption and MCD spectra in a nitrogen matrix at 20 K. The difference between the vapor spectra at room temperature and the matrix spectra at 20 K may be partially due to removal of hot bands, but matrix influences may also play a role. The absorption and MCD spectra exhibit vibronic progressions in two different false origin-based e_{2g} modes $\nu_6 = 521\ \text{cm}^{-1}$ and $\nu_9 = 1154\ \text{cm}^{-1}$ and the symmetric mode a_{1g} $\nu_1 = 923\ \text{cm}^{-1}$. Furthermore, the MCD shows positive and negative B terms as opposed to the positive A terms in the vapor spectra. The lack of A terms in the matrix spectra was attributed to broadening of the vibronic lines in the matrix environment. The assumption was made that the B terms would then dominate.

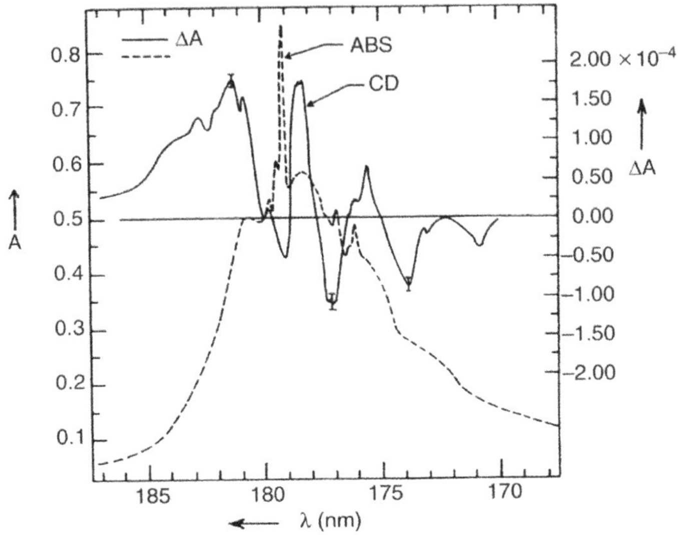

Figure 6.40. Absorption (lower curve) and 4.0–T MCD (upper curve) spectra for benzene vapor in the $^1E_{1u}$ region. Reprinted from ref. 40. Allen, S. D.; Mason, M. G.; Schnepp, O.; Stephens, P. J. *Chem. Phys. Letters* **1975**, *30*, 140. Copyright 1975, with permission from Elsevier.

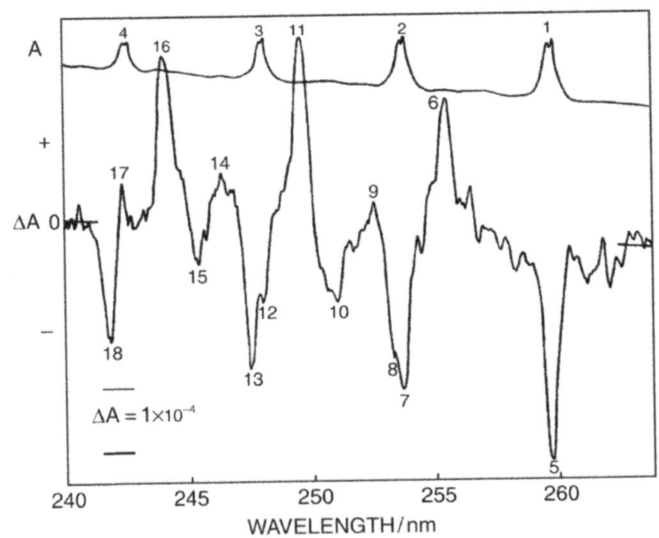

Figure 6.41. Absorption (upper curve) and 4.8 T MCD spectra of benzene in a nitrogen matrix at 20 K with a dilution of 1 : 250. Reprinted with permission from ref. 41. Barton, T. J.; Douglas, I. N.; Grinter, R.; Thomson, A. J. *Molec. Phys.* **1975**, *30*, 1677. Copyright Taylor and Francis Ltd. (http://www.tandf.co.uk/journals).

The B terms were ascribed to an interaction of the vibronic $^1B_{2u}$ states with the higher-energy vibronic $^1B_{1u}$ states, which also involve e_{2g} vibrational modes. The electric dipole intensity, as before, is assumed derived from mixing with the allowed $^1E_{1u}$ state via the e_{2g} vibrational modes.

A higher-energy matrix isolation study in the $^1B_{1u}$, $^1E_{1u}$, and $2R$ (Rydberg transition to $3p$) regions by using a synchrotron radiation source was reported [ref. 42]. The spectra are shown in Figure 6.42. The $^1A_{1g} \to {}^1E_{1u}$ spectra are notably different than shown in the vapor study (Figure 6.40). There appears to be a progression of positive A terms associated with corresponding absorption bands. The difference was explained by assuming a greater energy separation of the $^1E_{1u}$ and $2R$ states in the low-temperature matrix compared to the room temperature vapor. Evidence was found for the $^1A_{1g} \to 2R$ transition to higher energy in the 150- to 156-nm region (Figure 6.42, lower curves).

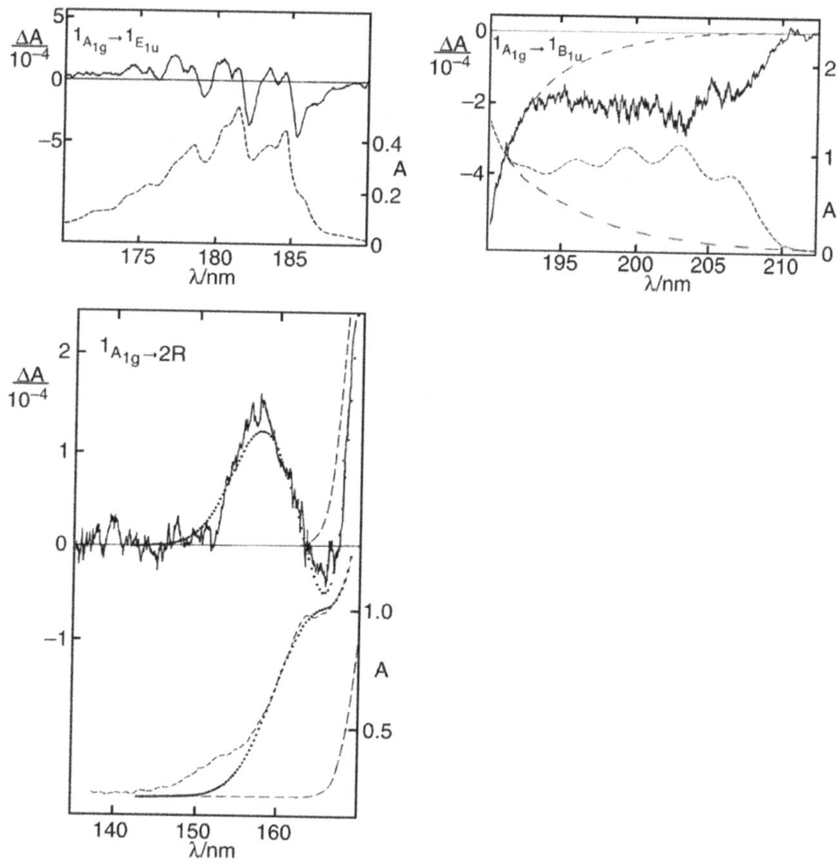

Figure 6.42. Absorption (lower curves) and 4.14-T MCD spectra for benzene in argon matrices at 5 K in the $^1E_{1u}$ (upper left), $^1B_{1u}$ (upper right), and $2R$ (lower) regions. Reprinted from ref. 42. Boyle, M. E.; Williamson, B. E.; Schatz, P. N.; Marks, J. P.; Snyder, P. A. *Chem. Phys. Letters* **1986**, *130*, 33. Copyright 1986, with permission from Elsevier.

6.10.2. Rydberg Transitions

By means of a synchrotron radiation source, MCD measurements at higher energy for benzene vapor were made. These spectra exhibit several Rydberg transitions in the 140- to 160-nm region, which are shown in Figure 6.43 [ref. 43]. The Rydberg transitions labeled nR, nR', nR'', and nR''' are transitions from the π HOMO $e_{1g}\pi$ level to $(n+1)p$, or $(n+1)f$ atomic orbitals localized on the carbon atoms. The most intense absorption labeled R' in Figure 6.43 is accompanied by a strong positive A term and is assigned as the $3R'$ band origin $^1A_{1g} \rightarrow {}^1E_{1u}$ arising from $\pi(e_{1g}) \rightarrow 4p_z(a_{2u})$. The absorption labeled $1 \rightarrow 1$ (101 cm^{-1} to the red of the $3R'$ band) and the accompanying positive A term in the MCD was interpreted as a hot band built upon the $3R'$ origin, which arises from a simultaneous absorption of one quantum of an e_{2u} (ν_{20}) vibration in both the ground and excited states. The calculated and observed A_1/D_0 values for the $3R'$ and the associated hot band assignment were found to be in reasonable agreement. The absorption labeled X was interpreted as another hot band from the $3R'$ origin, but one involving one of the nondegenerate vibronic components from the coupling of the e_{2u} vibration with $^1E_{1u}$, either $^1B_{1g}$ or $^1B_{2g}$. It was argued that neither of these states are expected to have either A- or B-term MCD intensity for lack of degeneracy or symmetry reasons. The bands labeled R'' and R''' in Figure 6.43 were both found to have positive A-term components, but the R'' band also exhibited a very large positive B term. These two bands were interpreted as the Rydberg transitions

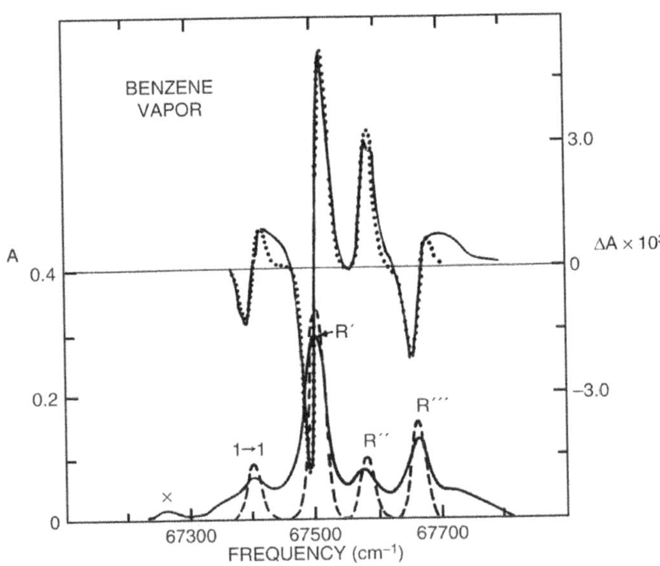

Figure 6.43. Absorption (lower curve) and 7.0-T MCD (upper curve) spectra for 13.2-cm path and 0.045-nm spectral slit width for benzene vapor at room temperature. The dotted and dashed curves are spectral fits. Reprinted from ref. 43. Snyder, P. A.; Lund, P. A.; Schatz, P. N.; Rowe, E. M. *Chem. Phys. Letters* **1981**, *82*, 546. Copyright 1981, with permission from Elsevier.

$\pi(e_{1g}) \rightarrow 4f(e_{1u})$, which are expected to give three dipole-allowed transitions: one $^1A_{1g} \rightarrow {}^1A_{2u}$ and two $^1A_{1g} \rightarrow {}^1E_{1u}$ transitions. The presence of the large B term for the R'' band may signal overlap between two of these transitions with R''' being the third. It was concluded, however, that a detailed assignment would require higher-resolution data [ref. 43]. This is almost always the case: The better the experimental data, the more precise the details of the interpretation. Clearly, MCD helps provide needed detail for understanding complex molecular spectra.

6.11. D_{4h} CYCLOBUTADIENE DIANION $Li_2[C_4(Me_3Si)_4]$

Cyclobutadiene dianion $C_4H_4^{2-}$ has been of interest from both synthetic and theoretical viewpoints. The ion is predicted to be planar and aromatic, but unfortunately the molecule is too unstable to isolate and therefore is not amenable to study. The tetrakis(trimethylsilyl)cyclobutadiene dianion, however, has been

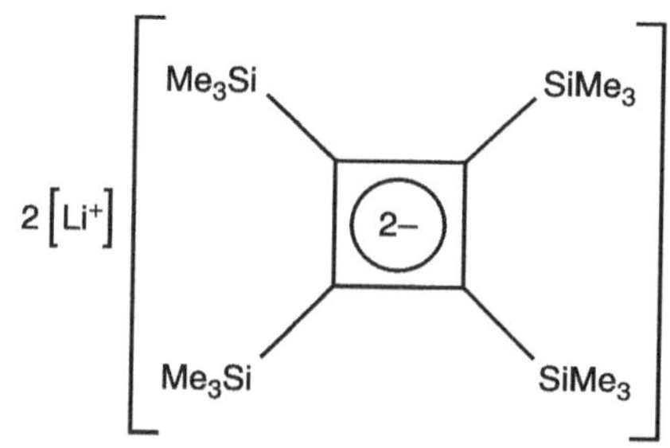

Figure 6.44. The lithium salt of tetrakis(trimethylsilyl) cyclobutadiene dianion. Reprinted with permission from ref. 44. Ishii, K.; Kobayashi, N.; Matsuo, T.; Tanaka, M.; Sekiguchi, A. *J. Amer. Chem. Soc.* **2001**, *123*, 5356. Copyright 2001 American Chemical Society.

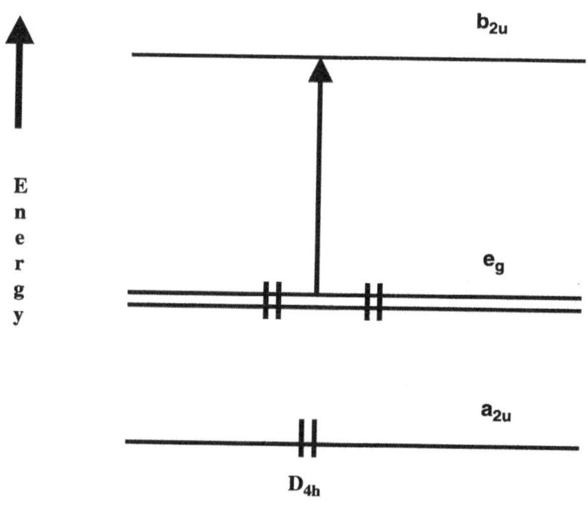

Figure 6.45. Molecular orbital energy levels for the cyclobutadiene dianion.

synthesized as the lithium salt, and X-ray crystalographic analysis showed that the C_4 ring is planar and approximates D_{4h} symmetry (Figure 6.44 [ref. 44]). ^{6}Li NMR spectroscopy showed the presence of strong ring current shielding consistent with aromaticity. The lowest-energy $\pi \rightarrow \pi^*$ transition observed at 289 nm is predicted to be the $e_g \rightarrow b_{2u}$ $^1A_{1g} \rightarrow {}^1E_u$ transition in accordance with the simple π-MO energy-level scheme shown in Figure 6.45. The MCD spectrum for a degassed *n*-hexane solution showed a clear negative *A* term for the lowest-energy intense band observed in the UV region (Figure 6.46 [ref. 44]). The sign of the *A*

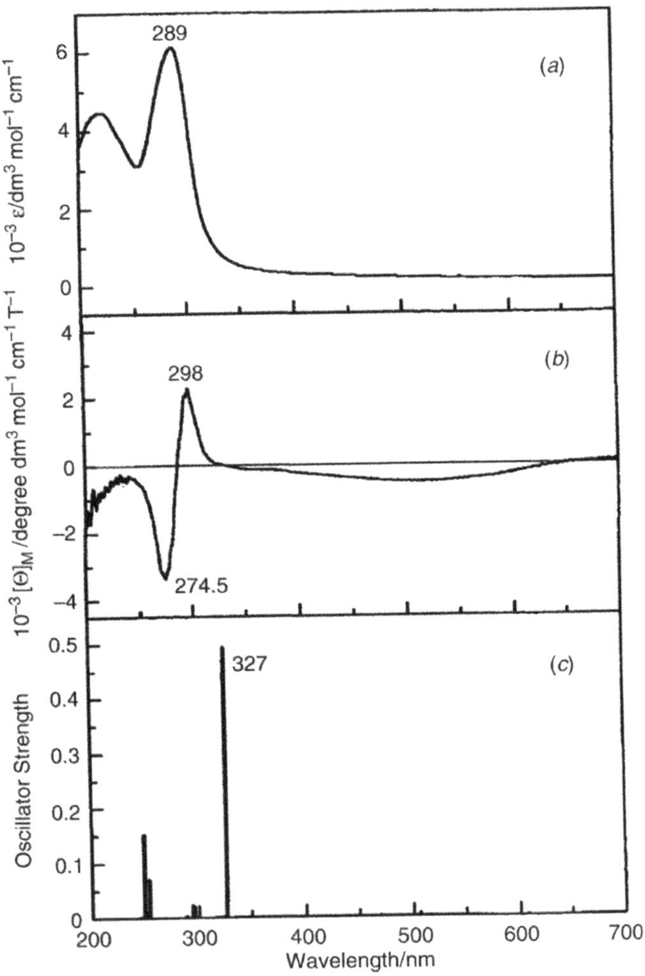

Figure 6.46. Absorption (*a*, upper curve) and MCD (*b*, middle curve) spectra for $Li_2[C_4(Me_3Si)_4]$ in *n*-hexane at room temperature and 1.1 T. The oscillator strengths and energies for the lowest-energy transitions from a ZINDO/S calculation are shown in the lower part of the figure (*c*). Reprinted with permission from ref. 44. Ishii, K.; Kobayashi, N.; Matsuo, T.; Tanaka, M.; Sekiguchi, A. *J. Amer. Chem. Soc.* **2001**, *123*, 5356. Copyright 2001 American Chemical Society.

term had been predicted by Michl [ref. 45] by utilizing a perimeter model for π electron systems of aromatic molecules, but at the time there were no examples available to test the prediction. The observation of the MCD *A* term is consistent not only with the planar D_{4h} structure, but also with the expectations of $C_4R_4^{2-}$ aromaticity for a $4n+2$ π electron system. Some ZINDO/S calculations, which included configuration interaction (CI), were also performed for $C_4(Me_3Si)_4^{2-}$ in order to give estimates of the transition energies and oscillator strengths of the lowest-energy transitions. The intense $\pi \rightarrow \pi^*$ band calculated at 327 nm was estimated to be 91% $e_g \rightarrow b_{2u}$. It may be noted here that Michl's perimeter model for aromatic annulenes has been useful in interpreting the MCD of a number of organic ring compounds [ref. 46].

6.12. ZINC PHTHALOCYANINE (TETRAAZOTETRABENZOPORPHYRIN) COMPLEX, ZnPc(2−)

Phthalocycyanine (Pc) complexes [refs. 47, 48] represent one of the most commercially important class of porphyrin molecules, with applications to dyestuffs, inks, and other pigments or coloring agents. The origin of their colors stems from the conjugated molecular structure, which has been of continuing interest to chemists. Figure 6.47 shows a sketch of the molecular structure for the diamagnetic Zn(II) complex, (ZnPc(2−)), a typical main group metal MPc(2−) complex. The Zn(II) ion is actually 48 pm above the plane of the four coordinating nitrogen donors forming a dome-shaped structure with Zn–N bond lengths of 206.1 pm. However, the electronic spectrum of the related MgPc(2−) complex where the Mg^{2+} resides within the plane of the 4 N donors is virtually identical to that of the ZnPc(2−) complex. This observation is taken to indicate that the spectrum is associated almost entirely with the highly conjugated 16-atom ring. Calculations

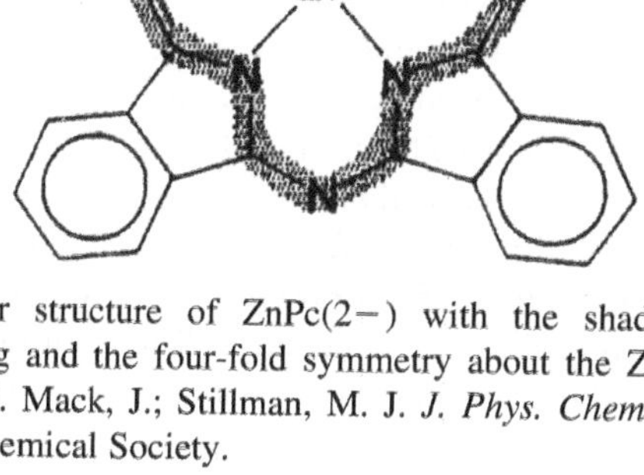

Figure 6.47. Molecular structure of ZnPc(2−) with the shaded path indicating the 16-atom conjugated ring and the four-fold symmetry about the Zn^{2+} ion. Reprinted with permission from ref. 47. Mack, J.; Stillman, M. J. *J. Phys. Chem.* **1995**, *99*, 7935. Copyright 1995 American Chemical Society.

and spectral assignments therefore are often referred to the idealized D_{4h} symmetry of the ring, even though the ZnPc(2−) complex symmetry is actually C_{4v} because of its dome-shaped structure.

The model most widely used to interpret the optical electronic spectra of MPc(2−) complexes was originally developed by Gouterman for porphyrin complexes (see ref. 48. for a synopsis of the electronic structural model development for porphyrin and Pc complexes). The Gouterman four-orbital LCAO model (based on D_{4h} symmetry) placed HOMO a_{1u} and a_{2u} orbitals close in energy (accidental degeneracy) for prophyrin complexes. The optical spectrum was then explained by *xy*-polarized $^1A_{1g} \rightarrow {}^1E_u$ $\pi \rightarrow \pi^*$ transitions to the e_g^* LUMO. The lowest-energy band, labeled *Q*, was of low intensity resulting from the forbidden character of the $a_{1u} \rightarrow e_g^*$ transition, while the higher-energy more intense band, labeled *B*, was consistent with the allowed character of the $a_{2u} \rightarrow e_g^*$ transition. The slightly lower symmetry of the Pc ligand due to the aza linkages and the fused benzene rings separates these two near-degenerate HOMO orbitals, but significant excited-state mixing between the *Q* and *B* states allows the lower-energy *Q* transition to gain intensity and become the dominant band in the Pc spectrum. Subsequently, another one-electron transition in the region of the $a_{2u} \rightarrow e_g^*$ transition was identified and, based on an MCD study, was assigned as $1b_{1u} \rightarrow e_g^*$; two primary absorption bands were then labeled B_1 and B_2. In addition, careful study revealed a *z*-polarized $^1A_{1g} \rightarrow {}^1A_{2u}$ transition between the *Q* and B_1 band systems ascribed to an $n \rightarrow \pi^*$ transition from the nonbonding aza N centers to the e_g^* LUMO. The relevant energy levels and electronic states are summarized in Figure 6.48 [ref. 48] (see also ref. 47) for ZnPc(2−).

6.12.1. Vitreous Solution Spectra at Low Temperature

The vis–UV absorption and MCD spectra for ZnPc(2−) in an isotropic solution-like environment was studied at low temperature in order to enhance resolution and give detail for the interpretation of the optical transitions. To accomplish this study, vitreous (glass-forming) solutions were prepared and cooled to low temperature, and CN^- ion was added to solutions to enhance the separation of the B_1 and B_2 bands. The absorption spectra were measured at 77 K and the MCD spectra were measured at 40 K for dimethylformamide/dimethyl acetamide (DMF/DMA) 5:2 v/v solutions [ref. 47]. The resulting spectra were analyzed by means of a simultaneous fit of both absorption and MCD spectra to Gaussian bands (the SIMPFIT method developed by Stillman and his group; see ref. 47 for appropriate references). The experimental spectra, together with the Gaussian deconvolution, for the *Q* and *B* band systems are displayed in Figure 6.49 [ref. 47]. In the region of the Q_{00} band at 671 nm (14,900 cm^{-1}) the fit required two oppositely signed *B* terms in addition to the large positive *A* term. These *B* terms are attributed to a Jahn–Teller distortion effect on the (near) degenerate excited state. A careful analysis of the region 560–660 nm in terms of vibrational structure built on the Q_{00} origin revealed also a negative *B* term at 604 nm (16,550 cm^{-1}), which was attributed to the origin of the $n \rightarrow \pi^*$ $b_{1g} \rightarrow e_g^*$ transition noted above.

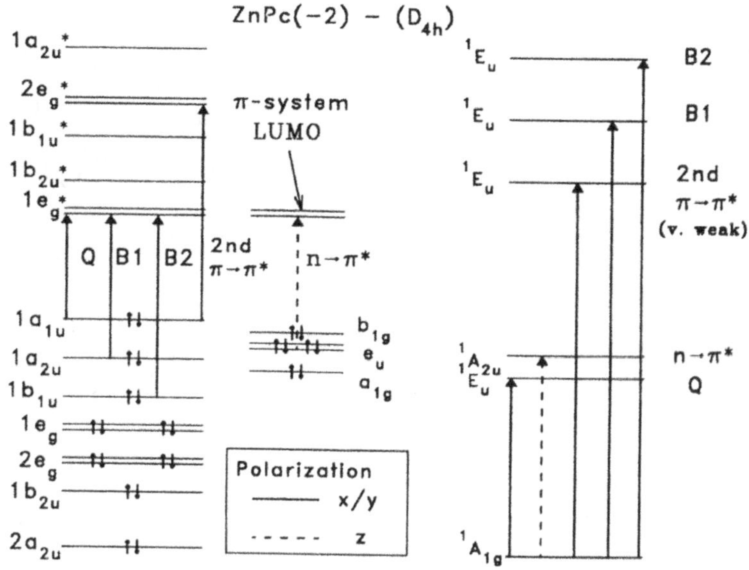

Figure 6.48. Molecular orbital energy levels (left side) and electronic states (right side) for ZnPc(2−). The one-electron energy-level diagram is based on the Gouterman model for the 16-atom cyclic polyene. Reprinted with permission from ref. 48. Mack, J.; Stillman, M. J. *Inorg. Chem.* **2001**, *40*, 812. Copyright 1995 American Chemical Society.

The weak bands to higher energy were then assigned as vibronic bands associated with this weak transition. Finally, the fit of the B_1 and B_2 bands required the presence of oppositely signed B terms rather than A terms. This observation was explained in terms of Jahn–Teller effects, similar to those presented by Q_{00}. It seems clear that from the deconvolution results the observed spectra possesses an underlying complexity that can be effectively probed by MCD. Perhaps this can be appreciated even more from an earlier study of ZnPc(2−) that was matrix-isolated in argon [ref. 49], which is presented in the Section 6.12.2.

6.12.2. Matrix-Isolated ZnPc(2−)

Absorption and MCD spectra in the Q and B band regions at ~5 K for ZnPc(2−) isolated in an argon matrix are shown in Figure 6.50 [ref. 49]. The spectra were sensitive to the isolation temperatures and annealing effects, leading to the conclusion that the complex occupies several different sites and is deposited with the molecular plane predominantly parallel to the deposition window. Thus the complex was assumed to be oriented with the four-fold axis parallel to the light propagation direction. The consequence of this orientation is to make z-polarized absorption much lower in intensity and dependent upon the orientation of the deposition window. Nevertheless, it is clear that the spectra reveal considerable structure. From an analysis of the Q_{00} band alone, it was concluded that there

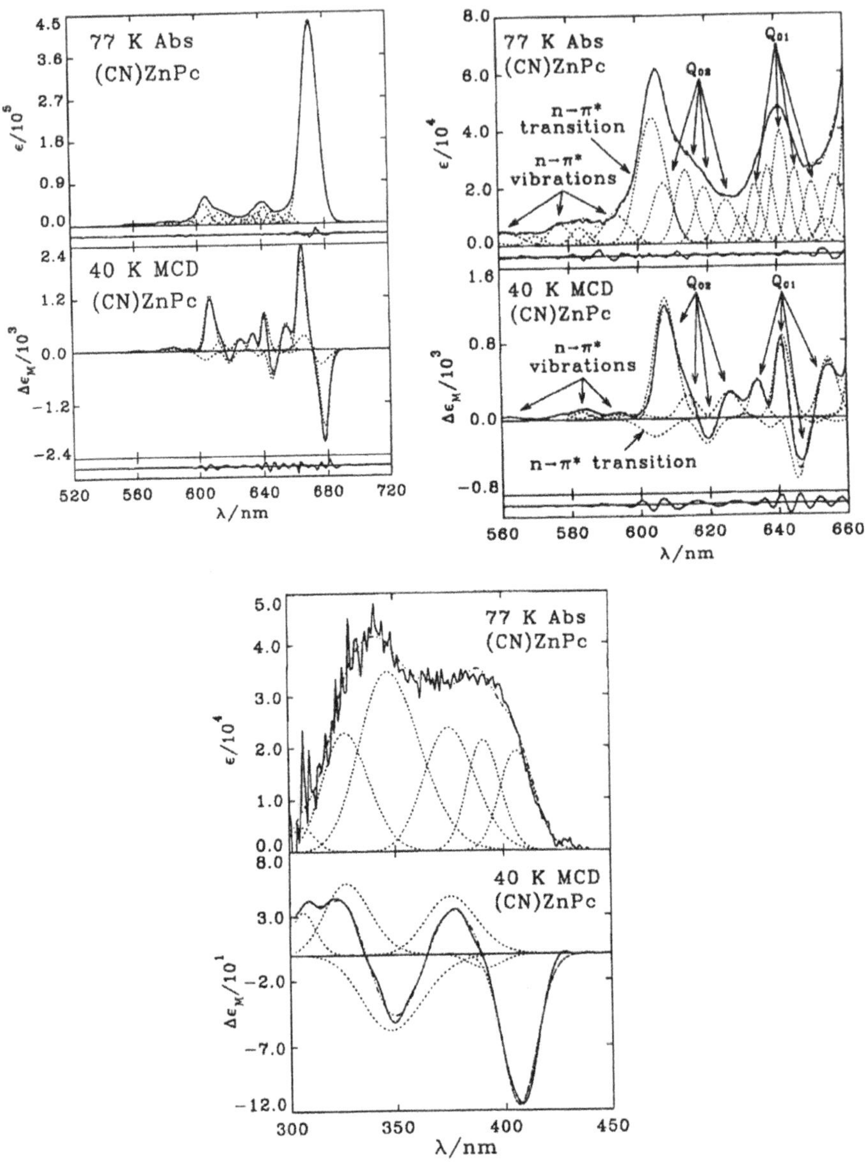

Figure 6.49. Absorption (solid lines upper curves) and MCD (solid lines, lower curves) for (CN)ZnPc(2−) in DMF/DMA 5:2 v/v glass at 77 K and 40 K, respectively. The Q band is given in the upper left panel, while a detail of the Q_{10} and Q_{20} are given in the upper right panel, which shows the $n \rightarrow \pi^*$ bands; the B band is presented in the lower panel. Dashed lines indicate the simultaneous absorption and MCD deconvolution. Reprinted with permission from ref. 47. Mack, J.; Stillman, M. J. *J. Phys. Chem.* **1995**, *99*, 7935. Copyright 1995 American Chemical Society.

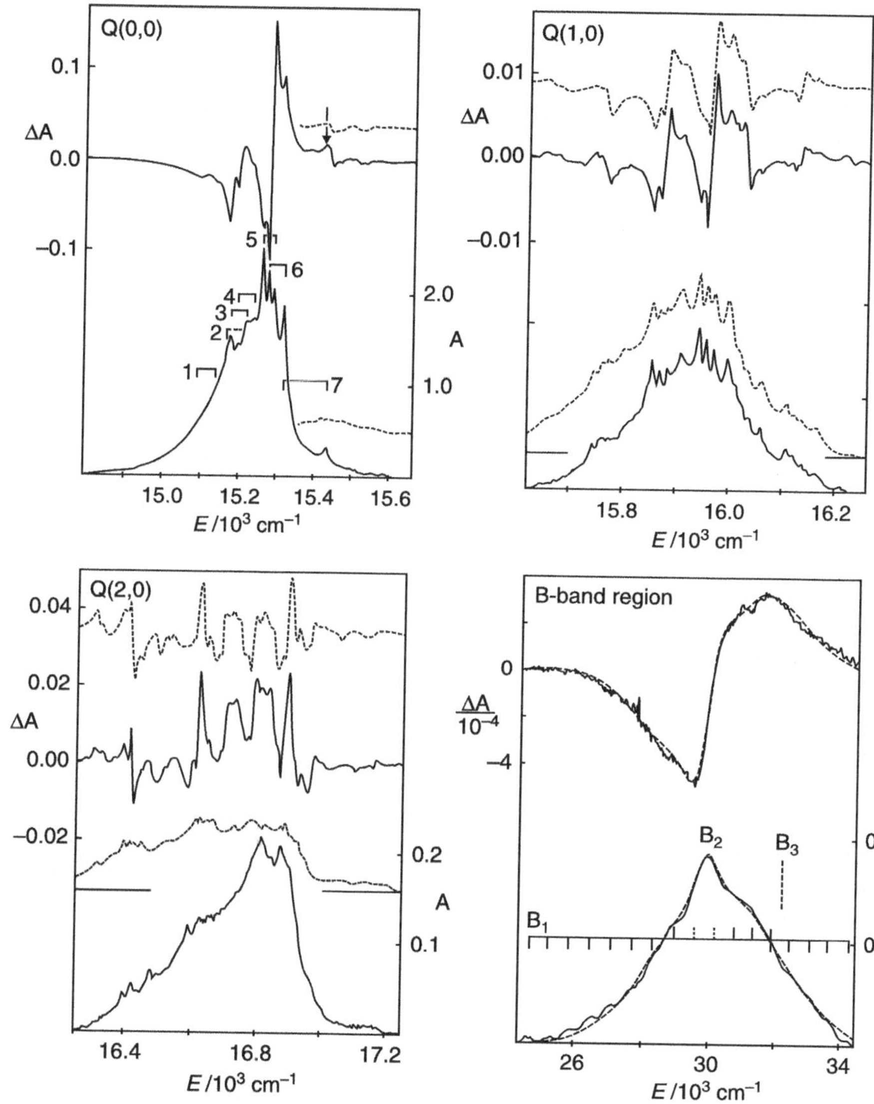

Figure 6.50. Absorption (solid lines, lower curves) and MCD (solid lines, upper curves) for ZnPc(2−) in an Ar matrix at 5 K. The dashed lines are simulated vibronic spectra or, in the lower right panel, a Gaussian fit. Reprinted with permission from ref. 49. VanCott, T. C.; Rose, J. L.; Misener, G. C.; Williamson, B. E.; Schrimpf, A. E.; Boyle, M. E.; Schatz, P. N. *J. Phys. Chem.* **1989**, *93*, 2999. Copyright 1989 American Chemical Society.

were seven different sites and that each exhibited a pair of oppositely signed *B* terms (pseudo *A* terms) corresponding to pairs of absorption peaks. The analysis of the spectra, in broad terms, was similar to the more recent glassy solution study, but the details of the vibronic interpretation of the Q_{10} and Q_{20} regions was somewhat different. In addition, a third band in the *B* region, B_3, was suggested as being another $n \rightarrow \pi^*$ transition. Based on its sensitivity to the orientation in the light beam, it was ascribed to a *z*-polarized $^1A_{1g} \rightarrow {}^1A_{2u}$ transition. The authors of the more recent study [ref. 47] acknowledged that such a third *B* band could be consistent with their analysis. The matrix isolated study also reported absorption and MCD spectra to higher energy (up to 74,000 cm^{-1} in the vacuum UV) obtained by means of synchrotron radiation measurements (see ref. 49 for discussion of the high-energy spectra and their measurement). Their spectra included the higher-energy bands labeled N, L, C, X_1, and X_2.

Some of these bands are certainly $\pi \rightarrow \pi^*$ transitions based on the polyene ring, but others may be related to Rydberg-type transitions with localization on the aza nitrogen atoms of the ring. One of the important conclusions reached from the analysis of the matrix-isolated spectra was that there must be extensive mixing of states of the same symmetry (configuration interaction or CI). Such mixing complicates a simple description of transitions based on a one-electron model. A detailed computation study would be necessary to describe the interaction of states in detail.

6.13. CHLOROPHYLL *Q* BAND, AN EXAMPLE USING *B* TERMS

The optical spectroscopic study of various chlorophyll (Chl) pigments in a variety of environments has been of interest over many decades because of their relevance to photosystems responsible for light-induced redox processes that form the basis of photosynthesis (for some MCD examples see refs. 50–53). The chromophore of Chl is based on a Mg^{2+} phorphyrin complex, which is attached to a protein structure of some type in the native state. An example of the porphyrin structure is shown in Figure 6.51. The absorption exhibited by Chl is primarily $\pi \rightarrow \pi^*$ type localized on the porphyrin ring. The lowest-energy HOMO → LUMO transition, which is called the *Q* band, occurs in the region of 550–750 nm. For an uncomplexed symmetrical porphyrin molecule of D_{4h} symmetry, this transition would correspond to $^1A_{1g}\ [(a_{1u})^2] \rightarrow {}^1E_u\ [(a_{1u})(e_g)]$. The degenerate excited state exhibits a negative *A* term. When the Chl complex is in a native environment and is attached to various protein structures, which often tend to be aggregated, the symmetry is low, and there are no longer degeneracies, so that only *B* terms are expected in the MCD spectra. In general terms the *Q*-band MCD spectra for a variety of Chl samples both free and protein bound are similar and consist of a positive *B* term for the lowest-energy component of the *Q* band, often labeled Q_y, and negative *B* term(s) to higher energy, ascribed to Q_x. The components of the *Q* band are separated from higher-energy absorptions and are therefore assumed to have a primary *B*-term interaction only with each other and

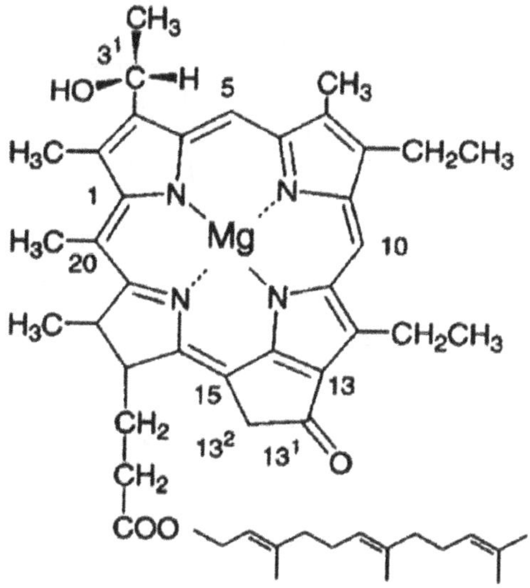

Figure 6.51. Structure of a chlorophyll porphyrin complex (3^1R)-[E,E] BChl c_F. Reprinted with permission from ref. 52. Umetsu, U.; Seki, R.; Wang, Z.-Y.; Kumagai, I.; Nozawa, T. *J. Phys. Chem. B* **2002**, *106*, 3987. Copyright 2002 American Chemical Society.

so must be polarized at right angles to each other and also perpendicular to the z-field direction (perpendicular to the plane of the complex). Hence the x and y designations for the bands. As an example, Figure 6.52 shows the Q-band absorption and MCD spectra, as well as the CD spectrum, for Chl a attached to a tetrameric assembly of water-soluble Chl-binding protein (WSCP) neighboring the Chl chromophores. It was argued that the narrowness (fwhm $\sim$180 cm^{-1}) implies only a small inhomogeneous broadening and therefore a well-defined Chl a-binding site in WSCP. The Q_y band in solvents are typically much broader (fwhm $\sim$400–500 cm^{-1}) and typically do not exhibit the low-energy shoulder. The middle section in Figure 6.52 labeled $Q_y(1, 0)$ shows resolved structure in the 610- to 660-nm range which is ascribed to vibration(s) built on $Q_y(0, 0)$. The higher-energy section from 565 to 610 nm is attributed to $Q_x(0, 0)$ from cauliflower WSCP, which contains two Chl molecules. In order to enhance resolution, the spectra were measured in a 40% 1 : 1 (v : v) ethylene glycol : glycerol glass at 1.7 K. The lowest-energy region is labeled $Q_y(0, 0)$ is assigned as the Q band origin at 672 nm and has a clear positive B term associated with the main absorption peak. There is also a low-energy shoulder on this band and on the MCD B term; the shoulder occurs at the CD maximum. This shoulder is assigned to molecular exciton coupling between Chl molecules and is split into two components (only one is evident in room temperature solutions of Chl a). The absorption in this region is particularly weak, making the assignment somewhat less precise. The negative B term, which is resolved into two features at 594 and 579 nm with roughly equal area as the $Q_y(0, 0)$ B term, is offered as evidence of the Q_x band in the higher-energy region [ref. 53]. The overall interpretation of the Q_y and Q_x

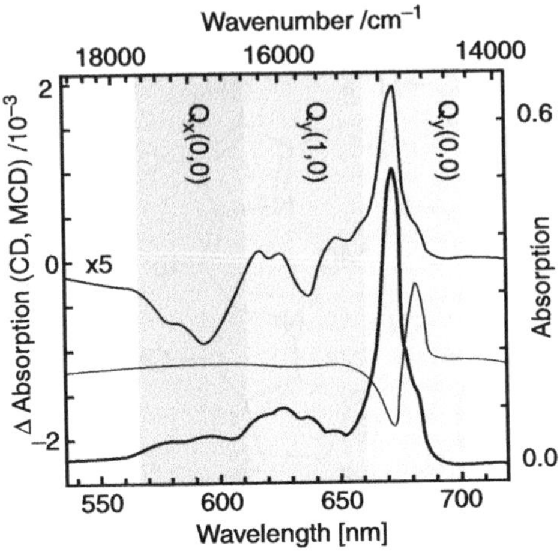

Figure 6.52. Absorption (lower curve), CD (middle curve), and MCD (upper curve) spectra for Chl *a*–WSCP in 40% 1 : 1 (v : v) ethylene glycol–glycerol glass at 1.7 K. The MCD spectrum (per Tesla) is multiplied by a factor of 5, and the CD spectrum is offset for clarity. Reprinted with permission from ref. 53. Hughes, J. L.; Razeghifard, R.; Logue, M.; Oakley, A.; Wydrzynski, T.; Krausz, E. *J. Amer. Chem. Soc.* **2006**, *128*, 3649. Copyright 2006 American Chemical Society.

positions agrees with earlier monomeric solution MCD studies [ref. 50] and MCD in liquid crystal matrices [ref. 51], but a recent room temperature solution study [ref. 52] placed the Q_x transition in the middle region ~650 nm for BChl *c* aggregates on the basis of a deconvolution analysis. Thus the assignment in this region is not entirely free of controversy. Finally, when a moment analysis was performed on the Q_y band MCD and absorption data [ref. 53], the magnitude of the $\bar{B}_0/\bar{D}_0$ values found for the Chl *a*–WSCP spectra at low temperature was ~60% higher than for Chl *a* in solution. The increase was surprising in view of ~50% reduction which has been observed in solvent-based aggregates and in photosynthetic centers. A simple exciton coupling between Chl centers in the tetramers was considered, but could not explain the size of the increase. The exciton coupling model is strongly supported by the observation of the CD effects in the region of the $Q_y(00)$ band which are more than ten times stronger than for Chl monomers. Exciton coupling can induce CD in systems where there is little or no signal for monomers; the CD results from molecular assymmetry of the dimeric coupled chromophores. However, neither the small energy shifts between the Q_y and Q_x components nor the estimated dipole interactions were sufficient to explain the MCD intensity effects. It was suggested that there must be specific Chl–protein interactions responsible for the effects, but the nature of the interactions remains unclear. Clearly, both Chl–Chl and Chl–protein interactions must be considered. MCD spectra can add an important dimension to spectroscopic measurements on chromophores embedded in natural protein systems, but such systems often derive complexity from the subtle aspects of protein structure and conformation and are therefore a challenge to understand. As protein structure becomes more defined, perhaps the spectral interpretation will become more precise.

6.14. SURFACE PLASMON BAND FOR COLLOIDAL GOLD NANOPARTICLES

The deep red to purple color of colloidal gold sols has fascinated chemists for years [ref. 54]. The color is due to a broad absorption band in the visible region of the spectrum (520–575 nm) which is dependent upon the size of the gold nanoparticles (~10–100 nm in spherical diameter). The band broadens and shifts to longer wavelength as the size of the particles increase, but the band disappears altogether for very small particles the size of a molecular complex or even a cluster complex (<~2 nm in diameter) and also for very large particles such as bulk gold metal particles (>~200 nm). The absorption has been attributed to a surface plasmon absorption due to the collective oscillations of free 6*s* Au electrons in the metal conduction band of the colloidal particles and has been described as a scattering phenomenon (classical Mie theory) [ref. 54]. In fact, the electronic origin of the surface plasmon band is not well understood. The observation of an MCD effect for the absorption band was therefore of some interest.

The MCD study [ref. 54] of colloidal gold in water at 298 K, and also in an isotropic solid silica xerogel at both 295 K and 5.5 K, revealed an unsymmetrical

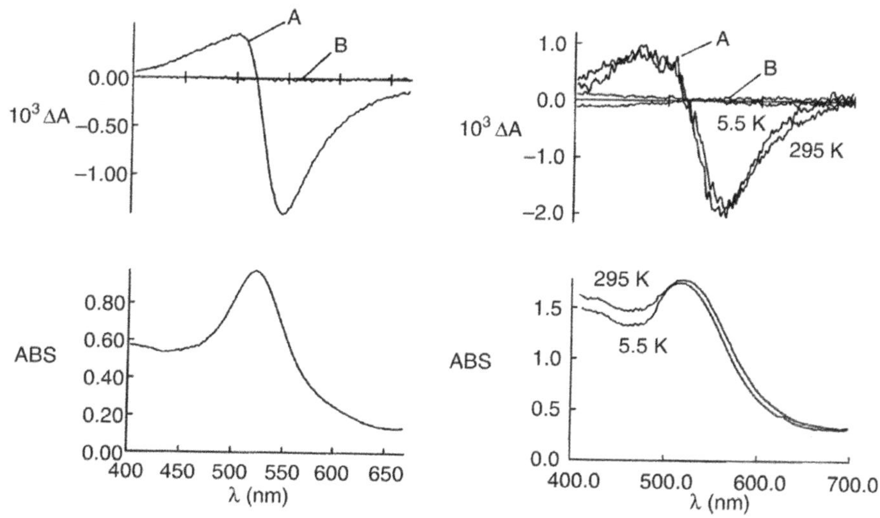

Figure 6.53. MCD (upper curves) and absorption (lower curves) spectra for colloidal gold in water at 298 K (left) and in an isotropic silica xerogel at 295 and 5.5 K (right). The MCD at 7.0 T (A) and at 0.0 T (B) are noted in each case. Reprinted with permission from ref. 54. Zaitoun, M. A.; Mason, W. R.; Lin, C. T. *J. Phys. Chem. B* **2001**, *105*, 6780. Copyright 2001 American Chemical Society.

positive *A* term as shown in Figure 6.53. In order to eliminate the possibility that a scattering effect might produce an anomolous CD effect, the measurements included a careful determination of the spectra in the absence of the field, as shown in the figure, which shows clearly that the spectra are due to MCD. The solid, isotropic xerogel also did not show any detectable static birefringence. Finally, the similarity of the spectra for the xerogel samples at 5.5 K and 295 K allowed the conclusion that the band is not temperature-dependent to within experimental error. The temperature independence of the MCD spectra eliminates the possibility of *C* terms, and therefore the MCD must consist of *A* and *B* terms. Thus, the origin of the MCD must be an "excited-state" phenomenon because any ground-state contribution would be expected to present temperature dependence and *C* terms. A moment analysis of the spectra showed that $\bar{A}_1/\bar{D}_0 = 0.77$, 0.65, and 0.70 for water at 298 K, xerogel at 295 K, and xerogel at 5.5 K, respectively, while $\bar{B}_0/\bar{D}_0 = -0.40 \times 10^{-3}$, -0.46×10^{-3}, and $-0.57 \times 10^{-3}\ \mathrm{cm}^{-1}$, respectively. The *A* term indicates that the excited state(s) must have spin and/or orbital angular momentum and therefore a magnetic moment. It was suggested [ref. 54] that the surface plasmon band probably owes its electronic origins to the heavy Au atom spin–orbit states of a $6(sp)^1$ conduction band, but the relationship between an atomic level electronic transitions and nanoparticle surface plasmon excitations is not presently understood.

7 Case Studies II. Paramagnetic Systems: *C* Terms

Paramagnetic molecules exhibit temperature-dependent *C* terms, which usually dominate the MCD spectrum, especially at low temperature. The temperature dependence of the MCD spectrum is a useful diagnostic for the *C* term, since *B* terms and *C* terms, which have the same band shape at room temperature and above, could be mistaken for each other. *B* terms, however, are temperature-independent. There are other temperature effects that could also be confused with the $1/T$ dependence of *C* terms. These include the narrowing of bands as temperature is reduced and matrix effects. Band narrowing will enhance the maximum values for all MCD spectra because signals are dependent upon the bandwidth, but such bandwidth effects do not usually follow a $1/T$ dependence. Matrix effects, including matrix phase changes at low temperature, likewise don't necessarily follow a $1/T$ dependence and are usually unpredictable in most cases. If there is doubt about the temperature dependence, then a careful temperature-dependence study is required to confirm the $1/T$ behavior, but usually the temperature dependence of MCD spectra containing *C* terms is striking.

7.1. PARAMAGNETIC CYANO COMPLEXES

7.1.1. LMCT Transitions for $Fe(CN)_6^{3-}$

An early example of *C* terms for an octahedral metal complex is given by the MCD for the three lowest-energy intense bands exhibited by the $Fe(CN)_6^{3-}$ ion (Figure 7.1 [ref. 1 Chapter 11; refs. 55, 56]). The $^2T_{2g}$ ground state for this ion derives from the low-spin t_{2g}^5 electron configuration for the complex, which has one hole in an otherwise filled, degenerate t_{2g} metal 3*d* orbital level and thus one unpaired electron. These bands were assigned as LMCT based on their intensity and relative energy as a function of metal oxidation state. The three intense bands 2.45, 3.27, and 4.05 μm^{-1} in the spectra of the Fe(III) complex $Fe(CN)_6^{3-}$ are all found lower in energy than the lowest energy intense band for the Fe(II) complex $Fe(CN)_6^{4-}$ found at 4.6 μm^{-1} (Figure 6.23). A red shift as the oxidation state of the metal increases is characteristic of LMCT bands. The bands were assigned as

A Practical Guide to Magnetic Circular Dichroism Spectroscopy, by W. Roy Mason

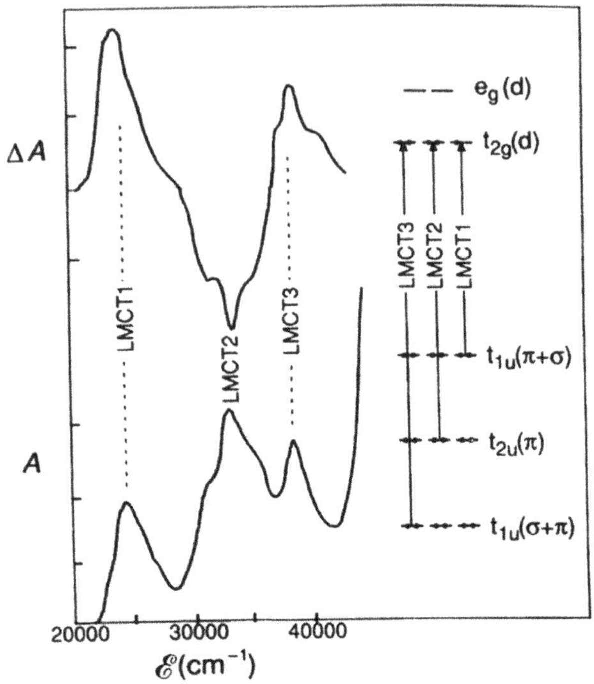

Figure 7.1. Absorption and MCD spectra for $Fe(CN)_6^{3-}$ in water solution. Reprinted with permission from ref. 56. Upton, A. H. P.; Williamson, B. E., *J. Phys. Chem.*, **1994**, *98*, 71. Copyright 1994 American Chemical Society.

$^2T_{2g} \rightarrow {}^2T_{1u}(\pi\sigma)$ $[t_{1u}(\pi\sigma)^5 t_{2g}^6]$, $^2T_{2g} \rightarrow {}^2T_{2u}(\pi)$ $[t_{2u}(\pi)^5 t_{2g}^6]$, and $^2T_{2g} \rightarrow {}^2T_{1u}(\sigma\pi)$ $[t_{1u}(\sigma\pi)^5 t_{2g}^6]$, respectively, as excitations from the filled ligand MOs $t_{1u}(\pi\sigma)$, $t_{2u}(\pi)$, and $t_{1u}(\sigma\pi)$ (see Figure 7.2). As discussed earlier, there is extensive intermixing between the σ and π ligand t_{1u} orbitals and, hence, the dual designation in the notation. The $t_{2u}(\pi)$ orbital has no σ counterpart, so mixing is not possible. The interpretation of the MCD spectra as *C* terms was based on the temperature dependence. A study [ref. 55] of the complex in isotropic solid KC1 crystals to 12 K confirmed the $1/T$ dependence, and therefore the *C*-term character of the observed MCD. Figure 7.3 shows a more recent study of $K_3[Fe(CN)_6]$ embedded in a thin poly(vinyl alcohol) (PVA) film [ref. 56]. It should be noted that *A* terms are also possible from the degenerate excited states, but the temperature-dependence studies showed that they are small compared to the *C* terms. The *C* terms are given by Eqs. (3.38) and (3.39) (Chapter 3), where $\alpha, \lambda = 1, 0, -1$; the ground-state degeneracy $|^2T_{2g}| = 3$; and $i = 1$ or 2 for $^2T_{iu}\lambda$, respectively:

$$C_0 = (1/(3\mu_B))\Sigma_{\alpha\lambda}[\langle^2T_{2g}\alpha|\mu_z|^2T_{2g}\alpha\rangle$$
$$\times (|\langle^2T_{2g}\alpha|m_{-1}|^2T_{iu\lambda}\rangle|^2 - |\langle^2T_{2g}\alpha|m_{+1}|^2T_{iu\lambda}\rangle|^2)]$$

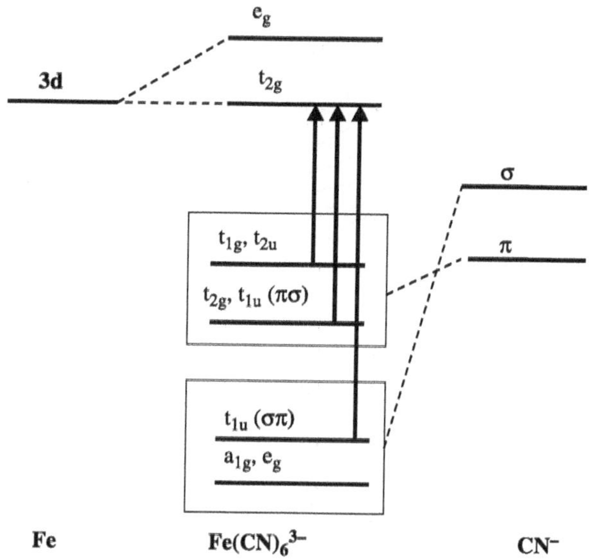

Figure 7.2. Schematic MO energy levels for $Fe(CN)_6^{3-}$ showing LMCT excitations from the occupied CN^- orbitals to the single hole in the t_{2g} metal orbital.

By assuming that spin–orbit coupling is small and thereby approximating $\mu_z^{t_{1g}} = -\mu_B L_z^{t_{1g}}$, the Wigner–Eckart theorem can be used to reduce this equation to RME form in the following equations:

$$C_0(^2T_{2g} \rightarrow {}^2T_{1u}) = -1/(9(6)^{1/2})\langle^2T_{2g}\|L_z^{t_{1g}}\|^2T_{2g}\rangle|\langle^2T_{2g}\|m^{t_{1u}}\|^2T_{1u}\rangle|^2$$
$$C_0(^2T_{2g} \rightarrow {}^2T_{2u}) = +1/(9(6)^{1/2})\langle^2T_{2g}\|L_z^{t_{1g}}\|^2T_{2g}\rangle(|\langle^2T_{2g}\|m^{t_{1u}}\|^2T_{2u}\rangle|^2)$$

If the t_{2g} orbital is approximated as a pure $3d$ atomic orbital, then the spin-independent $\langle^2T_{2g}\|L_z^{t_{1g}}\|^2T_{2g}\rangle$ RME can be reduced to one-electron form and evaluated at the Fe(III) ion center as (see ref. 1, Sections 13.2 and 13.4 for details)

$$\langle^2T_{2g}\|L_z^{t_{1g}}\|^2T_{2g}\rangle = \langle t_{2g}\|l_z^{t_{1g}}\|t_{2g}\rangle = -(6)^{1/2}$$

These equations together with the appropriate equations for D_0,

$$D_0 = (1/9)|\langle^2T_{2g}\|m^{t_{1u}}\|^2T_{1u}\rangle|^2 \quad \text{or} \quad D_0 = (1/9)|\langle^2T_{2g}\|m^{t_{1u}}\|^2T_{2u}\rangle|^2$$

respectively, lead to the values of C_0/D_0:

$$C_0/D_0(^2T_{2g} \rightarrow {}^2T_{1u}) = +1 \quad \text{and} \quad C_0/D_0(^2T_{2g} \rightarrow {}^2T_{2u}) = -1$$

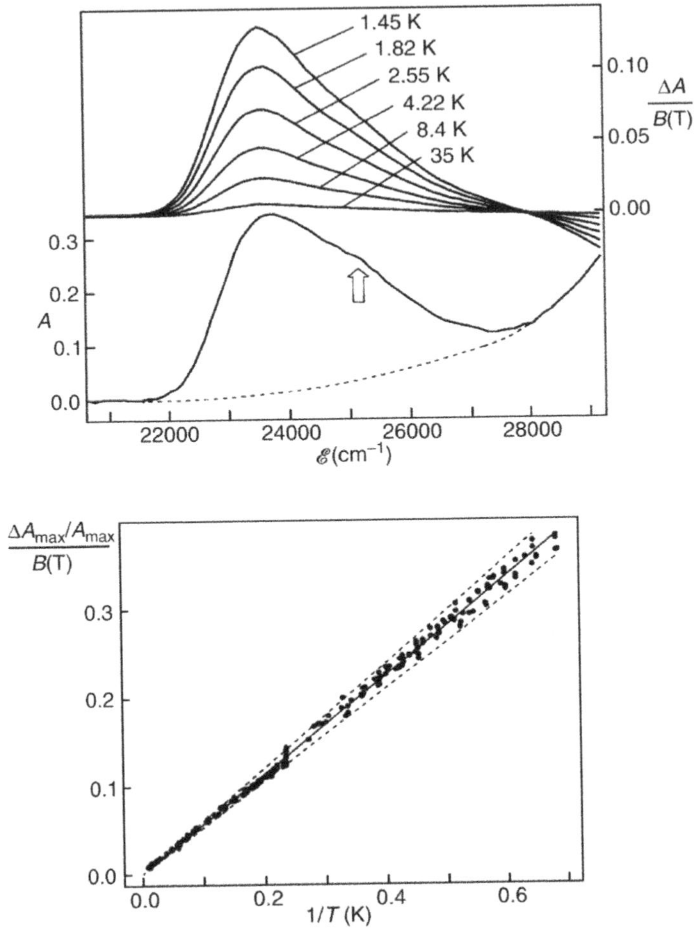

Figure 7.3. Temperature dependence of the first LMCT band for $K_3[Fe(CN)_6]$ in a poly(vinyl alcohol) film from 1.45 to 35 K. Reprinted with permission from ref. 56. Upton, A. H. P.; Williamson, B. E., *J. Phys. Chem.*, **1994**, *98*, 71. Copyright 1994 American Chemical Society.

The observed values of C_0/D_0 at room temperature for the three bands in the $Fe(CN)_6^{3-}$ spectrum are $+1.2$, -0.6, and $+0.6$, respectively, with some experimental uncertainty in the last two values [refs. 1, 54]. Thus the sign observed for the three C terms in the MCD spectra (positive, negative, and positive, respectively) establishes the energy ordering as ${}^2T_{2g} \rightarrow {}^2T_{1u}(\pi\sigma) < {}^2T_{2g} \rightarrow {}^2T_{2u}(\pi) < {}^2T_{2g} \rightarrow {}^2T_{1u}(\sigma\pi)$. Early assignments of the $Fe(CN)_6^{3-}$ spectrum had placed the ${}^2T_{2g} \rightarrow {}^2T_{2u}(\pi)$ transition at lowest energy because, it was argued, the $t_{2u}(\pi)$ MO was nearly nonbonding, while the two t_{1u} MOs were bonding [see ref. 27]. Thus, MCD C term signs provide strong experimental support for the energy ordering of the LMCT excited states. A more complete treatment of the $Fe(CN)_6^{3-}$ spectrum, including a consideration of *A* terms and spin–orbit effects, is described in Ref. 1, Sections 11.2 and 13.5, and also in ref. 56.

7.1.2. LMCT Spectra for $M(CN)_8^{3-}$, M = Mo(V), and W(V)

Another, more recent example of the temperature dependence of *C* terms for paramagnetic cyano complexes can be found in a study of the $M(CN)_8^{3-}$, [M = Mo(V) and W(V)] ions. These nd^1 octacyano complexes have a single unpaired electron in the HOMO, which is assumed to be the $d_{x^2-y^2}$ metal orbital. The molecular structure of these ions is either dodecahedral (D_{2d}) or square antiprismatic (D_{4d}),

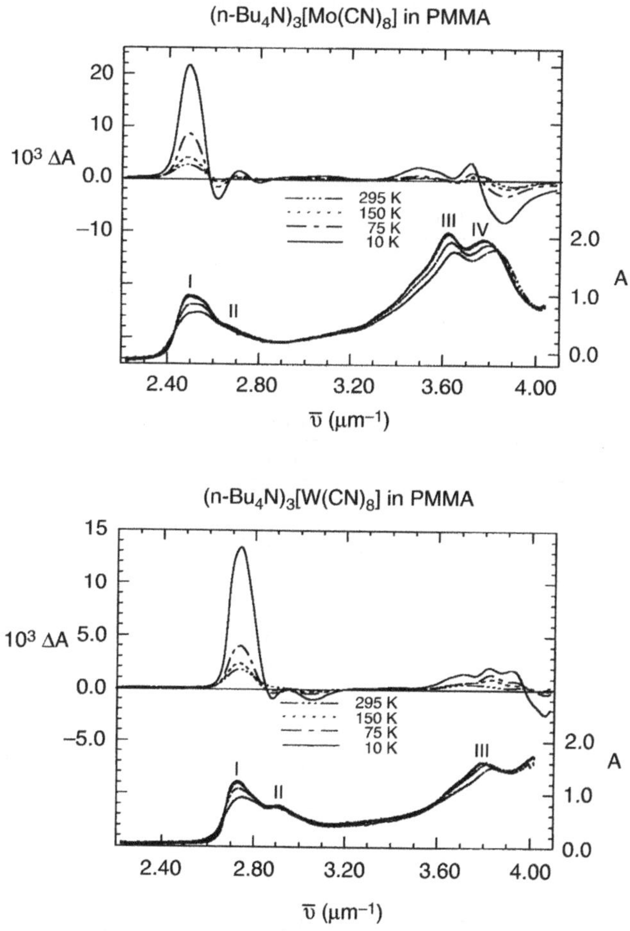

Figure 7.4. $M(CN)_8^{3-}$, [M = Mo(V) and W(V)], ions embedded in poly(methyl methacrylate) (PMMA) thin films at temperatures from 10 to 295 K. Reprinted from ref. 57. Isci, H.; Mason, W. R., *Inorg. Chim. Acta*, **2004**, *357*, 4065. Copyright 2004, with permission from Elsevier.

depending upon the crystalline environment. Their solution structure is not known for certain, but the D_{2d} and D_{4d} structures are believed to have nearly the same energy and therefore may be easily interconverted. The MCD spectra for the $M(CN)_8^{3-}$ ions embedded in thin poly(methyl methacrolate) (PMMA) films is shown in Figure 7.4 at various temperatures between 10 and 295 K. Band I in each case shows the strong temperature-dependence characteristic of a positive *C* term. Band IV for the Mo(V) complex also shows a combination of a negative *C* term and a negative *A* term; a similar pattern is also indicated for the W(V) complex.

The spectra have been interpreted within a D_{2d} structural model, which includes metal spin–orbit coupling, as LMCT from occupied mainly π or non-bonding CN^- orbitals to the partially filled HOMO $b_1(x^2 - y^2)$ metal-based *nd* orbital (see Figure 7.5). The assignment of band I was based on the *C*-term sign. The ground state, including spin–orbit coupling, for the b_1^1 configuration in D_{2d} is $E''(^2B_1)$. There are many closely spaced excited states, and states of the same

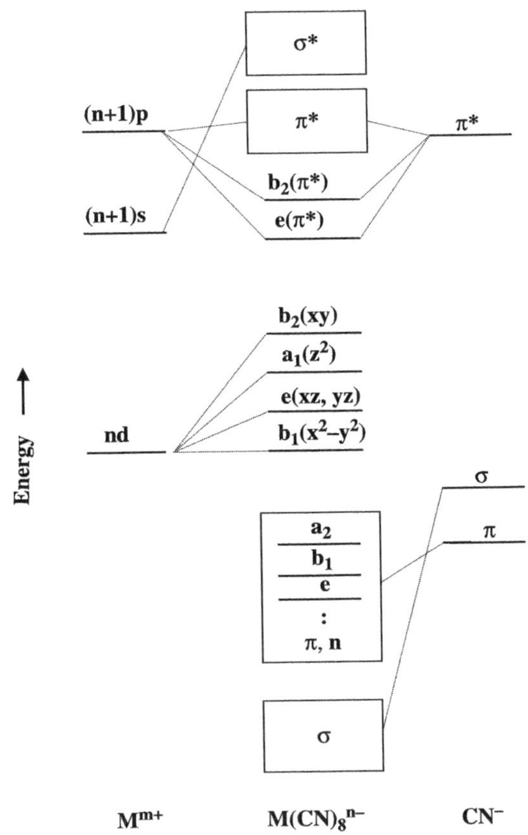

Figure 7.5. Energy levels for $M(CN)_8^{n-}$ ions assuming D_{2d} symmetry. For M(V) ions, the HOMO is $b_1(x^2 - y^2)$ with a single electron giving a ground state of $E''(^2B_1)$. Reprinted from ref. 57. Isci, H.; Mason, W. R., *Inorg. Chim. Acta*, **2004**, *357*, 4065. Copyright 2004, with permission from Elsevier.

symmetry can be intermixed. Therefore the composition of most of the excited states is complicated. However, a number of the symmetry types can be eliminated from consideration because they predict the wrong *C*-term sign when compared to the experimental spectrum. Band I was assigned as a combination of two close-lying transitions: $E''(^2B_1) \rightarrow E'(^2A_2)$ and $E''(^2B_1) \rightarrow E'(^2E)$ of the $a_2[b_1(x^2 - y^2)]^2$ and $e^3[b_1(x^2 - y^2)]^2$ excited configurations, respectively. Band IV was assigned as $E''(^2B_1) \rightarrow E'(^2B_1)$ of the $b_1[b_1(x^2 - y^2)]^2$ configuration on the basis of the negative *C* and the negative *A* terms predicted for the transition; ref. 57 should be consulted for more details and additional assignments.

7.2. GROUND-STATE MAGNETIZATION FOR UNSTABLE METALLOCENES

The saturation behavior of *C* terms at low temperatures and high fields (see Chapter 3, Section 3.10) offers the prospects of probing the magnetic properties of the ground state of paramagnetic molecules. This application is illustrated by an interesting study of the magnetization of several unstable matrix isolated metallocenes [ref. 58]. The unstable metallocenes Cp_2M, M = Mo, W, and Re; Cp $= \eta^5 - C_5H_5^-$ were prepared by photolysis of their corresponding hydrides Cp_2MH_n or other related Cp_2ML complexes and studied in a low-temperature argon matrix (Mo or W complexes) or a nitrogen matrix (Re complex) at 1.8–17 K and at fields from 0 to 8 T, in order to obtain their ground state magnetization properties. The normal way that ground-state magnetization is investigated is by EPR spectroscopy, but EPR measurements present problems for some heavy metal molecules, especially those that have an even number of unpaired electrons or large zero-field splittings so that the EPR signals are "silent," even though the molecules are paramagnetic. MCD *C*-term behavior provides an optical probe of ground states with orbital and spin degeneracy. The method was authenticated by *C*-term measurements for $(tol)_2V$ and Cp_2^*Re, tol $= \eta^6$-toluene and Cp* $= \eta^5-(CH_3)_5C_5^-$, in the argon matrix. In each of these cases the results could be compared with corresponding earlier EPR studies for frozen solutions at low temperature. The *C* terms observed for both complexes followed reproducible ΔA versus $\mu_B B/2kT$ curves. For $(tol)_2V$, where the ground state is $^2A_1$ (one unpaired electron and therefore spin-only angular momentum) and was described by a $\tanh(g_{iso}\mu_B B/2kT)$ function at 445.0 nm to give a value of $g_{iso} = 1.94$, Figure 7.6, compared favorably to EPR measurements on frozen solutions, where the *g* values were nearly isotropic with $g\|$ and $g\perp$ values of 2.00 and 1.98, respectively. The value of $g\|$ for the $^2E_2$ ground state (spin and orbital angular momentum) of Cp_2^*Re in the argon matrix was determined to be 5.07 from the *C* term at 588.0 nm, with the results from EPR for frozen solutions giving 5.08. The agreement between the EPR and matrix isolated MCD *C*-term studies is excellent and gives confidence that the *C*-term study provides comparable ground-state information.

The matrix isolated metallocenes were assumed to have axial symmetry (nominally D_5) with ground states $^3E_2$ for the d^4 Mo(II) and W(II) complexes and $^2E_2$

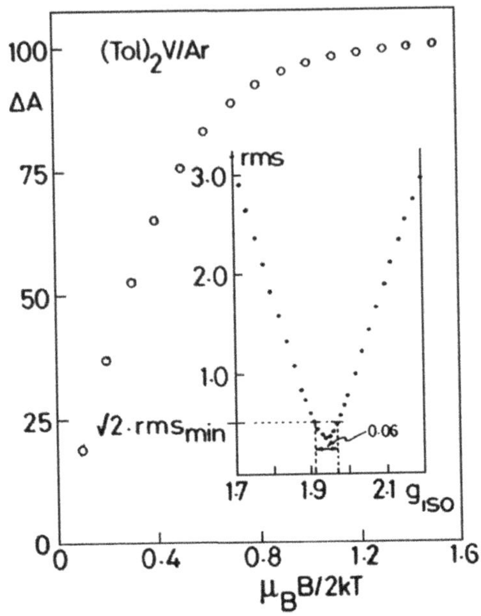

Figure 7.6. Magnetization curves for $(tol)_2V$ in an argon matrix at 445 nm and temperatures of 1.8 and 6.1 K. The small dots are the experimental points, while the open circles are the values calculated with $g_{iso} = 1.94$. The inset shows the rms deviation of $\Delta A(\text{calcd}) - \Delta A(\text{expt})$ plotted versus g_{iso}. Reprinted with permission from ref. 58. Graham, R. G.; Grinter, R.; Perutz, R. N., *J. Am. Chem. Soc.* **1988**, *110*, 7036. Copyright 1988 American Chemical Society.

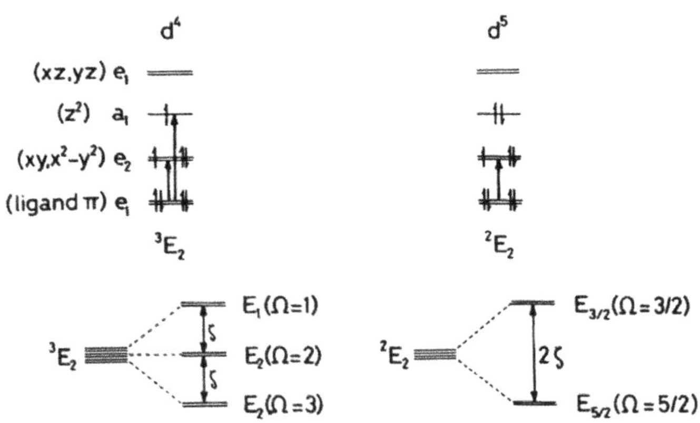

Figure 7.7. Ground-state configurations and energy levels for the d^4 Cp_2Mo and Cp_2W and d^5 Cp_2Re metallocenes, assumed to have axial *D*5 symmetry. The values of Ω give the total angular momentum quantum number. Reprinted with permission from ref. 58. Graham, R. G.; Grinter, R.; Perutz, R. N., *J. Amer. Chem. Soc.* **1988**, *110*, 7036. Copyright 1988 American Chemical Society.

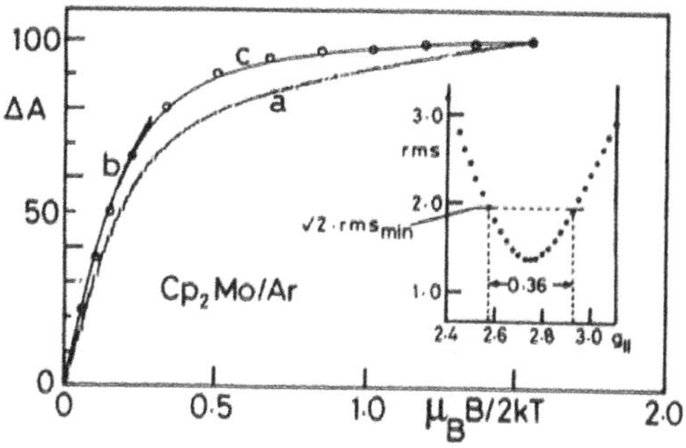

Figure 7.8. Magnetization curves for Cp_2Mo in an argon matrix: (a) experimental curve at 1.8 K and both 377 and 411.5 nm, superimposed; (b) experimental curve at 9.7 K from both wavelengths; (c) curve corrected for *B* terms, with open circles calculated for $k = 0.6$ and $\delta = 450\ cm^{-1}$ with spin–orbit coupling $\zeta = 600\ cm^{-1}$. Reprinted with permission from ref. 58. Graham, R. G.; Grinter, R.; Perutz, R. N., *J. Am. Chem. Soc.* **1988**, *110*, 7036. Copyright 1988 American Chemical Society.

for the d^5 Re((II) complex as shown in Figure 7.7. The *C*-term magnetization was interpreted by integration over orientation with respect to the field. The results for Cp_2Mo showed "nesting" in the magnetization curve (Figure 7.8) and required the inclusion of *B*-term contributions between Zeeman sublevels. The curves for Cp_2W and Cp_2Re did not show any measurable "nesting"; curves at different temperatures were practically superimposable. For all three complexes a complete interpretation required inclusion of spin–orbit coupling with an estimate of the orbital reduction parameter *k* and inclusion of quenching of orbital angular momentum by low-symmetry perturbations as given by a parameter δ. The results showed that the $g\perp$ values were close to zero and the $g\|$ values were 2.75, 3.07, and 5.34 for M = Mo, W, and Re, respectively.

7.3. DETERMINATION OF THE GROUND STATE FOR MATRIX-ISOLATED MANGANESE(II) PHTHALOCYANINE FROM *C* TERMS

Another interesting case study of ground-state properties is presented by a study of manganese(II) phthalocyanine, MnPc [ref. 59], where the Pc^{2-} ligand is the dianion of tetraazatetrabenzoporphyrin. The structure of the Pc^{2-} ligand with its planar idealized D_{4h} coordination is shown for the diamagnetic ZnPc complex in Figure 6.47 (Chapter 6, Section 6.12). A magnetic susceptibility study of MnPc on β-phase solid samples indicated a complex with an $S = 3/2$ ground state, and the quartet state $^4A_{2g}$ from a $d^5b_{2g}^2e_g^2a_{1g}$ configuration in D_{4h} symmetry was suggested. However, the complex was prone to interact with molecular oxygen O_2 in solutions and was suspected of intermolecular interactions in the solid state. By careful matrix isolation of the complex in a low-temperature argon matrix, the

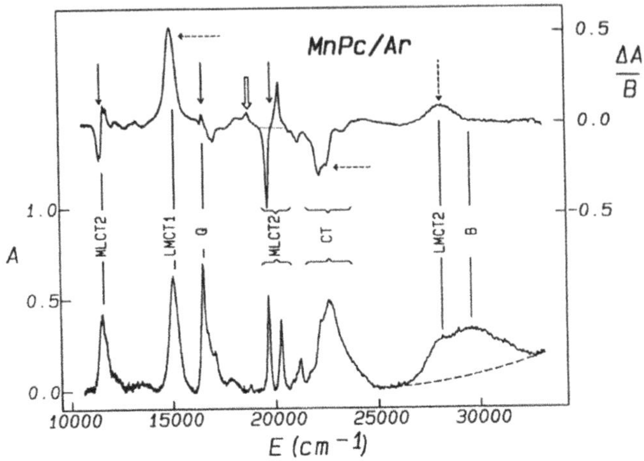

Figure 7.9. Absorption and MCD spectra of MnPc in an argon matrix at 5 K and $B =$ 0.15 T. Reprinted with permission from ref. 59. Williamson, B. E.; VanCott, T. C.; Boyle, M. E.; Misener, G. C.; Stillman, M. J.; Schatz, P. N., *J. Am. Chem. Soc.* **1992**, *114*, 2412. Copyright 1992 American Chemical Society.

spectra for the isolated complex were determined. The MCD spectrum (Figure 7.9) consisted of strong *C* terms that were both bisignate (pseudo *A* terms or temperature-dependent *A* terms) and single-signed positive and negative *C* terms. It was argued that an orbitally nondegenerate ground state could not produce single-signed *C* terms because the $^4A_{2g}$ manifold would produce both positive and negative *C* terms split only by the Zeeman energies in the applied field. This small energy difference would be expected to give bisignate *C* terms and not single-signed *C* terms such as the one observed at 1.5 μm^{-1} in Figure 7.9. An alternative ground state was assigned as ^4E_g arising from an alternative $e_g^3a_{1g}b_{2g}$ configuration. It was argued that the orbitally degenerate ground state would be split by spin–orbit coupling into four Kramers doublets that would be separated by energies approximately $\zeta_{3d}/3$, where ζ_{3d} is the spin–orbit coupling constant for Mn(II), about 300 cm^{-1}, as shown in Figure 7.10. This larger ground-state splitting together with differences in excited states governed by ligand field effects (Δ in Figure 7.10) would produce single-signed *C* terms separated by much larger energies. A magnetization curve determined carefully at 1.876 μm^{-1} further determined that the MnPc molecules, rather than being randomly oriented, were actually oriented with the *z* molecular axes parallel to the field by the matrix formation (*z* molecular axes perpendicular to the surface of the deposition window). The $g_{\parallel}$ value determined for the oriented molecules was 8.0, which was fully consistent with the ^4E_g ground state. If the molecules had been randomly oriented, the $g_{\parallel}$ value would have been 11.3 and therefore inconsistent with the ^4E_g ground state. It was hypothesized that the different magnetic susceptibility behavior observed for the β-phase solid was due to intermolecular interactions. Once the ground state of MnPc was formulated, the remainder of the spectrum in

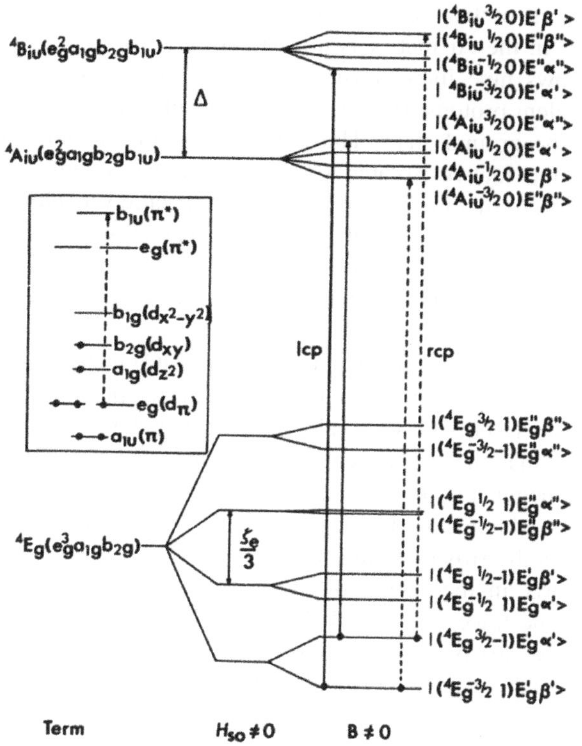

Figure 7.10. Energy levels for the $^4E_g(e_g^3a_{1g}b_{2g})$ ground state and transitions to two excited states. Reprinted with permission from ref. 59. Williamson, B. E.; VanCott, T. C.; Boyle, M. E.; Misener, G. C.; Stillman, M. J.; Schatz, P. N., *J. Am. Chem. Soc.* **1992**, *114*, 2412. Copyright 1992 American Chemical Society.

Figure 7.9 was assigned as a combination of LMCT, MLCT, and $\pi \rightarrow \pi^*$ transitions (see ref. 59 for details).

7.4. MONONEGATIVE CYCLOOCTATETRAENE ION (COT^-) MATRIX ISOLATED IN ARGON

The conjugated C_8H_8 cyclooctatetraene (COT) molecule has been of interest both structurally and theoretically for years. It has a nonplanar D_{2d} "tub"-like structure that is strain-free. The structure undergoes inversion and bond shift processes that are believed to involve planar transition states. In contrast, the dianion COT^{2-} has a planar D_{8h} structure with equal bond lengths and has the requisite $4n + 2$ π electrons for aromaticity; it also forms stable nonclassical η^8 metal complexes. The structure of the unstable radical monoanion of cyclooctatetraene COT^- has, therefore, been of considerable interest as an intermediate between two stable structures. The ion can be prepared by co-condensing COT and Cs in an argon matrix, and some theoretical studies along with esr measurements have suggested that the structure in the Ar matrix is most likely D_{8h} (or C_{8v}). The esr spectrum, for example, shows nine lines consistent with the unpaired electron interacting with eight equivalent protons. However, the esr results are limited in that they are exclusively for the ground state. Thus, there was considerable interest when it was found that the matrix isolated COT^- ion shows a temperature-dependent MCD and vibronic structure for the lowest-energy-allowed electronic transition [ref. 60]. Careful analysis of the MCD spectrum has allowed the development of a structural model for COT^- which is consistent with a symmetrical planar ground state. According to the model, both the ground state and the first allowed excited state are subject to Jahn–Teller (JT) effects.

Figure 7.11 shows Hückel molecular orbital (HMO) energies for a delocalized symmetrical planar D_{8h} structure, a less symmetrical planar D_{4h} structure with alternating long and short C–C bonds, and the nonplanar D_{2d} structure for the COT^- ion. Note that the highest-energy occupied orbital (HOMO) for the D_{8h} case is orbitally degenerate (e_{2u}), in contrast to the HOMO for either of the D_{4h} or D_{2d} structures which have nondegenerate HOMO's (b_{1u} or a_2, respectively). Figure 7.12 shows the structured absorption and MCD spectra obtained for COT^- in an argon matrix in the region of the lowest-energy-allowed electronic transition. The multiple traces for the MCD spectra were obtained at different temperatures and show clearly that the bands are C terms. The presence of C terms indicates ground-state degeneracy and is thus consistent with the $^2E_{2u}$ ground

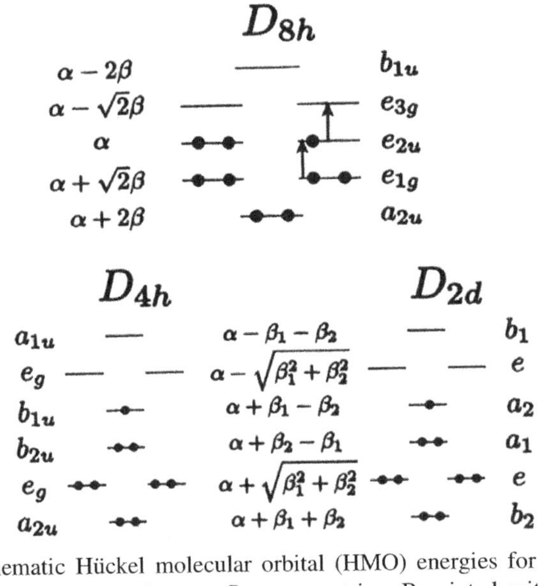

Figure 7.11. Schematic Hückel molecular orbital (HMO) energies for the π electrons of the COT^- ion assuming D_{8h}, D_{4h}, or D_{2d} symmetries. Reprinted with permission from ref. 60. Samet, C.; Rose, J. L.; Piepho, S. B.; Laurito, J.; Andrews, L. A.; Schatz, P. N. *J. Am. Chem. Soc.* **1994**, *116*, 11109. Copyright 1994 American Chemical Society.

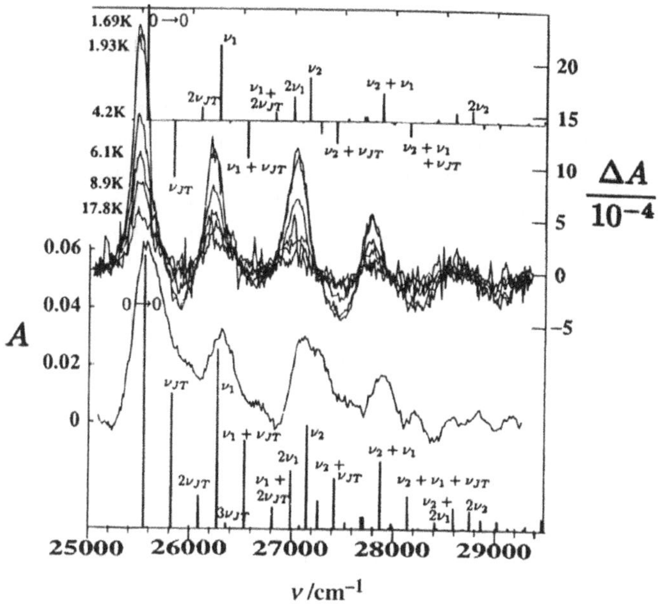

Figure 7.12. Absorption (lower curve) and MCD (upper curves) for the COT$^-$ ion in an argon matrix in the region of the $^2E_{2u} \rightarrow {}^2E_{1g}$ transition. The ΔA scale is per tesla, and the individual MCD curves were obtained at the Kelvin temperatures indicated to the left of the origin band. The "stick" features on horizontal axes for both absorption and MCD represent the relative intensities from the vibrational analysis of the structured band. Reprinted with permission from ref. 60. Samet, C.; Rose, J. L.; Piepho, S. B.; Laurito, J.; Andrews, L. A.; Schatz, P. N. *J. Am. Chem. Soc.* **1994**, *116*, 11109. Copyright 1994 American Chemical Society.

state for the D_{8h} planar structure and inconsistent with the $^2B_{1u}$ or $^2A_2$ ground states expected for the D_{4h} or D_{2d} possibilities, respectively. The assignment of the electronic transition is then given as $^2E_{2u}(e_{2u}^3) \rightarrow {}^2E_{1g}(e_{1g}^3)$. The C-term sign and magnitude in the HMO approximation for this transition can be determined by Eqs. (3.25) and (3.26) by assuming $2p\pi$ atomic orbitals on the C atoms:

$$\overline{C}_0 = (i/6)\langle {}^2E_{2u}(e_{2u}^3)|\mathbf{L}_z|{}^2E_{2_u}(e_{2u}^3)\rangle \cdot (\langle {}^2E_{2u}|\mathbf{m}|{}^2E_{1g}\rangle \times \langle {}^2E_{1g}|\mathbf{m}|{}^2E_{2u}\rangle)$$

$$\overline{D}_0 = (1/6)|\langle {}^2E_{2u}|\mathbf{m}|{}^2E_{1g}\rangle|^2$$

$$\overline{C}_0/\overline{D}_0 = i\langle {}^2E_{2u}(e_{2u}^3)|\mathbf{L}_z|{}^2E_{2u}(e_{2u}^3)\rangle = i\langle e_{2u}|\mathbf{L}_z|e_{2u}\rangle\rangle$$
$$\approx +1.33$$

where $\langle e_{2u}| = (1/\sqrt{8})(\phi_1 - \phi_3 + \phi_5 - \phi_7) - (i/\sqrt{8})(\phi_2 - \phi_4 + \phi_6 - \phi_8)$ is the HOMO with $\phi_i = 2p_\pi$ orbitals on the ith C atom. When a moment analysis (zeroth MCD moment divided by zeroth absorption moment) obtained from the experimental MCD and absorption spectra is plotted versus $1/kT$, the resulting ratio $\overline{C}_0/\overline{D}_0$ was estimated to be $\sim +0.02$, indicating a rather large quenching of angular momentum in the ground state (the contribution from B terms was found to be negligible). Such quenching is consistent with and strongly suggests the presence of Jahn–Teller (JT) effects in the ground state. Furthermore, the vibrational structure observed in the spectra does not exhibit simple, totally symmetric progression(s), which should show positive C terms for the vibronic bands just as for the origin, but rather the progression is a combination of both totally symmetric (a_{1g}) modes and a non-totally symmetric mode that shows negative C terms between the positive C terms. These negative C terms have been interpreted as due to a non symmetric mode attributed to JT distortions in the excited state. The vibrational analysis is consistent with a D_{8h} $^2E_{2u}$ ground state and a lower-symmetry excited state, either of D_{4h} or D_{2d} symmetry, that is subject to JT distortion by a nonsymmetric mode (b_{2g} or b_2) of $\sim$270 cm^{-1}. The symmetric progressions were interpreted as a $\sim$720-cm^{-1} ground-state mode (ν_1) and a $\sim$1600-cm^{-1} excited-state mode (ν_2). Comparison of the symmetric modes with those calculated for COT$^-$ lead to the conclusion that there are also substantial

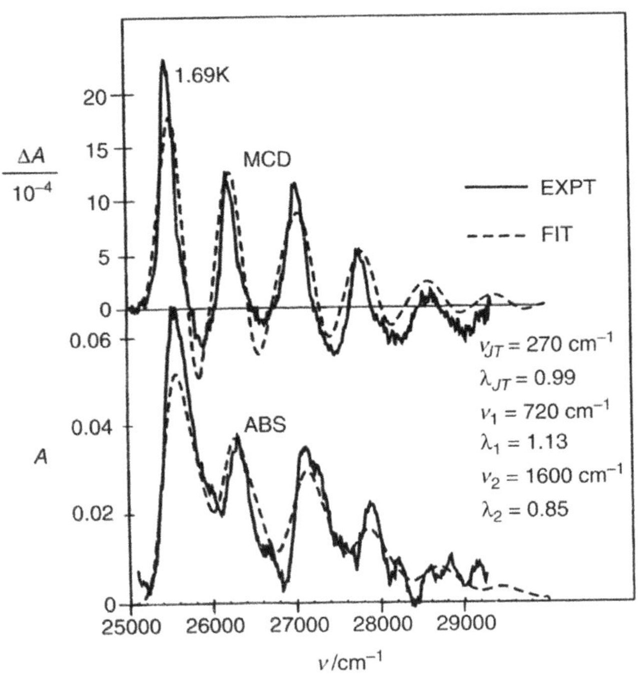

Figure 7.13. The "best fit" resulting from the vibronic analysis of the absorption and MCD spectra. Reprinted with permission from ref. 60. Samet, C.; Rose, J. L.; Piepho, S. B.; Laurito, J.; Andrews, L. A.; Schatz, P. N. *J. Am. Chem. Soc.* **1994**, *116*, 11109. Copyright 1994 American Chemical Society.

JT effects in the ground state as well, consistent with the observed angular momentum quenching. To fully account for the observed spectra, crystal field effects and spin–orbit effects in the ground state were also included (for the details of the analysis, see ref. 60). When the model is compared to experiment, the "best fit" follows the experimental spectra fairly well as shown in Figure 7.13; the "stick" spectra in Figure 7.12 illustrate the vibronic analysis. The ground-state structure can be visualized as two Kekulé-like structures separated by a JT barrier of $\sim$964 cm^{-1}, with a JT mode (b_{1g}) of $\sim$1140 cm^{-1}. The zero-point energies are estimated to be $\sim$60% of the barrier height, so that extensive tunneling between the two structures would be expected. The ground-state orbital angular momentum quenching (Ham effect) due to the combination of JT, crystal field, and spin–orbit effects was calculated to be by a factor of $\sim$0.18, somewhat less than the observed quenching, but close in view of all the approximations involved. The excited-state JT barrier and mode was $\sim$130 cm^{-1} and $\sim$270 cm^{-1}, respectively. This example illustrates how careful MCD spectroscopic measurement and detailed analysis can probe subtle dynamic structural features in unstable matrix isolated molecules.

7.5. GROUND-STATE ELECTRONIC AND STRUCTURAL INFORMATION FROM VARIABLE TEMPERATURE AND FIELD STUDIES OF *C* TERMS: APPLICATION TO NON-HEME-IRON ENZYMES

The behavior of *C* terms under conditions of variable temperature and field has been found to provide a useful probe of electronic structure and site geometry of metal ions in biologically important compounds [ref. 61]. In such compounds, the metal ion is often believed to be the active site for enzyme reactions. The method, developed by Solomon and his group, has been called "VTVH" and derives from the effect of low-symmetry ligand field at the metal site on the magnetically active levels of the paramagnetic ground state. A variety of different spin systems have been characterized, and these have been reviewed and discussed extensively, especially as they apply to active sites in non-heme-iron enzymes [refs. 61–64]. Here an example of the behavior of the high-spin $3d^6$ Fe(II) metal center with $S = 2$ is described in a qualitative way. When the ^5D free ion term for Fe(II) is subjected to a weak octahedral field, two high-spin states are produced by ligand field: $^5T_{2g}$ and ^5E_g. When the symmetry of the ligand field is lowered by axial or other distortions, these degenerate states split further, depending upon the type of distortion. Ligand field absorption bands for Fe(II) systems occur in the near IR and are generally weak because the transitions are Laporte forbidden. However, the MCD spectra for these transitions exhibit *C* terms that are observed to have high intensity at low temperature and high field and can be readily studied under a variety of conditions. For example, an Fe(II) ion in a five-coordinate environment approximating a square pyramid gives a MCD spectrum depicted in Figure 7.14*a*; shown there also is the *C* term at different field strength. A plot of MCD intensity versus $\beta H/2kT$ ($\beta = \mu_B =$ Bohr

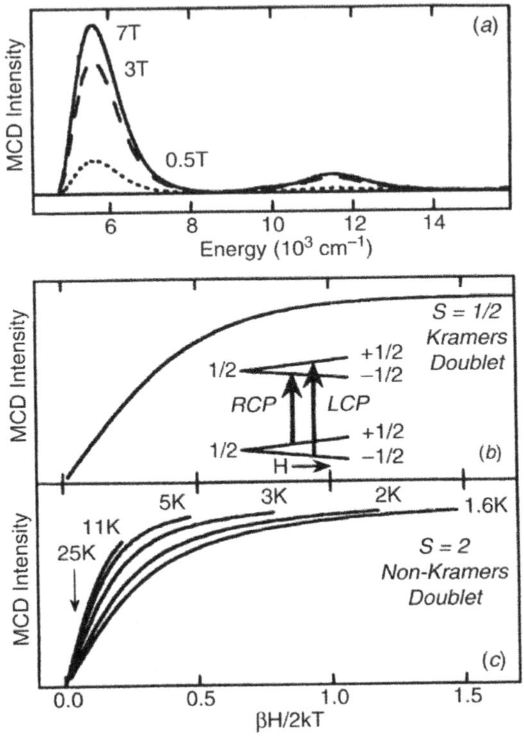

Figure 7.14. (*a*) MCD *C* terms at different field for a five-coordinate high-spin Fe(II) site. (*b*) Magnetization curve for a $S = 1/2$ Kramers doublet. (*c*) Magnetization curve for a non-Kramers doublet $S = 2$. Reprinted with permission from ref. 61. Solomon, E. I.; Brunold, T. C.; Davis, M. I.; Kemsley, J. N.; Lee, S.-K.; Lehnert, N.; Neese, F.; Skulan, A. J.; Yang, Y.-S.; Zhou, J. *Chem. Rev.* **2000**, *100*, 235. Copyright 2000 American Chemical Society.

magneton, $H \approx B$) can be constructed and will depend upon the type of spin state present for the ground state. As discussed above, a single unpaired electron will form a Kramers doublet with $S = 1/2$, which will be split by the magnetic field. This will lead to a $\tanh(g\beta H/2kT)$ magnetization saturation function in which curves at different temperatures will be superimposed (Figure 7.14*b*). If the ground state has more than one unpaired electron and/or has significant orbital angular momentum, the curves at different temperatures or at different fields will not superimpose and will give a "nested" appearance as shown in Figure 7.14*c* for an $S = 2$ system. Now the spin states are a function of metal ion geometry and, as shown in Figure 7.15, will change depending upon whether there is a positive or negative axial distortion from an octahedral (isotropic) environment as given by the *D* energy parameter. The case represented by the five-coordinate site in Figure 7.14 corresponds to the right-hand side of Figure 7.15 with a negative value of the *D* parameter and therefore negative zero field splitting (ZFS). A further rhombic distortion will split the $M_S = \pm 2$ levels of the ground state by an

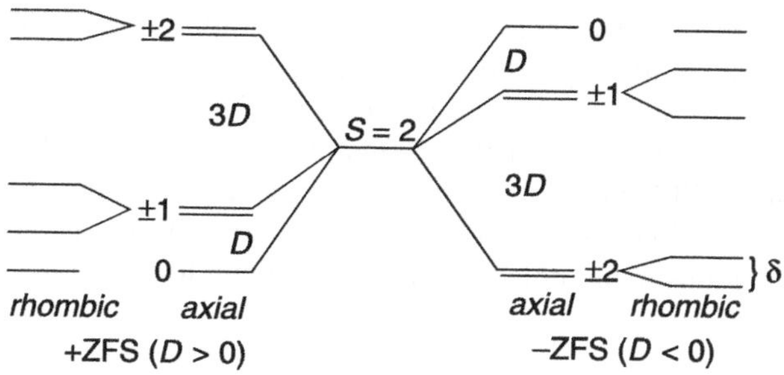

Figure 7.15. Ground-state energy levels for an $S = 2$ spin system: ZFS stands for zero-field splitting as measured by the D parameter; δ is the degree of rhombic distortion. Reprinted with permission from ref. 61. Solomon, E. I.; Brunold, T. C.; Davis, M. I.; Kemsley, J. N.; Lee, S.-K.; Lehnert, N.; Neese, F.; Skulan, A. J.; Yang, Y.-S.; Zhou, J. *Chem. Rev.* **2000**, *100*, 235. Copyright 2000 American Chemical Society.

energy amount δ in Figure 7.15. When δ is large, there will be no EPR signal (EPR silent) even though the system is clearly paramagnetic. The behavior of the Zeeman sublevels in the presence of a field leads to the nesting of the saturation magnetization curves as shown in Figure 7.16. The analysis of the nested variable temperature and field magnetization curves can then lead to experimental values of

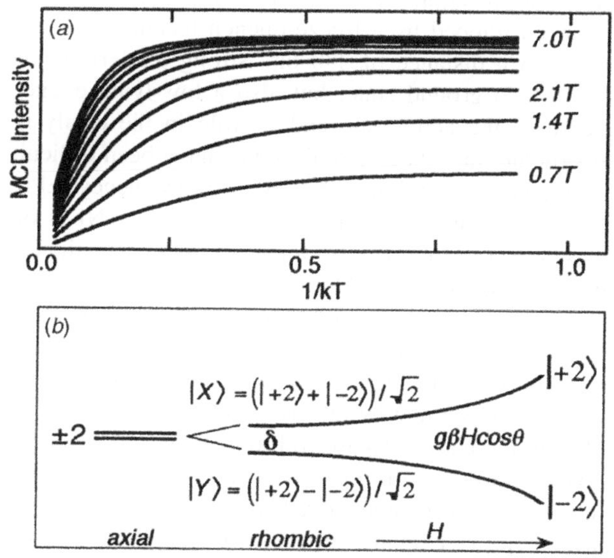

Figure 7.16. (*a*) Saturation magnetization as a function of $1/kT$ at different magnetic fields. (*b*) The behavior of the $M_S = \pm 2$ ground state as a function of axial, rhombic, and applied field. The angle θ is the angle between the z molecular axis and the field direction. Reprinted with permission from ref. 61. Solomon, E. I.; Brunold, T. C.; Davis, M. I.; Kemsley, J. N.; Lee, S.-K.; Lehnert, N.; Neese, F.; Skulan, A. J.; Yang, Y.-S.; Zhou, J. *Chem. Rev.* **2000**, *100*, 235. Copyright 2000 American Chemical Society.

the δ and $g\|$ parameters. The analysis includes integration over angular orientation of the molecular site with respect to the magnetic field, metal spin–orbit coupling, and corrections for *B* terms between Zeeman sublevels (see refs. 61–64 for details). The main utility of the method is that it allows a more or less direct probe of properties of the EPR inactive ground state by means of optical MCD measurements. Such information can be valuable for assessing the site geometry and spin state for the metal ions in biomolecules that are difficult to study otherwise.

7.6. MODELING THE STRUCTURE OF THE NATIVE INTERMEDIATE OF THE MULTICOPPER OXIDASE FROM *RHUS VERNICIFERA* LACCASE ISOLATED FROM THE JAPANESE LACQUER TREE

Multicopper oxidases such as laccases catalyze the four-electron reduction of O_2 to water:

$$O_2 + 4e^- + 4H^+ \xrightarrow{4Cu^{2+}} 2H_2O$$

The reduced form of the enzyme produces the so-called "native intermediate" (NI) in which a trinuclear copper site is mutually bridged by the product of the O_2 reduction has been trapped by a freeze-quench technique. The structure exhibits at least one Cu–Cu distance of 330 pm, strong antiferromagnetic coupling, and a doublet ($S = 1/2$) ground state, but also a low-lying excited state at 150 cm^{-1}. The strucure of the active site and its role in the catalysis is clearly of interest, but the structure and its electronic states have been difficult to characterize by physical methods including EXAFS, multifrequency EPR, and VTVH MCD [ref. 65]. However, a very characteristic feature of NI is observed in the MCD spectrum as shown in Figure 7.17: a pair of oppositely signed, temperature-dependent *C* terms at 316 nm (31,600 cm^{-1}) and 364 nm (27,500 cm^{-1}) which form a negative pseudo *A* term. The negative pseudo *A* term was interpreted as due to O $\rightarrow$ Cu(II) LMCT within the structure of the trinuclear site [ref. 65]. Therefore in order to model the site, two structurally different trinuclear Cu complexes were proposed and studied by means of MCD spectroscopy for comparison with NI isolated from the native laccase [ref. 66]. Figure 7.18 shows the structures of the tris-hydroxy- and μ3-oxo-bridged trinuclear Cu(II) complexes (TrisOH or μ_3O, respectively) which were considered as a models for NI. TrisOH has D_3 symmetry with three μ_2-OH ligands in the trigonal plane. The Cu(II) pairs are antiferromagnetically coupled, giving a ^2E ground state ($J = -105\ cm^{-1}$). Ground-state zero-field splitting (ZFS) is observed to be 65 cm^{-1}. The μ_3O model complex has C_3 symmetry and has a ferromagnetic ^4A ground state ($J = +54.5\ cm^{-1}$) and a ZFS of $-5.0\ cm^{-1}$. The ferromagnetic interaction is a result of the μ-O ligand protruding out of the Cu_3 plane. When the μ_3-oxo ligand is closer to the Cu_3 plane, an antiferromagnetic contribution dominates, resulting

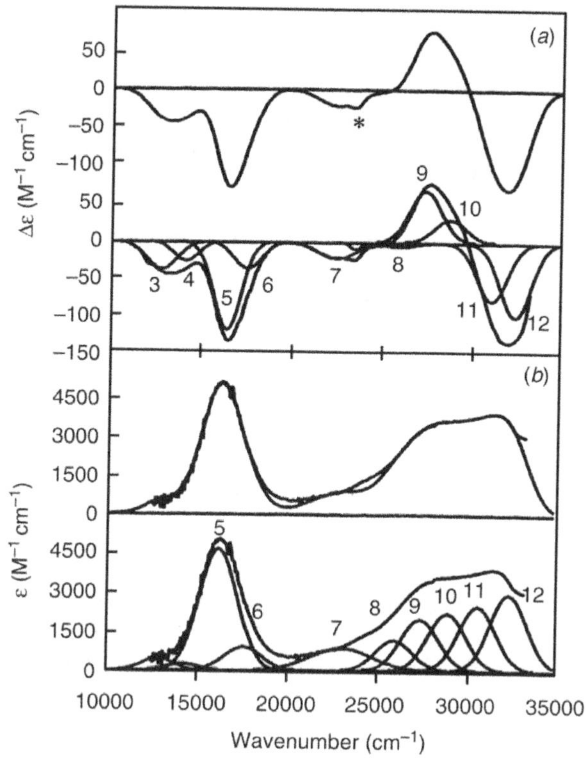

Figure 7.17. (*a*) MCD at low temperature (5–40 K) and (*b*) Absorption at room temperature spectra for the trinuclear copper NI laccase, together with their simultaneous fits to a minimum number of Gaussian bands. Reprinted with permission from ref. 65. Lee, S.-K.; George, S. D.; Antholine, W. E.; Hedman, B.; Hodgson, K. O.; Solomon, E. I. *J. Am. Chem. Soc.* **2002** *1247*, 6180. Copyright 2002 American Chemical Society.

in a ^2E ground state (see ref. 66 and structural references therein for details of the structures and properties of the model complexes).

The absorption and MCD spectra for TrisOH and μ_3O complexes are shown in Figures 7.19 and 7.20, together with a deconvolution from a simultaneous fit of both the absorption and MCD spectra to a minimum number of Gaussian bands. The bands labeled 1–4 in both spectra are assigned to ligand field transitions; their energies and intensities are similar to those found in mononuclear Cu(II) complexes. Bands 5–11 for TrisOH are interpreted as O → Cu(II) LMCT, while band 12 is ascribed to a N → Cu(II) LMCT transition. For μ_3O, bands 5–7 and 9 are O → Cu(II) LMCT, while bands 8, 10, and 11 are N → Cu(II) LMCT. The LMCT assignments were aided by band polarizations from careful VTVH measurements. The *xy*-polarized O → Cu(II) LMCT transitions did not show "nesting" behavior in the magnetization curves for TrisOH; the magnetization for μ_3O was more complicated due to the low ZFS, and simulations were required to predict the polarization (see ref. 66 for details). The nature of the orbitals

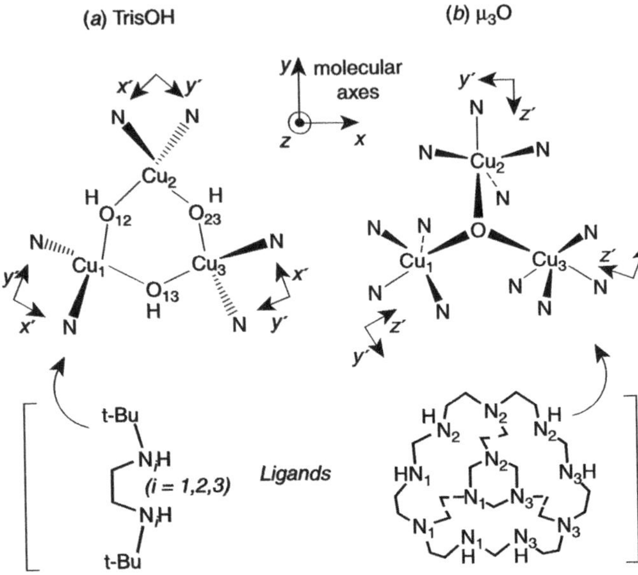

Figure 7.18. TrisOH and μ_3O model trinuclear Cu(II) complexes. Reprinted with permission from ref. 66. Yoon, J.; Mirica, L. M.; Stack, T. D. P.; Solomon, E. I. *J. Am. Chem. Soc.* **2005**, *127*, 13680. Copyright 2005 American Chemical Society.

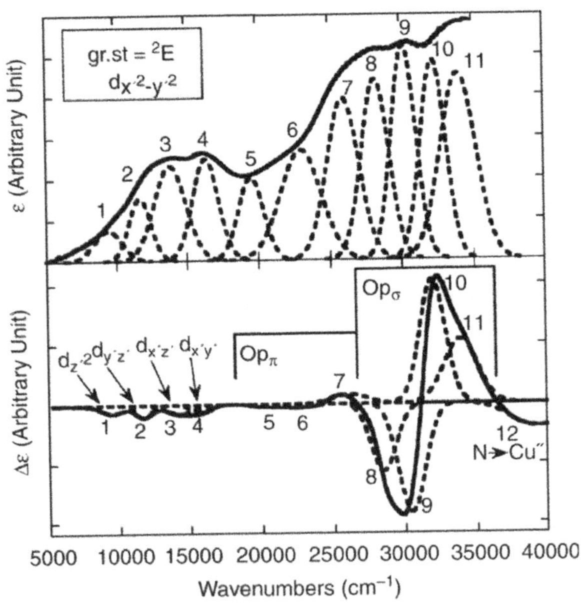

Figure 7.19. TrisOH mull absorption (at 10 K, upper curve) and MCD (at 5.0 K and 7.0 T, lower curve) spectra (solid lines). The Gaussian fit is indicate by dashed lines. Reprinted with permission from ref. 66. Yoon, J.; Mirica, L. M.; Stack, T. D. P.; Solomon, E. I. *J. Am. Chem. Soc.* **2005**, *127*, 13680. Copyright 2005 American Chemical Society.

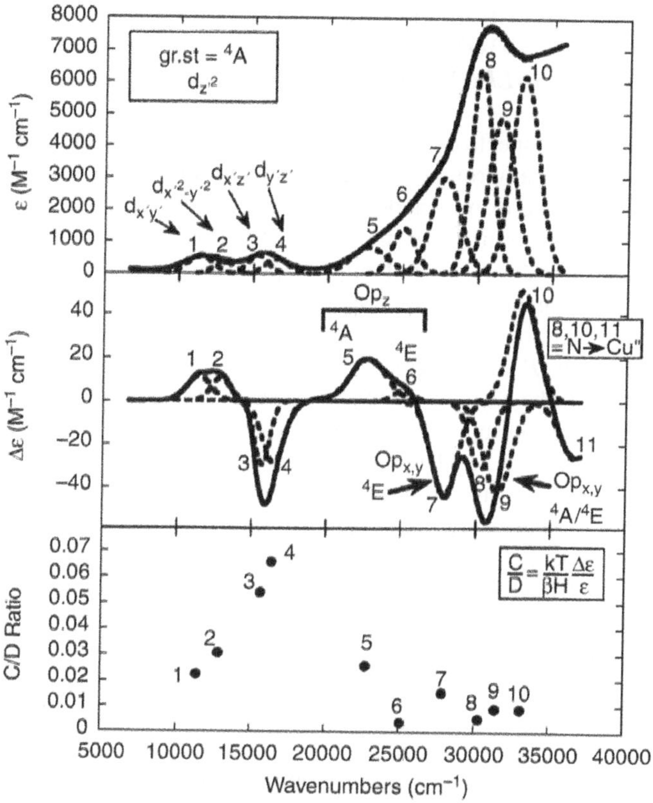

Figure 7.20. μ_3O solution absorption (at 10 K, upper curve) and MCD (at 5.0 K and 7.0 T, middle curve) spectra and C/D ratios between MCD and absorption intensities (bottom points). Dashed lines indicate Gaussian fit. Reprinted with permission from ref. 66. Yoon, J.; Mirica, L. M.; Stack, T. D. P.; Solomon, E. I. *J. Am. Chem. Soc.* **2005**, *127*, 13680. Copyright 2005 American Chemical Society.

involved in the transitions for both complexes was explored by inclusion of Cu(II) spin–orbit coupling and rather extensive DFT calculations. In particular, the origin of the *C*-term intensities that give rise to the pseudo *A* term were explored. It was found that the two component *C* terms for TrisOH were associated with O → Cu(II) LMCT (bands 5–11, Figure 7.19) to the same Cu center coupled by spin–orbit effects. In the case of the μ_3O complex, the *C* terms arise from O → Cu(II) LMCT (bands 5–7 and 9, Figure 7.20) to different Cu centers coupled by oxo-centered excited state spin–orbit coupling. The TrisOH complex was found to present a *positive* pseudo *A* term for the ^2E ground state, whereas a shift of the μ_3-oxo ligand into the Cu_3 plane predicted a *negative* pseudo *A* term. On this basis, the TrisOH model could be ruled out, and the μ_3O model was considered the best descriptor of the NI Cu_3 site. This study, though rather involved, provides a good example where MCD spectra can provide significant insight to a very challenging enzyme active site description.

7.7. HIGH-SPIN METAL CENTERS WITH GROUND-STATE SPIN > 1/2; APPLICATION TO THE PROTEIN-OXIDIZED RUBREDOXIN, *DESULFOVIBRIO GIGAS*

The potential of MCD spectroscopy to probe high-spin ground-state properties for metal ions in complex environments such as sites in biological systems was early recognized as the above examples demonstrate. For spin states >1/2 or for orbitally or spin-forbidden transitions, the analysis of the spectra has not been very easy to accomplish and has the potential to provide misinformation. One of the problems with the standard approach to the analysis of MCD spectra (see ref. 1) is that the treatments are valid only for the so-called linear field limit at temperatures where $kT \gg \mu_B B$, the Zeeman interaction. Perturbation theory is applied only to first order in the Zeeman interaction. At low temperature and at high field, these approaches are appropriate only for an isolated Kramers doublet $S = 1/2$. The approach taken by A. J. Thomson and co-workers [ref. 67] to correct these problems has been to consider spin–orbit coupling and Zeeman interaction to second order for ground states described by spin-Hamiltonian formalism and to assume that the optical transition bandwidths are much greater than Zeeman shifts and zero-field splitting (ZFS). Their approach is a rigid-shift approximation as described in Chapter 3. Their analytical method features the simultaneous fit of the absorption and MCD spectra and, at the same time, the fit of low-temperature magnetization curves at several different wavelengths. This is made possible by taking advantage of the symmetry aspects of the forms of the MCD bands and their magnetization curves. Group theory, using irreducible tensor methods as described in ref. 1, is used in the development of analytical equations for the spectral and magnetization fits. This is a challenging task at fields where the results are nonlinear and would otherwise require the diagonalization of large Hamiltonian matrices of excited states, a task that is computationally demanding. However, by setting up a plausible transition model with a finite number of excited states and making use of the symmetry aspects of the states and operators involved in the transitions, with a minimum of adjustable parameters that are dependent upon the structural requirements of the problem, such as the zero field splitting and its axial *D* and rhombic *E* components, simultaneous fits of experimental spectra and magnetization curves can be achieved. The model is symbolically challenging and readers are referred to ref. 67 for details. The computational approach, however, is rather straightforward. Absorption bands are assumed to be Gaussian, and MCD spectra are assumed to be a combination of Gaussians of either + or − sign (*B* and *C* terms) and their derivatives (*A* terms). Of course, only the *C* terms are temperature-dependent. The observed spectra are considered as a combination of overlapping components.

An application of this simultaneous analytic approach is given by the high-spin $3d^5$ Fe(III), with $S = 5/2$ in a tetrahedral site, in the protein oxidized rubredoxin from the bacterium *Desulfovibrio gigas* [ref. 67]. The analysis allows the determination of the sign of the axial ZFS parameter *D*, which cannot be accomplished by EPR, for example. Figure 7.21 presents the one-electron energy levels for

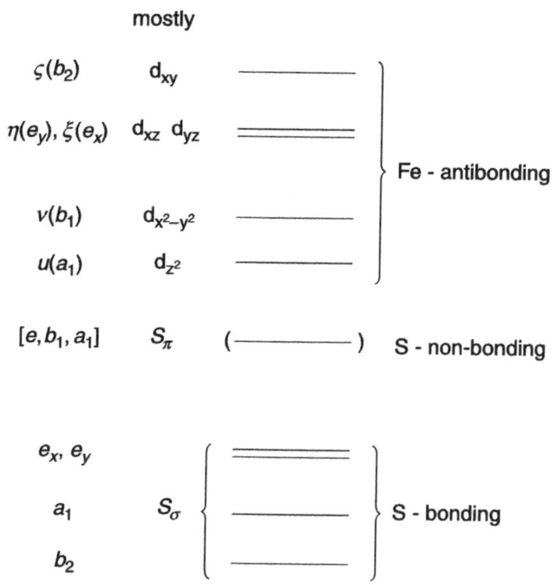

Figure 7.21. One-electron energy levels for high-spin $3d^5$ Fe(III) assuming D_{2d} symmetry. The S_σ and S_π ligand orbitals are doubly occupied, while the antibonding Fe orbitals are all singly occupied. Reprinted with permission from ref. 67. Oganesyan, V. S.; George, S. J.; Cheesman, M. R.; Thomson, A. J. *J. Chem. Phys.* **1999**, *110*, 762. Copyright 1999 American Institute of Physics.

high-spin Fe(III) with four thiol ligands from cysteine residues in a distorted pseudotetrahedral $[Fe(S\text{-}cys)_4]^-$ complex (D_{2d} symmetry assumed). The ground state is $S = 5/2$, ${}^6A_1$, and therefore spin-allowed excited states will be ${}^6B_2$ (z-polarized) and ${}^6E$ (xy-polarized). The observed spectra are shown in Figure 7.22 and have been assigned to LMCT transitions from the thiol S-donors to the Fe(III) metal based $3d$ orbitals. Based on intensity expectations from orbital overlaps, the bands in the 600–350-nm region were assigned as the five $S_\sigma \rightarrow Fe_{\sigma^*}$ type transitions from the ${}^6A_1$ ground state to the following excited states:

${}^6B_2(1)$: $S_\sigma(e) \rightarrow Fe_{\sigma^*}(e)$(570 nm)	${}^6E(3)$: $S_\sigma(b_2) \rightarrow Fe_{\sigma^*}(e)$(391 nm)
${}^6B_2(2)$: $S_\sigma(a_1) \rightarrow Fe_{\sigma^*}(b_2)$(396 nm)	${}^6E(4)$: $S_\sigma(a_1) \rightarrow Fe_{\sigma^*}(e)$(494 nm)
	${}^6E(5)$: $S_\sigma(e) \rightarrow Fe_{\sigma^*}(b_2)$(465 nm)

When the effective site symmetry was assumed to be lower, S_4 instead of D_2d, then an additional transition becomes allowed to the ${}^6E(6)$ state: S_σ(e) $\rightarrow Fe_{\sigma^*}$ (e) (489 nm). The overall fit to the observed spectra are also shown in Figure 7.22. Figure 7.23 shows the Gaussian deconvolution of the absorption and MCD spectra in terms of the transitions 1–5 given above and ^{6}E(6) for an improved fit. Figure 7.24 shows the calculated magnetization curves for 1.60, 4.19, and 9.80 K at four wavelengths, together with the experimental points

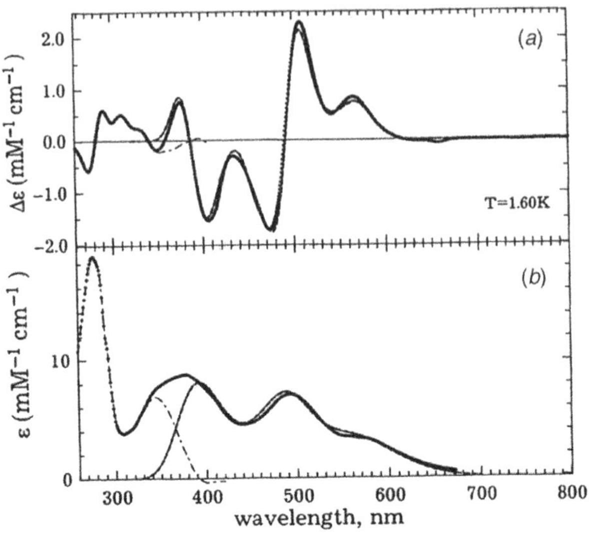

Figure 7.22. Absorption (lower curve) and MCD (upper curve) spectra (solid lines) for the protein oxidized rubredoxin from the bacterium *Desulfovibrio gigas* at 1.6 K in a 50% water–glycerol (v:v) glass. Calculated spectra (dashed lines) using $E/D = 0.25$ and $D = -0.605$ cm^{-1} and parameters obtained from simultaneous fit of magnetization curves (Figure 7.24). Reprinted with permission from ref. 67. Oganesyan, V. S.; George, S. J.; Cheesman, M. R.; Thomson, A. J. *J. Chem. Phys.* **1999**, *110*, 762. Copyright 1999 American Institute of Physics.

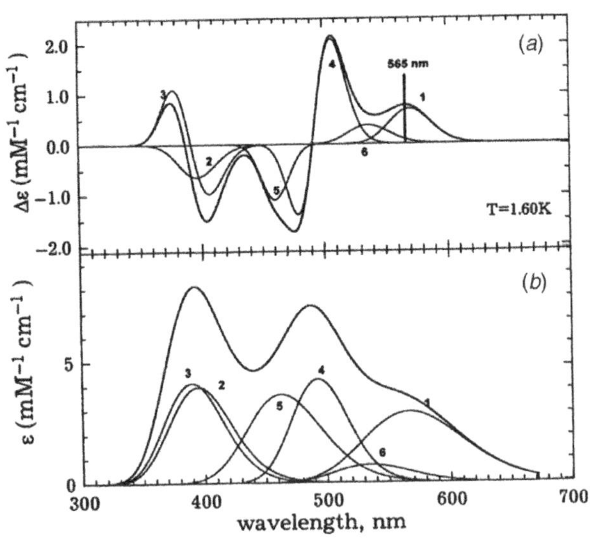

Figure 7.23. Deconvolution of the absorption (lower curves) and MCD (upper curves) using Gaussians and their derrivative forms. The curve numbers refer to the excited states 1–6. Reprinted with permission from ref. 67. Oganesyan, V. S.; George, S. J.; Cheesman, M. R.; Thomson, A. J. *J. Chem. Phys.* **1999**, *110*, 762. Copyright 1999 American Institute of Physics.

(circles). The magnetization fit is also very good. As can be seen from Figures 7.22 and 7.23, the overall fits are very good. Any time that spectra are fit with adjustable parameters or selections from a number of possible transition choices, there is concern that the fit is an artifact of the selection or parameters (a fit of *any* spectrum can be made if there are enough parameters!). In this case, the minimum number of parameters and transitions are used and the fit is simultaneous to two different experimental spectra and the magnetization curves, which approach greatly reduces the possibility of artifactual fits. This is an important conclusion from this study. To further validate the method, the MCD spectrum and magnetization curves were also calculated by diagonalizing the complete 54×54 matrix of excited states using the Hamiltonian $H = H_0 + H_{SO} + H_{Zeeman}$, where H_0 is for energies and wavefunctions in the lower S_4 point group. The results were virtually identical to those achieved some 100 times faster by the simpler analytical method [ref. 67]. Furthermore, the best fits were for a negative

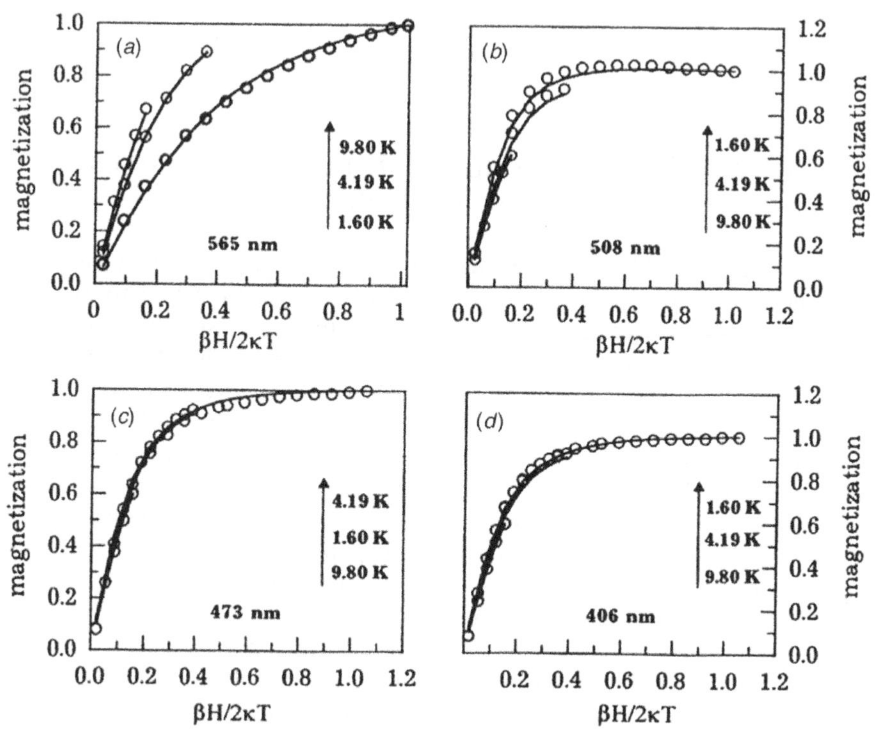

Figure 7.24. Magnetization curves for MCD (see Figure 7.2) at 1.6, 4.19, and 9.80 K (nested subplots from bottom to top) and wavelengths indicated, normalized to MCD recorded at 5 T. Experimental data are indicated by circles, while solid lines are the simulated magnetization curves with $E/D = 0.25$, and $D = -0.605\ \text{cm}^{-1}$. Reprinted with permission from ref. 67. Oganesyan, V. S.; George, S. J.; Cheesman, M. R.; Thomson, A. J. *J. Chem. Phys.* **1999**, *110*, 762. Copyright 1999 American Institute of Physics.

value for the *D* ZFS parameter. When a positive value for *D* was used, the magnetization curves could be reproduced satisfactorily, but the fit to the observed MCD was not acceptable. It was also noted [ref. 67] that fitting parameters based on the linear polarization of the transitions is an uncertain procedure, especially for cases where the ground state has no orbital angular momentum (such as the $^6A_1$ state discussed here), because of the importance of spin–orbit coupling in determining the MCD spectrum. Thus, linearly *z*-polarized transitions in an axially symmetric complex can gain MCD intensity only by second-order mixing with nearby *xy*-polarized transitions. The magnetization curves are highly dependent upon this type of interaction.

The analytical method described above has been extended to symmetry- and spin-forbidden transitions for high-spin metal ions and includes weak low-energy ligand field transitions [ref. 68]. The extension considers spin–orbit coupling perturbation to third order and also uses the irreducible tensor method to take full advantages of symmetry. The method was applied to the lowest-energy spin-forbidden ligand field transitions for the pseudotetrahedral high-spin Fe(III) $S = 5/2$, in the protein rubredoxin from *Methanobacterium thermoautotrophicum* (*strain Marburg*), with the prediction of MCD signs, intensities, and magnetization curves.

7.8. MCD OF Co^{2+} AS A PROBE OF METAL-BINDING SITES IN *E. COLI* METHIONYL AMINOPEPTIDASE

An interesting use of MCD to characterize two different metal-binding sites at the active site of methionyl aminopeptidase (MetAP) was reported recently [ref. 69]. MetAP catalyzes the hydrolysis of N-terminal methionine residues from proteins and is believed to be important in natural protein maturation and degradation processes. The aminopeptidase was isolated from *Escherichia coli*, *Ec*MetAP, and was found to bind Co^{2+} in two different sites, one five-coordinate, Co(1), and the other six-coordinate, Co(2), illustrated in Figure 7.25. The Co^{2+}-bound *Ec*MetAP was used to model the active sites within this important peptidase by studying the MCD of the two different paramagnetic Co^{2+} centers. It is not known whether Co^{2+}-bound MetAP is a natural form of the peptidase because of the low natural abundance of cobalt in biological systems, or more likely whether some other metal ion present in greater concentration, such as Zn^{2+} or Fe^{2+}, might be present *in vivo*. Nevertheless, there seems to be no evidence that peptidase activity is dependent upon which metal ion occupies the two metal sites. The MCD study of the two types of Co^{2+} environments were of interest because they present very different magnetization when investigated by variable temperature-variable field (VTVH) studies [ref. 69].

Figure 7.26 shows the absorption and MCD spectra for Co^{2+}-bound *Ec*MetAP and illustrates the different temperature dependence of the negative *C* terms at 495 nm and 567 nm in the inset. The negative *C* term at 495 nm was attributed to the 6 coordinate Co(2), while the *C* term at 567 nm was ascribed to the

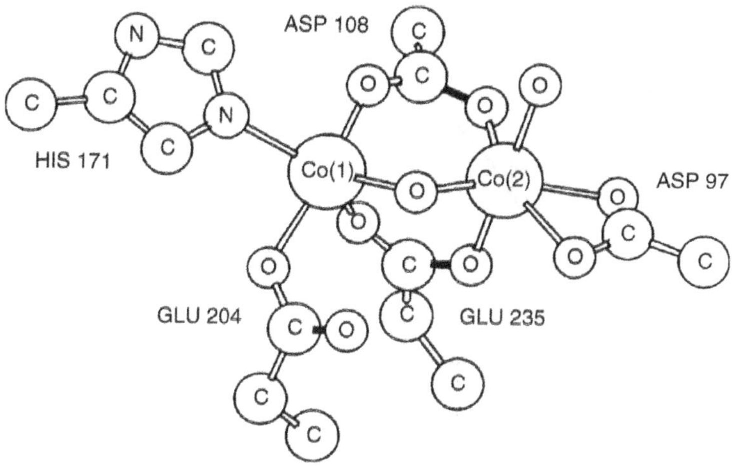

Figure 7.25. Active site of *Ec*MetAP with only amino acid side chains that serve as ligands for the Co^{2+} sites shown. Co(2) has in addition a bound water molecule. Reprinted with permission from ref. 69. Larrabee, J. A.; Leung, C. H.; Moore, R. L.; Thamrong-nawasawat, T.; Wessler, B. S. H. *J. Am. Chem. Soc.* **2004**, *126*, 12316. Copyright 2004 American Chemical Society. (Originally published in ref. 70.)

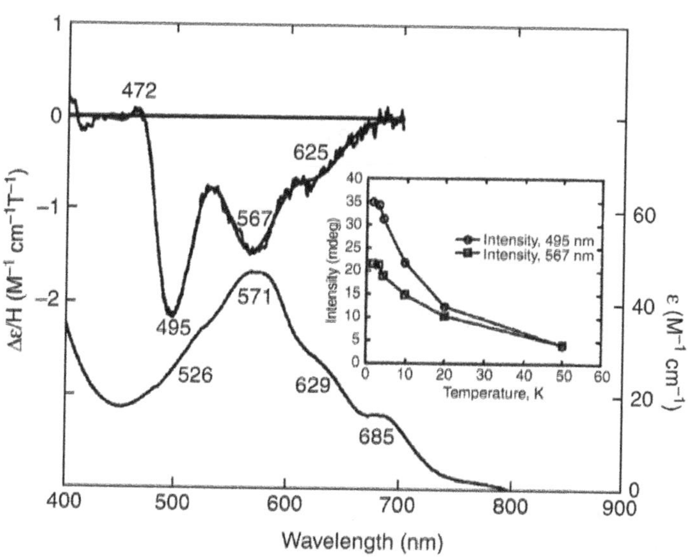

Figure 7.26. Absorption (at room temperature in a buffer solution, lower curve) and MCD (at 1.6 K with 1:2 buffer/glycerol v/v glassy solution, upper curve) for *Ec*MetAP at pH 7.5. The inset shows the MCD signal intensity at 495 nm (top curve) and 567 nm (bottom curve) taken at 3.5 T versus Kelvin temperature. Reprinted with permission from ref. 69. Larrabee, J. A.; Leung, C. H.; Moore, R. L.; Thamrong-nawasawat, T.; Wessler, B. S. H. *J. Am. Chem. Soc.* **2004**, *126*, 12316. Copyright 2004 American Chemical Society.

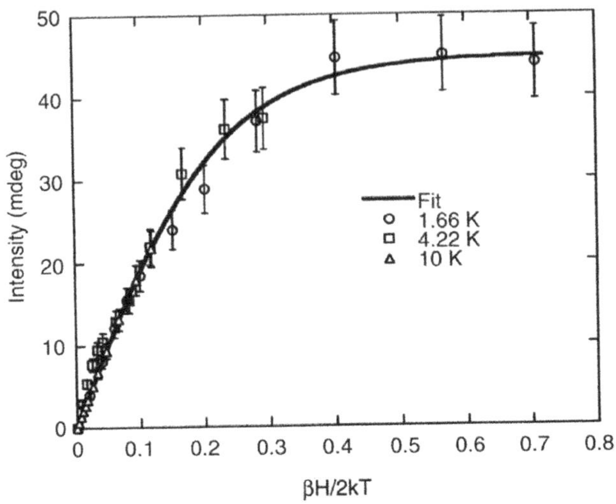

Figure 7.27. Magnetization curve for the *C* term at 495 nm at 1.66, 4.22, and 10 K at fields up to 3.5 T. Reprinted with permission from ref. 69. Larrabee, J. A.; Leung, C. H.; Moore, R. L.; Thamrong-nawasawat, T.; Wessler, B. S. H. *J. Am. Chem. Soc.* **2004**, *126*, 12316. Copyright 2004 American Chemical Society.

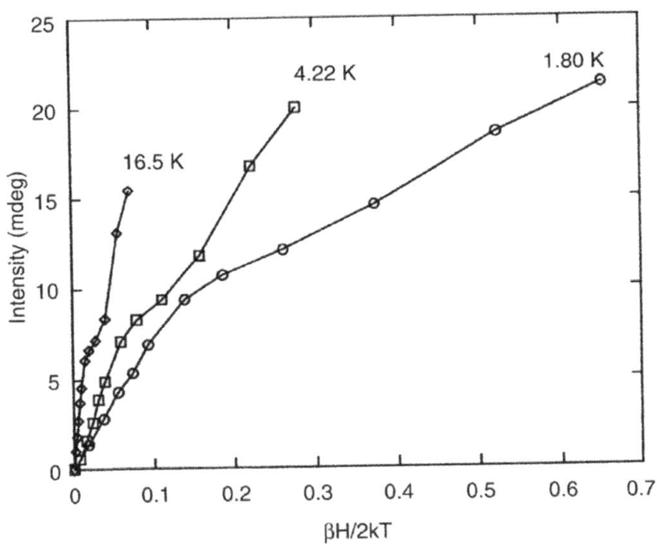

Figure 7.28. Magnetization curves for the *C* term at 567 nm at 1.8, 4.22, and 16.5 K at fields up to 3.5 T. Reprinted with permission from ref. 69. Larrabee, J. A.; Leung, C. H.; Moore, R. L.; Thamrong-nawasawat, T.; Wessler, B. S. H. *J. Am. Chem. Soc.* **2004**, *126*, 12316. Copyright 2004 American Chemical Society.

five-coordinate Co(1) site. The magnetization study of the 495-nm C term is shown in Figure 7.27. The curve was fit to a typical tanh function:

$$\Delta\varepsilon = I \tanh[g\beta\mathbf{H}/(2kT)] + B\mathbf{H} \quad (7.1)$$

where I is a constant, g is the isotropic g factor, β = Bohr magneton, $\mathbf{H}$ = magnetic field strength, k = Boltzmann's constant, T = Kelvin temperature, and B is the temperature-independent B term, taken to be zero. The tanh function is expected for the pseudo-Kramer's doublet ground state for an octahedral CoO_6 environment with strong spin–orbit coupling in a $^4T_{1g}$ Co^{2+} ground state with a substantial zero-field splitting (ZFS) (see Figure 7.29 [ref. 69]). Data at three different temperatures were within experimental error of the expected curve. In contrast, the magnetization curves for the 567-nm C term is much more complicated as shown in Figure 7.28, where the curves at different temperatures are not overlaid, but are "nested" as a result of $J = 1/2$ and $3/2$ interaction (see Figure 7.29 [ref. 69]).

The magnetization curves were interpreted by means of the ground-state energy-level schemes for six-coordinate and five-coordinate Co^{2+} shown in Figure 7.29. These schemes assume that the two different coordination environments result in different ground-state symmetries: $^4T_{1g}$ for the octahedral CoO_6 environment for Co(2), and ^4A for the idealized five-coordinate CoO_4N for Co(1) with D_{3h} or C_{4v} symmetry. The octahedral scheme features both spin–orbit coupling and zero-field splitting (ZFS) of the lowest ($J = 1/2$) level. The five-coordinate scheme assumes extensive ZFS, which would intermix the $J = 1/2$ and $3/2$ levels. Support for the octahedral scheme resulted from the isotropic g value determined to be 4.45 from the fit of the experimental magnetization curve

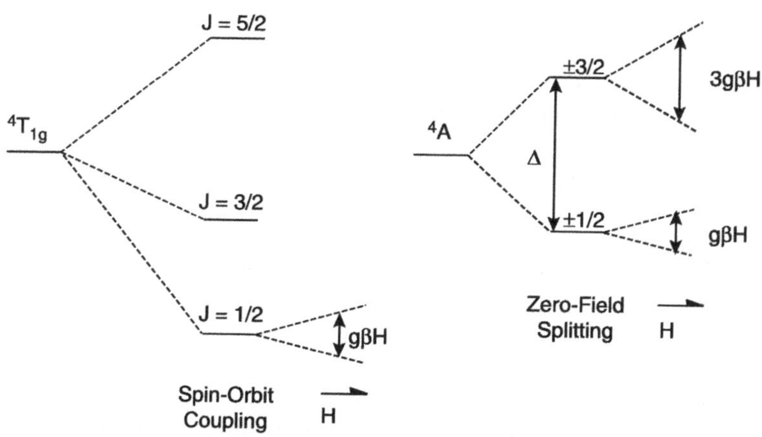

Figure 7.29. Ground-state spitting for six-coordinate (left) and five-coordinate (right) Co^{2+} complexes. Reprinted with permission from ref. 69. Larrabee, J. A.; Leung, C. H.; Moore, R. L.; Thamrong-nawasawat, T.; Wessler, B. S. H. *J. Am. Chem. Soc.* **2004**, *126*, 12316. Copyright 2004 American Chemical Society.

by means of Eq. (7.1); this value is typical for six-coordinate Co^{2+}. The ZFS for the five-coordinate Co^{2+} was estimated to be 10 ± 5 cm^{-1} by means of Eq. (7.2), where Δ is the ZFS (see ref. 69 for details):

$$\text{Intensity} = [1 + \exp(-\Delta/(kT))]^{-1} + \exp(-\Delta/(kT))[1 + \exp(-\Delta/(kT)]^{-1} \quad (7.2)$$

It was also noted that the fit of the 495-nm *C* term to a simple Kramer's doublet model at low-temperature rules out any strong magnetic coupling between the five- and six-coordinate Co^{2+} centers. The study represents an interesting probe of a complex site that features two metal ions in different sites that are not strongly coupled but within the same active site of an enzyme.

7.9. OPTICALLY DETECTED ELECTRON PARAMAGNETIC RESONANCE (ODEPR) BY MICROWAVE-MODULATED MCD: APPLICATION TO THE OPTICAL ANISOTROPY OF THE BLUE COPPER PROTEIN *PSEUDOMONAS AERUGINOSA* AZURIN

A rather different approach to MCD measurements of paramagnetic systems has been developed by A. J. Thomson and co-workers. The approach involves the use of a transverse laser beam to optically detect electron paramagnetic resonance (ODEPR) in a sample placed in a microwave (MW) field [refs. 71, 72]. In this experiment, the sample is placed in a variable magnetic field (0–1 Tesla) oriented in the z direction, B_0, and is then bathed with a transverse MW field, B_1, of suitable frequency, while a laser beam modulated with a PEM to be alternately lcp and rcp is passed through the sample in the transverse x direction. The experimental layout is shown in Figure 7.30 [ref. 71]. In a conventional MCD experiment, circularly polarized light along a longitudinal field interacts with a magnetic moment in the ground state, and the differential absorption is described by the MCD *C* term, illustrated in Figure 7.31*a* for light parallel to B_0. However, for a magnetic moment from unpaired electron spin(s), the transverse MW field B_1 will causes the spin magnetic moment to precess about B_0. When transverse laser light is introduced (Figure 7.31*b*), the interaction of the laser light with the perpendicular (xy) component of the precessing moment (MCD *C* term in the xy plane) will cause the beam to undergo modulation at the MW frequency. The frequency of the modulated laser light that emerges from the sample is mixed with the MW frequency by means of coherent Raman scattering. The scattered Raman radiation from a paramagnetic center can be detected with relatively high sensitivity by optical heterodyne detection of the "beat" (difference) frequency between the laser and MW frequencies [ref. 73]. By using a fixed-frequency laser and scanning the magnetic field B_0, the magnetization in the direction of the laser beam is measured, and electron spin resonance can occur. The signal from the laser light is detected and, after suitable demodulation, gives the resonance(s) and thus the appearance of a conventional EPR spectrum. Both absorption and dispersion EPR signals may be detected separately by adjusting the phase (by means of the phase

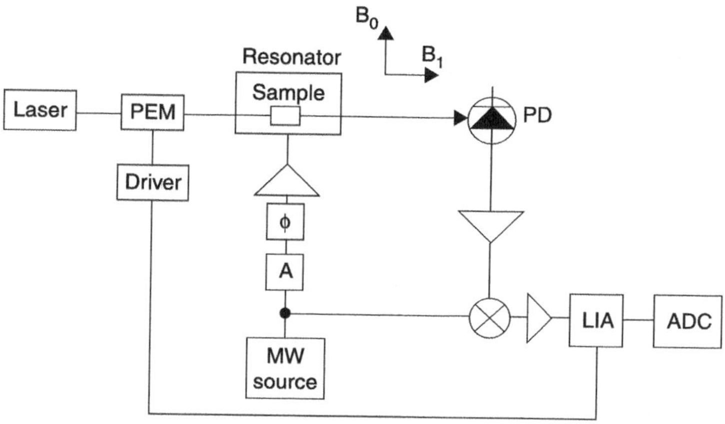

Figure 7.30. Layout of the ODEPR experiment. PEM, photoelastic modulator; PD, photodiode detector; A, attenuator; ϕ, phase shifter; LIA, lock-in amplifier; ADC, analog-to-digital (A/D) converter. Reprinted from ref. 71. Bingham, S. J.; Suter, D.; Schweiger, A.; Thomson, A. J. *Chem. Phys. Lett.* **1997**, *266*, 543. Copyright 1997, with permission from Elsevier.

shifter ϕ in Figure 7.30) of the MW field. This ODEPR spectrum, however, is dependent upon the laser frequency and will reproduce the longitudinal MCD spectrum if the signal is determined as a function of laser frequency at the constant applied B_0 corresponding to the EPR resonance [ref. 72]. This is a consequence of the proportionality of the magnetization in the z direction M_z to that in the xy direction M_{xy} at the resonance condition, in the absence of saturation:

$$M_{xy} = \omega_{\text{Rabi}} T_2 M_z$$

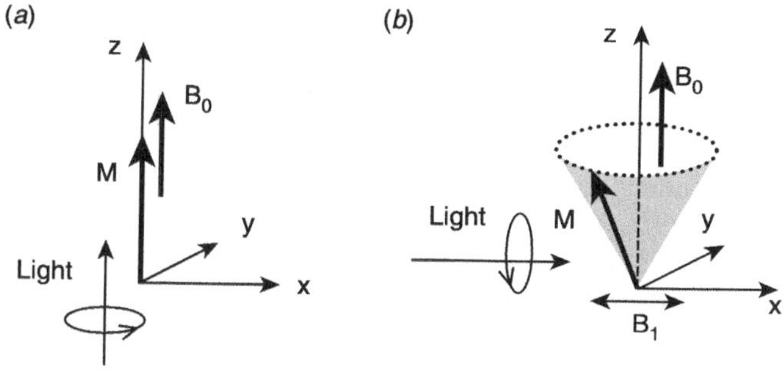

Figure 7.31. (*a*) A conventional MCD experiment with circularly polarized light propagated parallel to B_0; (*b*) A transverse circularly polarized beam interacting with a magnetic moment precessing about B_0 caused by a microwave field B_1. Reprinted with permission from ref. 72. Borger, B.; Bingham, S. J.; Gutschank, J.; Schweika, M. O.; Suter, D.; Thomson, A. J. *J. Chem. Phys.* **1999**, *111*, 8565. Copyright 1999 American Institute of Physics.

where ω_{Rabi} is the Rabi frequency and T_2 is the phase memory time, both parameters dependent upon the laser power and saturation effects. This, in turn, leads to a proportionality between the MCD differential absorptions ΔA_{xy} and ΔA_z:

$$\Delta A_{xy} = \omega_{Rabi} T_2 \Delta A_z$$

assuming that the proportionality of the MCD to the magnetization is the same in both the xy and z directions.

The ODEPR experiment was verified and characterized by examining prototypical systems including the metalloprotein cytochrome c-451 (low-spin Fe(III) $S = 1/2$) and single-crystal ruby (Cr^{3+} : Al_2O_3) ($3d^3$, $S = 3/2$) [ref. 72]. To illustrate, Figure 7.32*a* shows the ODEPR for ruby at a 13.66-GHz MW frequency and at 1.6 K. Figure 7.32*b* shows the ODEPR signal as a function of laser wavelength and shows a comparison with the relevant portion of a conventional MCD spectrum for ruby measured at 4.9 T and 1.6 K.

The application of ODEPR to the optical anisotropy of the blue copper protein *Pseudomonas aeruginosa* azurin is a particularly good example of the application of the technique. Since the single crystals of azurin and many other metalloproteins are not readily available, spectroscopic measurements must rely on non-oriented samples, usually frozen solutions at low temperature. ODEPR studies provide a useful method to characterize optical anisotropy, to discover the polarization of individual transitions relative to the molecular axis system, and to establish a basis for deconvolution of conventional broad band absorption and MCD spectra [ref. 74]. The results from conventional EPR spectroscopy indicate that azurin has an axially symmetric g tensor, with g_z approximately parallel with the bond from the copper to a sulfur from a methionine ligand. Since the optical and magnetic anisotropy are fixed to the same molecular axis sytem, the modulated MCD ΔA_x along the laser beam propagation direction was calculated to be given by the proportionality

$$\Delta A_x \propto T(\theta) f(\theta, \sigma)[C_z g_z (g_\perp^2/g^2)\sin^2(\theta) + C\perp\ g\perp(g_z^2/g^2)\cos^2(\theta) + 1) \qquad (7.3)$$

where $T(\theta) = \tanh(g(\theta)\mu_B B_0/(2kT))$, the Boltzmann factor with $\theta =$ the angle between the molecular z axis and B_0; $f(\theta, \sigma) =$ lineshape function with width σ; and C_z and $C_\perp = (C_x + C_y)/2$ are C-term magnetization components. The relationship in Eq. (7.3) can provide the parameters for fitting the ODEPR spectra determined with different laser wavelengths. An example of the dispersion ODEPR at different laser wavelengths and their fit is given in Figure 7.33. From such data the ratio of C parameters as a function of wavelength can be deduced. These parameters can then be used as the basis of a fit of the experimental longitudinal MCD spectrum shown in Figure 7.34. For an axial system the MCD proportionality in Eq. (7.4) is assumed:

$$\Delta A_z \propto g_z C_z + 2g_\perp C_\perp \qquad (7.4)$$

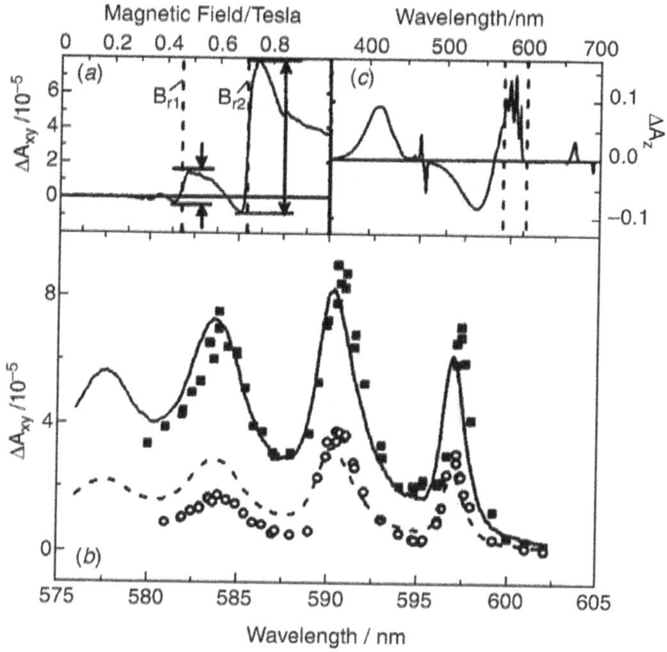

Figure 7.32. (*a*) ODEPR for ruby. MW frequency = 13.66 GHz, $T = 1.6$ K; along the crystal *c* axis with a path length of 2 mm and nominal concentration of 2×10^{-3} Cr:Al. (*b*) ODEPR signal with resonance fields (noted in (*a*) by vertical dashes) B_{r1} (circles) and B_{r2} (squares) as a function of laser wavelength, together with scaled conventional longitudinal MCD measured at 4.9 T and 1.6 K (solid and dashed lines). (*c*) Longitudinal MCD of ruby recorded at 4.9 T and 1.6 K (dashed vertical lines shows the wavelength range in (*b*)). Reprinted with permission from ref. 72. Borger, B.; Bingham, S. J.; Gutschank, J.; Schweika, M. O.; Suter, D.; Thomson, A. J. *J. Chem. Phys.* **1999**, *111*, 8565. Copyright 1999 American Institute of Physics.

The complex overlapping band structure presented by the MCD is then decomposed into individual bands by means of Eq. (7.5):

$$\Delta A_z(\nu) = \Sigma \Delta A_{zi} \exp((\nu - p_i)^2/2w_i^2) \qquad (7.5)$$

where ν indicates the photon energy, p_i and w_i are the position and width of each band component with the averaged MCD amplitude ΔA_{zi}. The best simultaneous fit was with six Gaussian bands as shown in Figure 7.34. The six bands, together with their MCD *C* terms, were interpreted within a D'_{2d} double group. The three lowest energy bands with *C* terms centered at 10,542 cm^{-1} (negative), 12,594 cm^{-1} (positive), and 13,766 cm^{-1} (negative) were assigned to Cu(II) ligand field transitions from occupied 3*d* orbitals to the half-filled $3d_{xy}$, based on their *C*-term signs (see ref. 74 for detailed arguments). Band 4 at 15,592 cm^{-1} (negative C term), the strongest in the MCD spectrum, was assigned to a Cys S$\pi \rightarrow$ Cu $3d_{xy}$

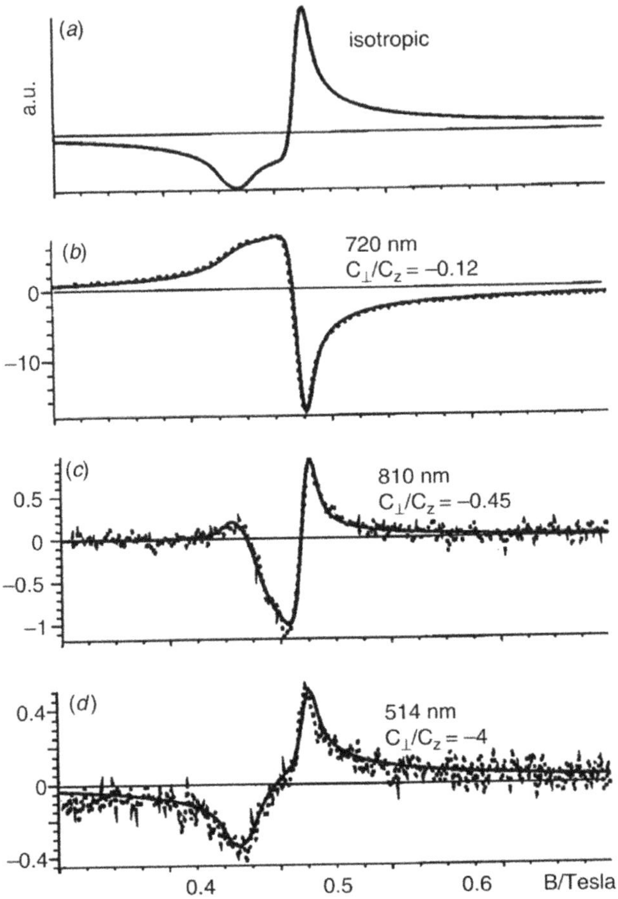

Figure 7.33. ODEPR spectra at different laser wavelengths. The fit curves (solid lines) for the spectra were determined by means of Eq. (7.3) with the parameter ratios noted in the figure. Reprinted with permission from ref. 74. Borger, B.; Gutschank, J.; Suter, D.; Thomson, A. J.; Bingham, S. J. *J. Am. Chem. Soc.* **2001**, *123*, 2334. Copyright 2001 American Chemical Society.

LMCT charge-transfer transition. Crystal field calculations, together with spin–orbit coupling mixing $3d_{x^2-y^2}$ into the ground state, dictate that the transition dipoles are predominantly in the *xy* plane, leading to almost pure C_z and a negative *C* term. Bands 5 and 6 at 19,907 cm^{-1} (positive *C* term) and 21,400 cm^{-1} (negative *C* term) were attributed to other protein donor atoms → Cu $3d_{xy}$ transitions, perhaps from a His N donor. The interpretation was found to be consistent with other blue copper proteins such as plastocyanin, which had been previously studied by means of single-crystal absorption spectra. The authors state that the ODEPR approach to interpreting the MCD is more accurate than alternative MCD saturation curve approaches because the different components of the optical

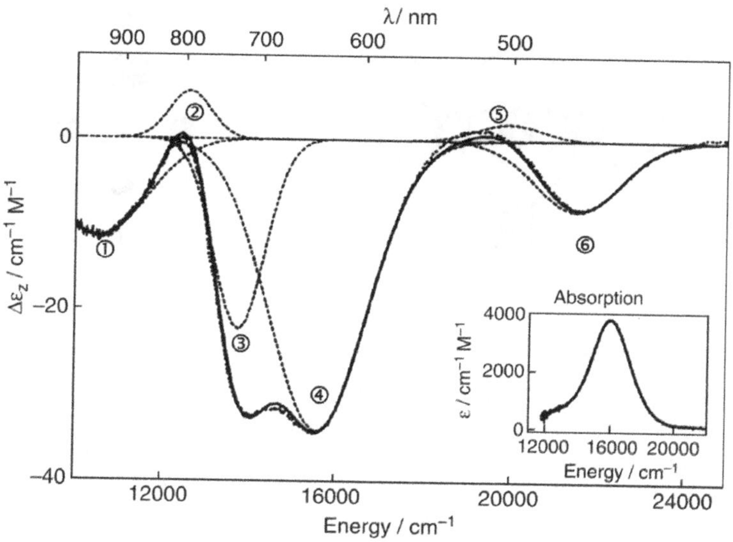

Figure 7.34. MCD spectrum of azurin at 1.8 K and 5 T with the corresponding broad absorption shown in the inset. The MCD spectrum is fit to six individual bands (*C* terms) shown as the dotted lines. Reprinted with permission from ref. 74. Borger, B.; Gutschank, J.; Suter, D.; Thomson, A. J.; Bingham, S. J. *J. Am. Chem. Soc.* **2001**, *123*, 2334. Copyright 2001 American Chemical Society.

anisotropy contribute to different parts of the ODEPR spectrum, an advantage when there is only small *g*-value anisotropy as in the copper proteins [ref. 74].

7.10. SINGLE-MOLECULE MAGNETS (SMMs)

Single-molecule magnets (SMMs) are paramagnetic complexes, usually cluster complexes, that possess a barrier to relaxation (reorientation) of spin which gives rise to a spin polarization and exhibit hysteresis in their molecular magnetic susceptibility versus an external field [ref. 75]. Such SMMs are of interest in the development of small magnetic particles capable of storing data at high densities and then being interrogated by fast (optical) means.

7.10.1. Spin Polarization in the $S = 10$ Mixed Valence Mn(IV)–Mn(III) Cluster Complex $[Mn_{12}O_{12}(O_2CR)_{16}(H_2O)_4]$

A widely studied example of an SMM is the mixed valence Mn cluster complex $[Mn_{12}O_{12}(O_2CR)_{16}(H_2O)_4]$, where $RCO_2^- = CH_3CO_2^- =$ acetate. The structure of this cluster consists of four Mn(IV) ions ($S = 2$) in a Mn_4O_4 cubane-like core surrounded by eight Mn(III) ions ($S = 3/2$) bridged by carboxylate oxygens and is illustrated in Figure 7.35. The complex has an overall ground-state spin $S = 10$.

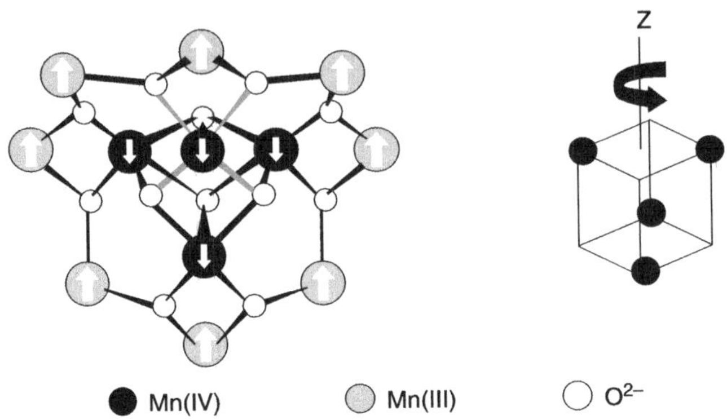

Figure 7.35. Structure of $[Mn_{12}O_{12}(O_2CCH_3)_{16}(H_2O)_4]$ (ligand framework removed) showing the $Mn_4^{IV}O_4$ cubane-like core. The direction of the S_4 axis is ahown on the right. (Reprinted with permission from ref. 75. McInnes, E. J. L.; Pidcock, E.; Oganesyan, V. S.; Cheesman, M. R.; Powell, A. K.; Thomson, A. J. *J. Am. Chem. Soc.* **2002**, *124*, 9219. Copyright 2002 American Chemical Society.

The cluster framework has axial symmetry along a unique *z*-axis (approximately S_4) of the cubane Mn(IV) core. This structure induces a zero-field splitting (ZFS) that is -0.5 cm^{-1}. This leads to a splitting of the 21 M_S sublevels of the $S = 10$ ground state at zero magnetic field, with the $M_S = \pm 10$ pair lying lowest at approximately 50 cm^{-1} below the $M_S = 0$. This ZFS creates an energy barrier to the interconversion of + and − spin states between two potential wells as illustrated in Figure 7.36 [ref. 75]. At high temperatures the interconversion occurs rapidly, but at very low temperatures, below 3 K, the so-called "blocking temperature," the spin polarization can be trapped. If, for example, a magnetic field parallel to the *z* axis is applied, spin polarization occurs because the degeneracy between the $\pm n$ sublevels is removed, and the lower-energy $-n$ sublevels are preferentially populated. Below the blocking temperature the spin polarization relaxation is very slow and remains even after the applied field is removed, leading to magnetic hysteresis. Thus, information could be stored as up or down spins and would remain as long as the spin polarization persists. The applied field essentially provides a method of inducing or erasing the spin polarization, and therefore the information. Interest in optical methods to determine the state of polarization has quite naturally focused on MCD, since the applied field is part of the experiment [ref. 75]. This has been made possible by MCD measurement for low-temperature glassy solutions containing the acetato complex $[Mn_{12}O_{12}(O_2CCH_3)_{16}(H_2O)_4]$, or $Mn_{12}Ac$, and by preparing an analogous complex complex with long-chain carboxylate bridging ligands $C_{14}H_{29}CO_2^-$ in the analogous complex $[Mn_{12}O_{12}(O_2CC_{14}H_{29})_{16}(H_2O)_4]$, or $Mn_{12}C_{15}$. This latter complex has higher solubility in organic solvents and allows the complex to be cast into thin films of poly(methyl methacrylate), PMM. The thin PMM films are transparent at wavelengths longer than 300 nm and can be conveniently cooled to low temperatures for spectroscopic measurements.

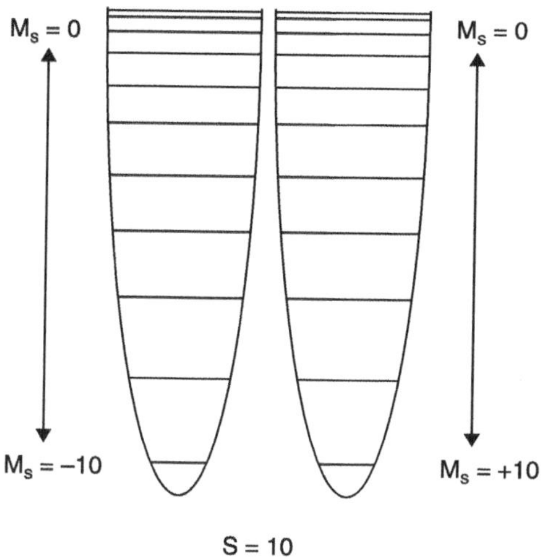

Figure 7.36. Ground-state spin sublevels for the $S = 10$ $[Mn_{12}O_{12}(O_2CCH_3)_{16}(H_2O)_4]$, which is subject to an axial ZFS. The potential well depths are approximately 50 cm^{-1} at zero applied field. Reprinted with permission from ref. 75. McInnes, E. J. L.; Pidcock, E.; Oganesyan, V. S.; Cheesman, M. R.; Powell, A. K.; Thomson, A. J. *J. Am. Chem. Soc.* **2002**, *124*, 9219. Copyright 2002 American Chemical Society.

The absorption and MCD spectra for $Mn_{12}C_{15}$ is shown in Figure 7.37. The absorption spectrum is broad and featureless in the 12,500- to 28,000-cm^{-1} region, while the MCD spectra show considerable detail. As expected, the MCD features change sign, but not shape, when the field is reversed, so that the spectra are symmetric about the zero ΔA or $\Delta \epsilon$ axes. However, the MCD spectral pattern persists at zero field after either a +5- or −5-T field is applied just prior to a zero-field scan. This observation illustrates clearly the persistent spin polarization induced by the applied field when the temperature (in this case, 1.8 K) is lower than the blocking temperature of 3 K. Careful magnetization studies at 4.2 K (above the blocking temperature) using MCD signals as a function of temperature and field were interpreted in terms of a ground-state spin $S = 10$ and an axial ZFS parameter $D = -0.6$ K, which agreed favorably with magnetization results from single crystals of $Mn_{12}Ac$ and high-field EPR measurements [ref. 75]. Polarization of the optical transitions responsible for the C terms relative to the axis of zero-field distortion were also determined from the magnetization experiments. While results for $Mn_{12}Ac$ were fit in a straightforward way, the magnetization for $Mn_{12}C_{15}$ in PMM films required the additional assumption of a nonrandom distribution of the molecular z axes relative to the applied field direction. This effect was rationalized by assuming that some stress occurred from the polymer during the casting of the film; this stress influenced the molecular orientation within the film. At temperatures below the blocking temperature, clear hysteresis loops of MCD intensity versus field were observed for the clusters. Figure 7.38 shows examples for both $Mn_{12}Ac$ and $Mn_{12}C_{15}$. The shape of the loops were dependent upon the particular

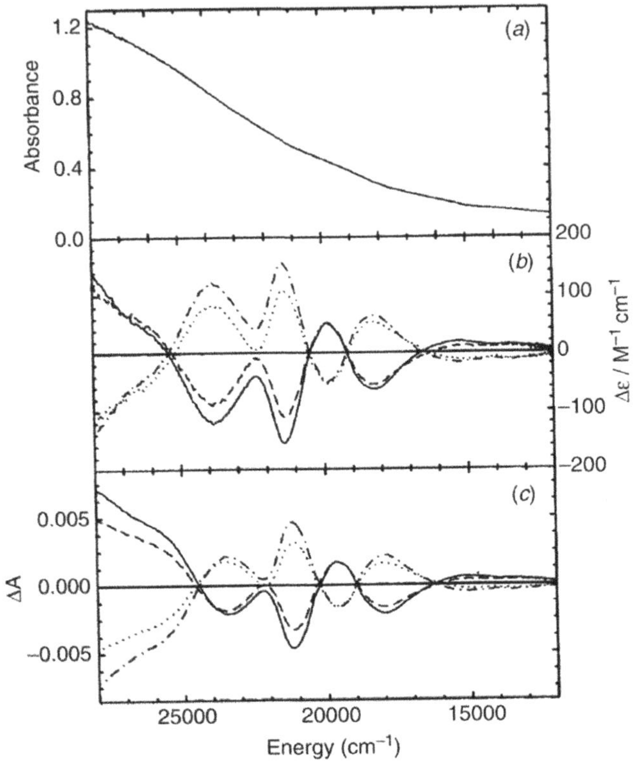

Figure 7.37. (*a*) Absorption spectrum of $Mn_{12}C_{15}$ in a PMM film at room temperature. (*b*) MCD spectra of $Mn_{12}C_{15}$ in $CHCl_3$/toluene (1 : 1 v/v) glassy solution at 1.8 K. (*c*) MCD spectra for $Mn_{12}C_{15}$ in a PMM film at 1.8 K. For parts b and c the solid curves are for +5-T field, the dash dot curves are for −5-T field, the dashed curves are for 0 T after application of a +5-T field, and the dotted curves are for 0 T after application of −5-T field. Reprinted with permission from ref. 75. McInnes, E. J. L.; Pidcock, E.; Oganesyan, V. S.; Cheesman, M. R.; Powell, A. K.; Thomson, A. J. *J. Am. Chem. Soc.* **2002**, *124*, 9219. Copyright 2002 American Chemical Society.

transition monitored, as well as on the medium. The loops clearly demonstrate the retention of spin polarization after reduction of the polarizing field to zero. The maximum intensity for this effect appears to be for transitions *xy*-polarized relative to the original polarizing field (see ref. 75 for details). Thus MCD is a sensitive optical method of determining the spin polarization in these Mn SMMs.

7.10.2. Single Ion and Cluster Spin in the $S = 6$ $[Cr_{12}O_9(O_2CCMe_3)_{15}]$ Cluster Complex

The success of MCD as a probe of magnetic properties of the Mn cluster complexes, which were of interest as SMMs, has also prompted experimental studies concerning the relationship between single-ion spin properties and that of a

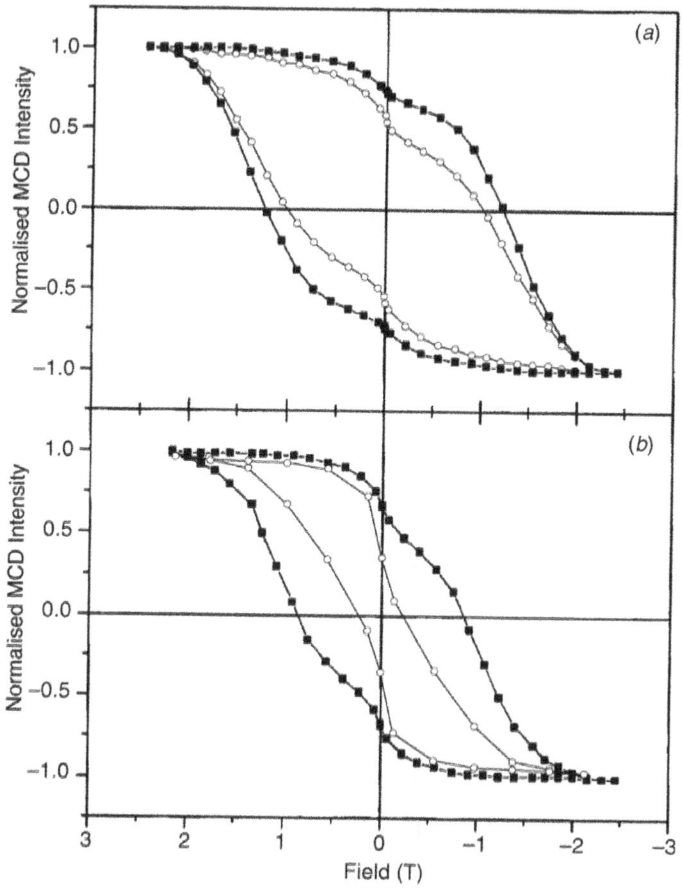

Figure 7.38. Hysteresis loops for (*a*) $Mn_{12}C_{15}$ in PMM at 1.7 K for *C* terms at 21,200 cm^{-1} (filled squares) and 19,600 cm^{-1} (open circles) and (*b*) $Mn_{12}Ac$ at 1.7 K and 21,200 cm^{-1} in PMM film (filled squares) and DMF/MeCN (3:1 v/v) glassy solution (open circles). Reprinted with permission from ref. 75. McInnes, E. J. L.; Pidcock, E.; Oganesyan, V. S.; Cheesman, M. R.; Powell, A. K.; Thomson, A. J. *J. Am. Chem. Soc.* **2002**, *124*, 9219. Copyright 2002 American Chemical Society.

complicated cluster. The Cr(III) cluster complex $[Cr_{12}O_9(O_2CCMe_3)_{15}]$, the structure of which is shown in Figure 7.39, was characterized by multifequency EPR spectroscopy, together with susceptibility and magnetization studies, as having a ground spin state of $S = 6$ with $g_{ZZ} = 1.965$ and $g_{XX} = g_{YY} = 1.960$ along with the axial and rhombic zero-field splitting parameters of $D_{S=6} = +0.088\ cm^{-1}$, $E_{S=6} = 0$, respectively [ref. 76]. It was noted that this cluster would not be a good candidate for an SMM because the axial ZFS should be negative like the Mn clusters. However, an analysis of high-resolution MCD spectra at low temperature showed that the *D* value for a single Cr(III) ion with CrO_6 coordination was $-1.035\ cm^{-1}$. This result was used together with a vector coupling model to

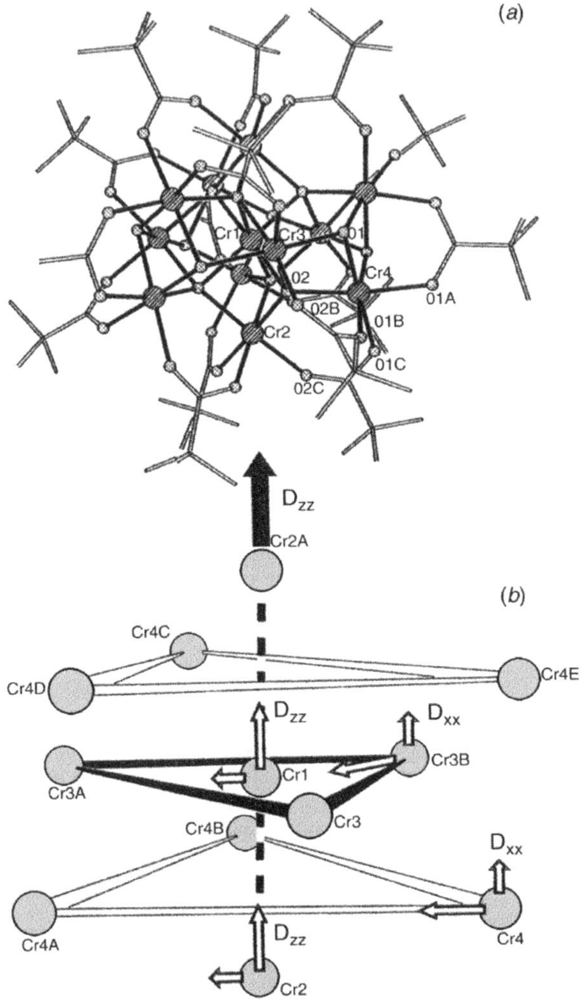

Figure 7.39. Crystal structure and view of the Cr_{12} skeleton in the Cr(III) cluster. D_{ZZ} represents the ZFS for the cluster, while D_{zz} and D_{xx} represent single ion ZFS axes. Reprinted with permission from ref. 76. Collison, D.; Murrie, M.; Oganesyan, V. S.; Piligkos, S.; Poolton, N. R. J.; Rajaraman, G.; Smith, G. M.; Thomson, A. J.; Timko, G. A.; Wernsdorfer, W.; Winpenny, R. E. P.; McInnes, E. J. L. *Inorg. Chem.* **2003**, *42*, 5293. Copyright 2003 American Chemical Society.

show that the cluster ZFS was due almost entirely to the single-ion components and that the relative orientations of these components can lead to a cluster ZFS of opposite sign.

The absorption (at 293 K) and MCD (at 1.8 K) spectra for $[Cr_{12}O_9(O_2CCMe_3)_{15}]$ are shown in Figure 7.40. The MCD spectra reveal considerably more detail than the absorption spectrum; this detail allows a relatively complete interpretation

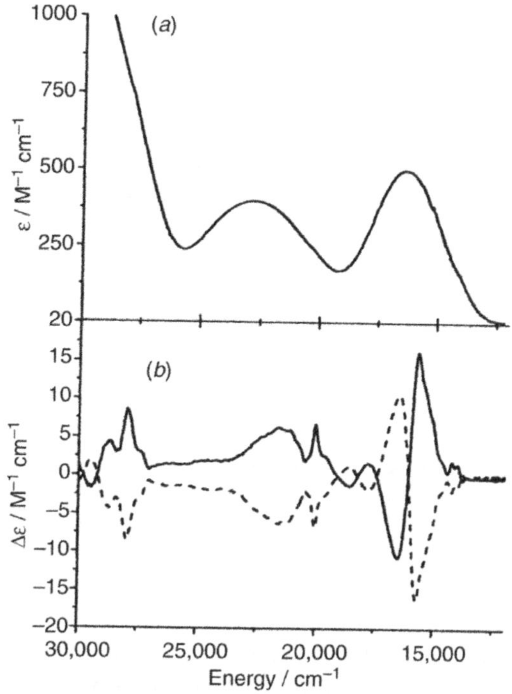

Figure 7.40. (*a*) Absorption (at 293 K) and (*b*) MCD (at 1.8 K, solid line at +5 T and the dashed line at −5 T) spectra for $[Cr_{12}O_9(O_2CCMe_3)_{15}]$ in $CHCl_3$/toluene (1 : 1 v/v) solution. Reprinted with permission from ref. 76. Collison, D.; Murrie, M.; Oganesyan, V. S.; Piligkos, S.; Poolton, N. R. J.; Rajaraman, G.; Smith, G. M.; Thomson, A. J.; Timko, G. A.; Wernsdorfer, W.; Winpenny, R. E. P.; McInnes, E. J. L. *Inorg. Chem.* **2003,** *42*, 5293. Copyright 2003 American Chemical Society.

based on a single-ion model. The assignments of the MCD features are based on the energy levels in Figure 7.41 for a monomeric $Cr^{III}O_6$ center, as originally discussed in ref. 77. The lowest-energy absorption $^4A_2 \rightarrow {}^4T_2$ band consists of two transitions at 16,502 and 15,673 cm^{-1} assigned to $^4A_2 \rightarrow {}^4A_1$ and ^4E, respectively. The field dependence of the MCD spectra at 1.8 K was simulated to give the polarization of the optical transitions. The $^4A_2 \rightarrow {}^4E$ transition was found to be predominantly (91%) XY polarized, while the $^4A_2 \rightarrow {}^4A_1$ transition was a mixture of XY (65%) and XZ (35%) polarized, assuming an $S = 6$ ground state with $D_{S=6} = +0.088$ cm^{-1}. The assignment of the broad band at 22,800 cm^{-1}, along with the narrow and weaker MCD features to doublet excited states, allowed a fairly complete description of a trigonally distorted $Cr^{III}O_6$ center, including inter-state splitting and Racah parameters. From these parameters the single-ion ZFS was determined as $|D| = 1.035$ cm^{-1}. Consideration of the g values for the single-center calculation showed $g_{zz} = 1.957$ and $g_{xx} = 1.961$ and therefore $g_{zz} < g_{xx}$, which is expected to lead to a *negative* ZFS (from ref. 77). The ZFS for the cluster was then interpreted by using vector coupling of individual single ion ZFS within the Cr_{12}

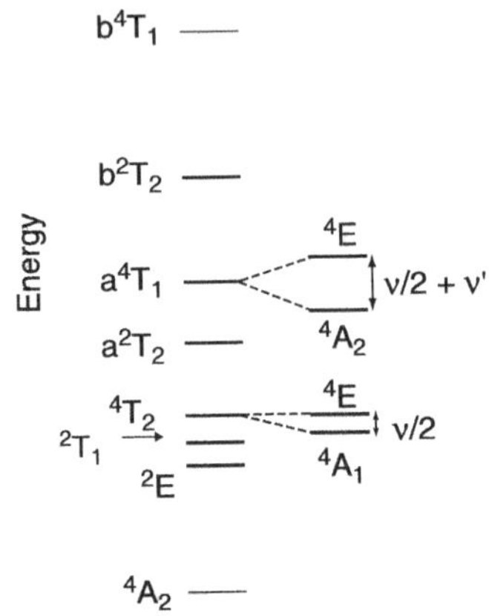

Figure 7.41. Electronic states for a $Cr^{III}O_6$ center (after ref. 77). Reprinted with permission from ref. 76. Collison, D.; Murrie, M.; Oganesyan, V. S.; Piligkos, S.; Poolton, N. R. J.; Rajaraman, G.; Smith, G. M.; Thomson, A. J.; Timko, G. A.; Wernsdorfer, W.; Winpenny, R. E. P.; McInnes, E. J. L. *Inorg. Chem.* **2003,** *42*, 5293. Copyright 2003 American Chemical Society.

skeleton. Several coupling schemes were considered [ref. 76]. The model that emerged noted that all but three of the 12 Cr(III) ions project the positive in-plane components of their **D** tensors onto the unique cluster axis; the three along the axis project negative components. Thus even though the individual ion ZFS values are negative, the vector sum gives a positive value as a result of relative orientation. Good quantitative agreement was obtained for both D_{ZZ}, the cluster ZFS, and $g_{ZZ} > g_{XX}$, the cluster g values. This study emphasises that cluster properties need not be restricted to metal ions that give the same sign ZFS as that of the overall cluster.

7.11. LANTHANIDE IONS IN CRYSTALLINE ENVIRONMENTS

A number of interesting MCD measurements on lanthanide ions in crystalline matrices have been reported by Görller-Walrand and co-workers [refs. 78–81]. The noteworthy aspects of these measurements include the use of an ordered crystalline matrix for the spectroscopic study of $4f \rightarrow 4f$ transitions, which are more atomic-like than most molecular transitions discussed earlier and have complex angular momentum characteristics. MCD measurements for crystalline samples require either an isotropic crystal (cubic for example) or crystal system with a well-defined optic axis where the index of refraction is the same for all perpendicular directions to the axis. If these requirements are not met, then there will be large static birefringence and magnetically induced birefringence that will create signals many orders of magnitude larger than the MCD effects of interest. In this section, several examples will be described.

7.11.1. $Na_3Ln(ODA)_3 \cdot 2NaClO_4 \cdot 6H_2O$ (LnODA, Ln = Eu and Nd)

The oxydiacetato ligand, $O_2CH_2OCH_2O_2^{2-}$, forms complexes with a number of lanthanide Ln^{3+} ions that are isostructural and crystallize in the trigonal space group $R32$. In the unit cell the environment of the LnO_9 coordination polyhedron is a distorted tricapped trigonal prism in a site that is approximately D_3, with the three-fold axis parallel to the crystal optic axis. A view down the trigonal axis is shown in Figure 7.42. The top and bottom triangles of the O donors are twisted so that two enantiomeric crystalline forms are possible and therefore a chiral environment is also found about the metal ion. Crystal samples were grown from aqueous solution and then polished perpendicular to the trigonal (optic) axis. Since the Ln^{3+} is in a chiral environment, samples will exhibit natural circular dichroism (CD) in addition to MCD. The MCD spectra were corrected by subtracting the CD measured at zero field. The spectral analysis included an evaluation of the CD rotatory strength in addition to MCD intensities [refs. 78, 79].

The EuODA and NdODA complexes represent very different electronic configurations. Eu^{3+} has a $4f^6$ ground configuration with a $^7F_0$ ground state that has non degenerate A_1 symmetry in D_3. With the z axis along the C_3 trigonal crystal axis, the dipole-allowed axial spectra consists only of $A_1 \rightarrow E$-type transitions and therefore only A and B terms in the MCD. In contrast, the Nd^{3+} ion has $4f^3$ and a $^4I_{9/2}$ ground state, so that A, B, and C terms are observed. In both cases a parametric crystal field model [ref. 82] is used to calculate the energies and intensities of the $f \rightarrow f$ transitions. The Hamiltonian is given by

$$H = H_0 + H_{CF}$$

where H_0 is for the free ion and H_{CF} results from the crystal field imposed upon the ion by the environment. The model has 20 parameters for H_0 and 6 parameters

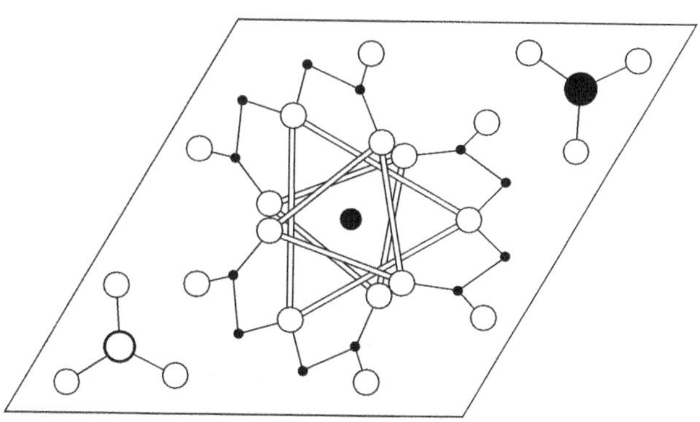

Figure 7.42. The crystal environment about the lanthanide ion in the $Ln(ODA)_3^{3-}$. The view is down the trigonal axis for a λ configuration. Reprinted with permission from ref. 78. Görller-Walrand, C.; Verhoeven, P.; D'Olieslager, J.; Fluyt, L.; Binnemans, K.; *J. Chem. Phys.* **1994**, *100*, 815. Copyright 1994 American Institute of Physics.

for H_{CF}, and the calculated energies are determined by a simultaneous diagonalization of free ion and crystal field Hamiltonian terms (26 parameters in all). The axial absorption, CD, and MCD spectra are then used to refine the intensity parameters; the MCD intensities are found to be very sensitive to the sign and values of these parameters. The MCD spectra are then fit or simulated for comparison with experiment [refs. 78, 79].

The measurements for EuODA were in the range 4500–25,500 cm^{-1} at 150 K (spectra were collected but not analyzed at lower temperatures because of concerns of a crystal-phase transition that occurs between the *R*32 space group and a subgroup *P*321 in which the symmetry at the metal center is only C_2).

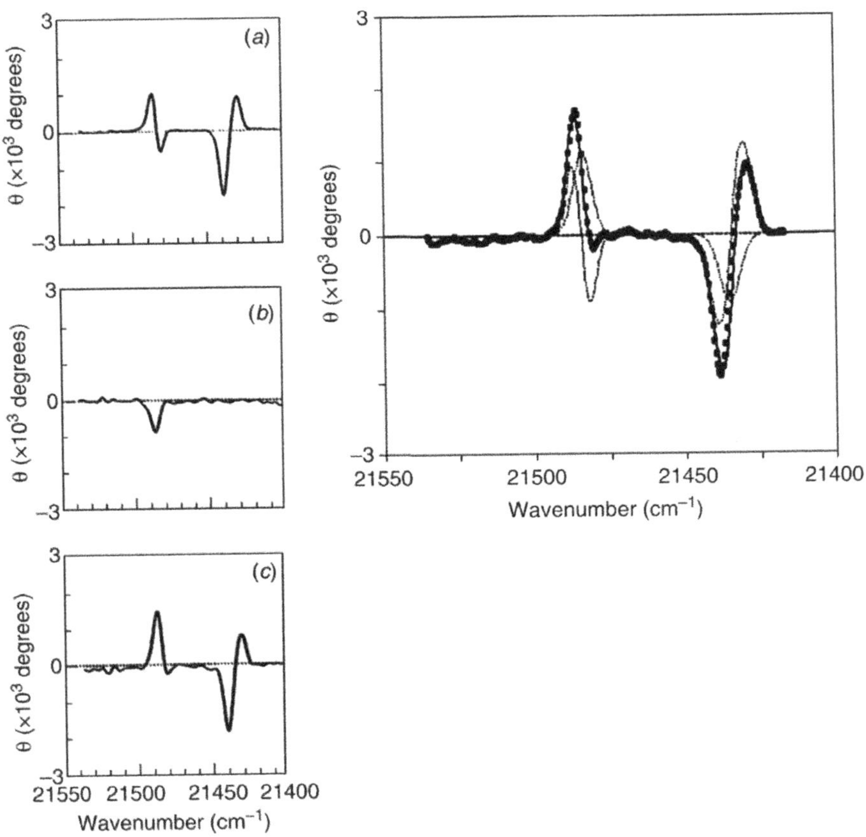

Figure 7.43. MCD + CD (*a*, upper left), CD (*b*, center left), MCD with CD subtracted (*c*, lower left), and the intensity fit for the MCD (right) spectra for the $^7F_0 \rightarrow {}^5D_2$ transition for EuODA at 150 K and 6.2 T; the crystal path length was 0.09 mm. $\theta = 33000\ \Delta A$ Reprinted with permission from ref. 78. Görller-Walrand, C.; Verhoeven, P.; D'Olieslager, J.; Fluyt, L.; Binnemans, K.; *J. Chem. Phys.* **1994**, *100*, 815. Copyright 1994 American Institute of Physics.

Figure 7.43 shows an example for the $^7F_0 \rightarrow {^5D_2}$ transition. Eleven transitions were fit in the spectral range considered. The final set of parameters provided an acceptable fit for the energies and signs of the MCD and CD features in the spectra. The relative intensities paralleled those observed, but the magnitudes were not always very close. However, in general the MCD spectra provided a sensitive test for the parametric model (see ref. 78 for details).

In the case of NdODA, the measurements were in the range 5000–30,000 cm^{-1} and the temperature was 4.2 K. Even though there is a phase transition at 74 K, it was assumed that the results could still be interpreted within the D_3 symmetry assumption. The parametric model was modified by adding the Zeeman term $H_{Zeeman} = -\mu_z B_z = \mu_B(L_z + 2S_z)B_z$:

$$H = H_0 + H_{CF} + H_{Zeeman}$$

Transitions between Zeeman levels were treated directly. This contrasts with the traditional *A*-, *B*-, and *C*-term description used conventionally. The electric dipole strengths of the lcp and rcp light absorptions were calculated for each transition, and then the overall MCD spectrum was simulated by the relationship

$$\Delta\varepsilon(\nu,T) = 327\nu_{Zeeman} N_i(T)(D_{lcp} - D_{rcp})\rho_{Zeeman}(\nu)$$

where $D_{lcp} - D_{rcp}$ represents the difference between the calculated dipole strengths of lcp and rcp light, $N_i(T)$ is the Boltzmann fractional population of the Zeeman component of the $^4I_{9/2}$ ground state for the particular transition, and $\rho_{Zeeman}(\nu)$ is the Zeeman band-shape function (see ref. 79 for details). The energies and dipole strengths of 99 absorption transitions from the $^4I_{9/2}$ ground multiplet were calculated for NdODA. Figure 7.44 illustrates two of the transition multiplets: $^4I_{9/2} \rightarrow {^4I_{15/2}}$ and $^4F_{9/2}$, respectively, which are typical of the results. Figure 7.45 shows the temperature dependence of the $^4I_{9/2} \rightarrow {^2P_{1/2}}$ transition; the simulation included the Boltzmann populations as a function of temperature for the Zeeman split $I_{9/2}$ multiplet. It is clear that the simulations are not perfect, but the MCD, CD, and axial absorption spectra all complement one another in describing the electronic transitions. The MCD and CD simulations are sensitive to the parameter values of the model, and as such they provide a good test for parameter reliability. The atomic-like $f \rightarrow f$ transitions thus provide a good basis for this model, which has value in describing the fine structure of the electronic spectra.

7.11.2. $LiErF_4$, $LiYF_4/Eu^{3+}$, and KY_3F_{10}/Eu^{3+} Crystals

Lanthanides Er^{3+} and Eu^{3+} have also been studied in single crystals where the metal ions are surrounded by F-ligands [refs. 80, 81]. The $LiErF_4$ and the Eu^{3+}-doped $LiYF_4$ feature an eightfold coordination environment about the lanthanide that is approximately D_{2d}, while the Eu^{3+} doped KY_3F_{10} is cubic and exhibits a

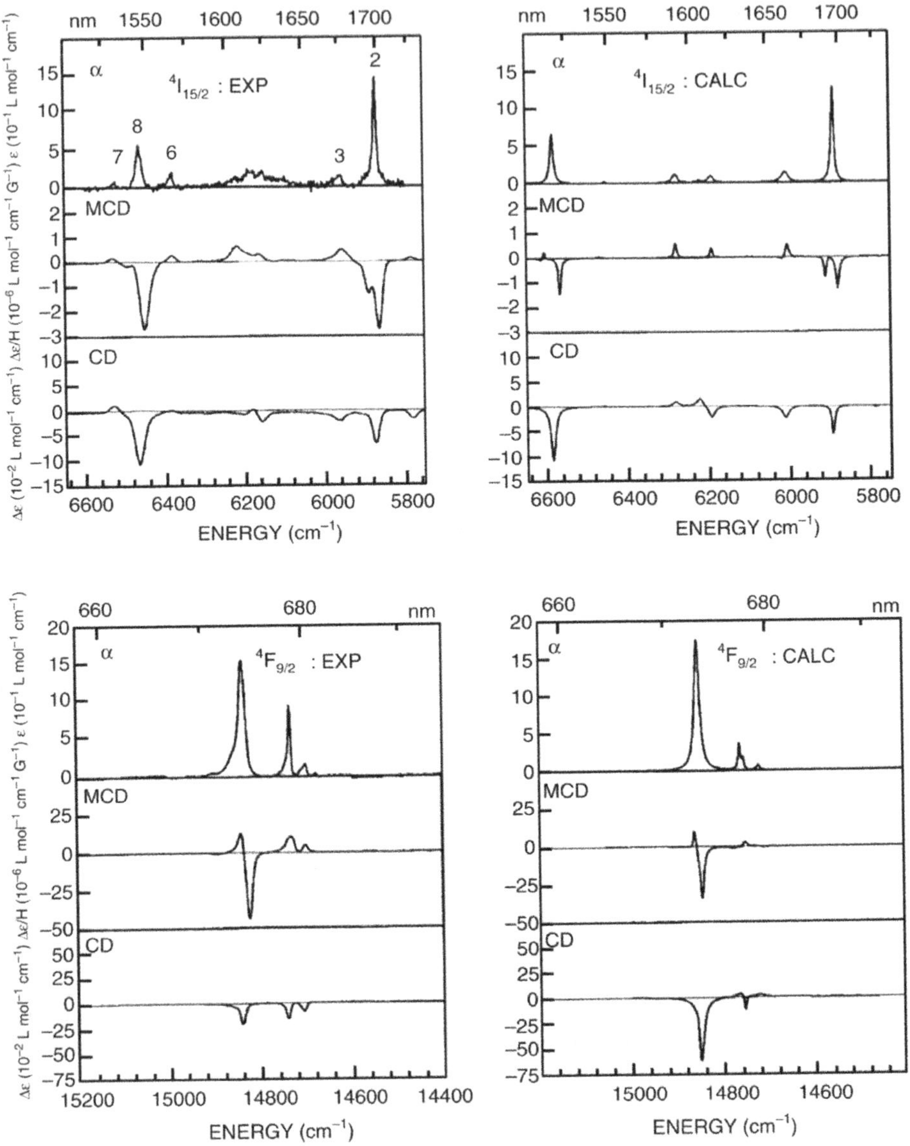

Figure 7.44. Experimental (left) and simulated (right) axial absorption (α, top curves). MCD (center curves) and CD (bottom curves) spectra for NdODA at 4.2 K and 6.2 T. Reprinted with permission from ref. 79. Fluyt, L.; Couwenberg, I.; Lambaerts, H.; Binnemans, K.; Görller-Walrand, C.; Reid, M. F. *J. Chem. Phys.* **1996**, *105*, 6117. Copyright 1996 American Institute of Physics.

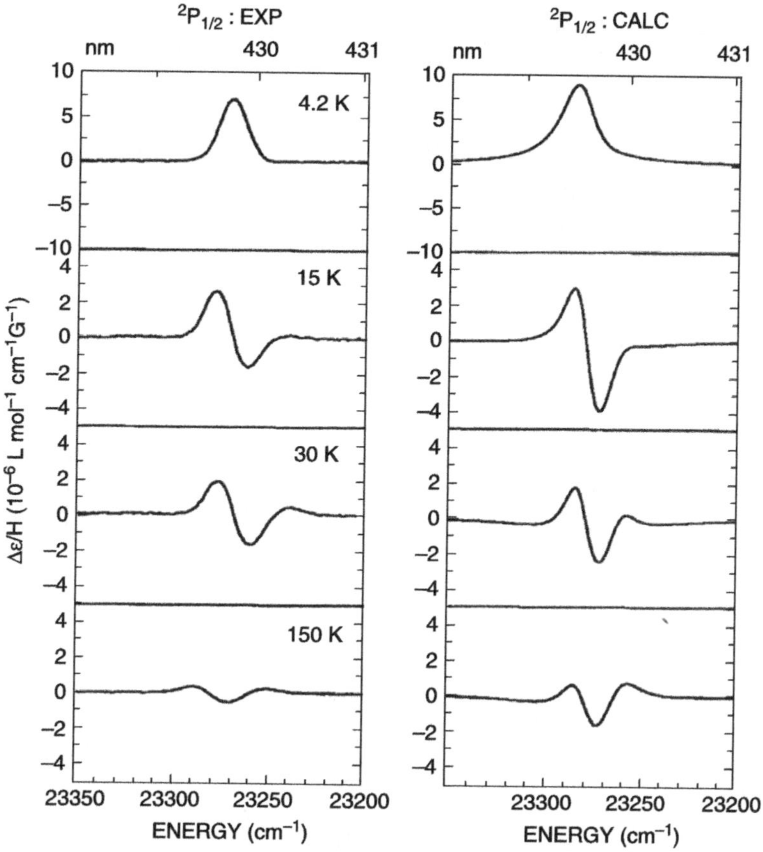

Figure 7.45. Temperature dependence for the $^4I_{9/2} \rightarrow {}^2P_{1/2}$ transition for NdODA at 6.2 T. Reprinted with permission from ref. 79. Fluyt, L.; Couwenberg, I.; Lambaerts, H.; Binnemans, K.; Görller-Walrand, C.; Reid, M. F. *J. Chem. Phys.* **1996**, *105*, 6117. Copyright 1996 American Institute of Physics.

square antiprismatic C_{4v} coordination of the Eu^{3+} ion. Figure 7.46 shows sketches of the site symmetries.

The study of $LiErF_4$ crystals was similar to that described above for NdODA. Parametric crystal field calculations for the $4f^{11}$ Er^{3+} in a D_{2d} site were performed and compared with absorption spectra; 12 different transitions were measured and calculated. The MCD spectra were then graphically simulated as a test of the validity of the parameter set [ref. 80]. The success of this approach is in the comparison of the experimental and simulated spectra. As for NdODA, the comparison is not perfect, but the simulated spectra follow the shape (and for the MCD, also the sign) of the experimental spectra to a fair degree. Figure 7.47 shows an example of one of the observed and simulations for LiErF4, that of the $^4I_{15/2} \rightarrow {}^2H(2)_{11/2}$ transition. Overall, the absorption measurements extended from 6000 to

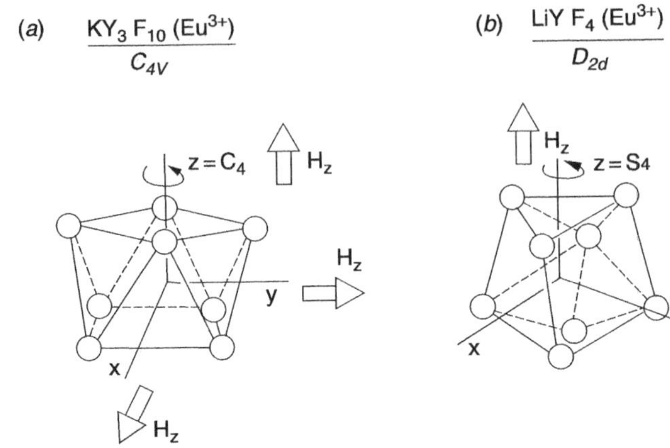

Figure 7.46. Site symmetries in $LiYF_4/Eu^{3+}$ and $LiErF_4$ (D_{2d}, right) and KY_3F_{10}/Eu^{3+} (C_{4v}, left). (Adapted from ref. 81. Reprinted with permission from ref. 81. Görller-Walrand, C.; Behets, M.; Porcher, P.; Laursen, I. *J. Chem. Phys.* **1985**, *83*, 4329. Copyright 1985 American Institute of Physics.)

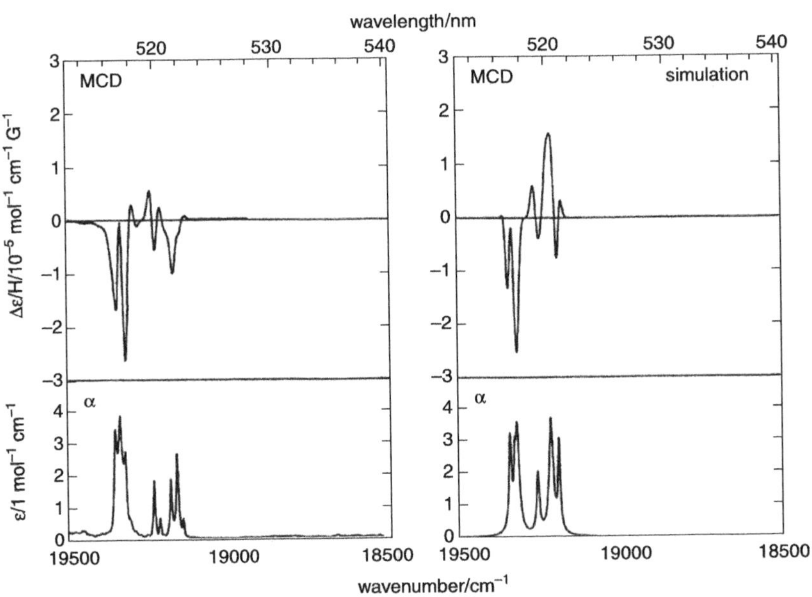

Figure 7.47. Observed axial α absorption (left lower curve) and MCD (left upper curve) spectra of the $^4I_{15/2} \rightarrow {}^2H(2)_{11/2}$ transition for $LiErF_4$ and their simulation (right curves). The observed spectra were measured at 4.2 K; the MCD spectra was determined at 0.9 T. From ref. 80. Heyde, K.; Binnemans, K.; Görller-Walrand, C. *J. Chem Soc. Faraday Trans.* **1998**, *94*, 843. Reproduced by permission of the Royal Society of Chemistry.

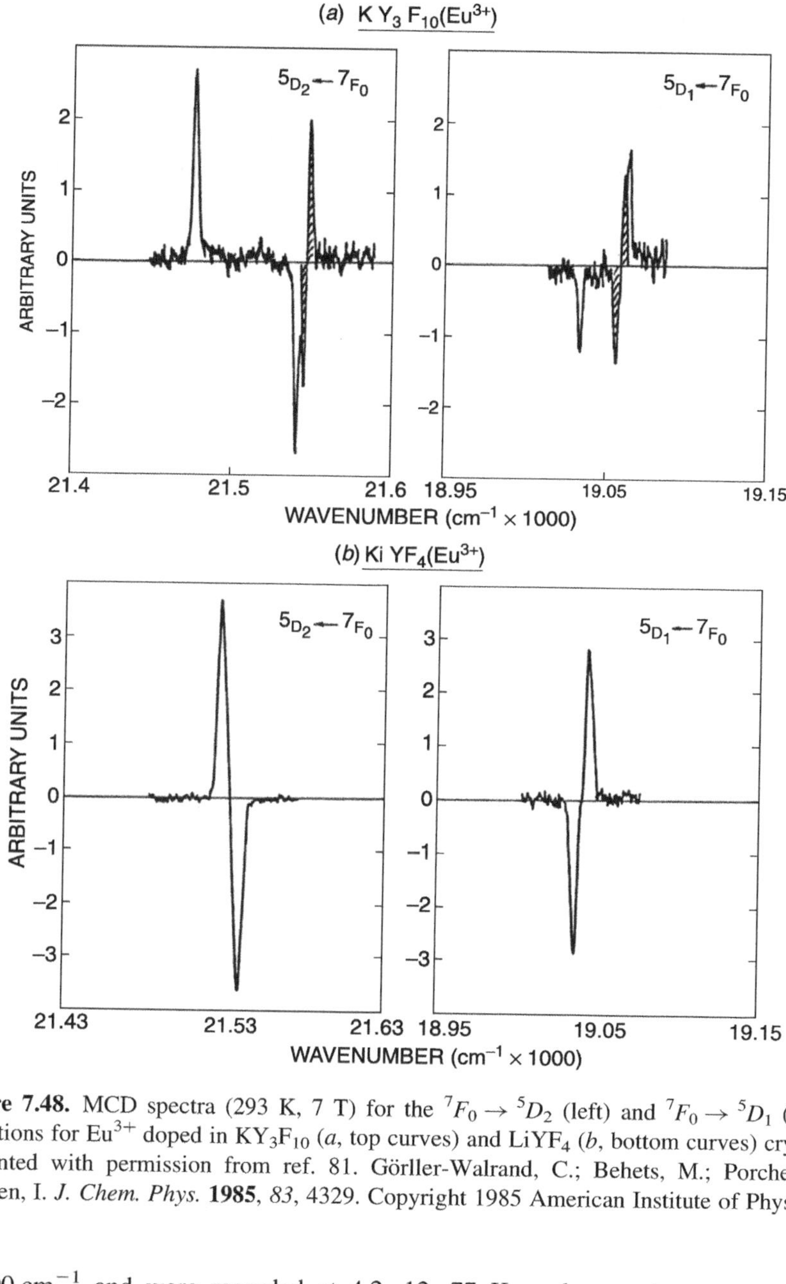

Figure 7.48. MCD spectra (293 K, 7 T) for the $^7F_0 \rightarrow {}^5D_2$ (left) and $^7F_0 \rightarrow {}^5D_1$ (right) transitions for Eu^{3+} doped in KY_3F_{10} (*a*, top curves) and $LiYF_4$ (*b*, bottom curves) crystals. Reprinted with permission from ref. 81. Görller-Walrand, C.; Behets, M.; Porcher, P.; Laursen, I. *J. Chem. Phys.* **1985**, *83*, 4329. Copyright 1985 American Institute of Physics.

32,000 cm^{-1} and were recorded at 4.2, 12, 77 K, and room temperature. The MCD spectra were measured at 0.9 T and at the same temperatures as the absorption, but only over the range 14,000–32,000 cm^{-1}. The original paper [ref. 80] can be consulted for some of the additional transitions showing observed and simulated spectra. The MCD specrta of Eu^{3+} in $LiYF_4$ and in KY_3F_{10} crystals were found to be sensitive to the site symmetry. The negative *A* term for the $^7F_0 \rightarrow {}^5D_2$ transition for the $LiYF_4$ crystal changes to a positive *A* term for the KY_3F_{10} crystal, while the $^7F_0 \rightarrow {}^5D_1$ transition exhibits a positive *A* term for both crystalline environments [ref. 81]. This is shown in Figure 7.48. The authors suggest that in certain cases, such as this one, MCD spectra can serve as a site-symmetry probe because neither polarized absorption nor fluorescence spectra can detect a difference [ref. 80].

8 Magnetic Vibrational Circular Dichroism (MVCD) and X-Ray Magnetic Circular Dichroism (XMCD)

In Chapters 4–7 the description of MCD assumes intravalence electronic excitations, electronic excited states, and circularly polarized light in the visible–ultraviolet spectral regions. MCD has also been observed for vibrational or rotation–vibrational excitations in the infrared, MVCD, and for core → valence level absorption edges in the X-ray region, XMCD. The application of MCD techniques in these regions is described briefly in this chapter, along with a few representative examples. It must be obvious that the experimental setup for these measurements must be appropriate to the spectral region involved, and therefore the apparatus necessary will depend upon the wavelength of the light used for the measurement. Infrared or X-ray polarizers and optics are quite different from than those used in the vis–UV region. For XMCD particularly, intense X-ray photon sources are required, which means proximity to synchrotron laboratories is necessary for measurements. Nevertheless, the measured differential absorption between lcp and rcp light by a sample in a strong magnetic field oriented along the propagation direction is entirely analogous to measurements in the vis–UV, and the origin of MVCD and XMCD likewise results from the Zeeman interaction of the field with magnetic moments within the sample. MVCD has been measured for samples in the gas phase and in condensed phases; XMCD has largely involved solid-state samples, often at low temperature.

8.1. MVCD

MVCD was first reported in 1981 [ref. 83] and has been developed largely by Keiderling and co-workers, and it was applied initially to vibrational studies of molecules with high symmetries in the condensed phase. More recently, with improvements in instrumental methods for higher resolution, MVCD was used to study the molecular Zeeman effect of rotationally resolved vibrations of small

A Practical Guide to Magnetic Circular Dichroism Spectroscopy, by W. Roy Mason

molecules in the gas phase. The infrared measurements necessitated the use of sources and detectors appropriate for the IR region and grid polarizers, lenses, and photoelastic modulator (PEM) elements that have high transmission in the IR. The application of Fourier transform IR (FTIR) methods to MVCD were also of note because the range of measurement and resolution were greatly enhanced over dispersive spectrometers. However, the differential signals for lcp and rcp light are often very small and instrumental artifacts have been, and are still, problematic because they can be large compared to the signals of interest.

8.1.1. Instrumentation

MVCD instrumentation requires a sensitive vibrational CD (VCD) spectrometer that operates in the IR or near-IR region, and the sample in a strong magnetic field parallel to the light beam, just as for vis–UV. The magnet considerations are essentially the same as for the vis–UV as described in Chapter 4. The sensitivity and measurement quality of the VCD spectrometer represents the main difference between MVCD and MCD. VCD spectrometers (see ref. 84 for a descriptive review of VCD instrumentation) utilize a grid polarizer and a PEM to provide modulated lcp and rcp light at high acoustic frequencies much the same way as for vis–UV measurements. The PEM element must transmit IR radiation in the region of interest. Quartz elements can be used in the near-IR down to about 6250 cm^{-1}, but crystal materials such as CaF_2 (to about 1250 cm^{-1}), BaF_2 (to 800 cm^{-1}), or ZnSe (to 500 cm^{-1}) are more suited to the vibrational region. Anti-reflective coatings are often used, and crystal lenses, rather than focusing mirrors, are better for reducing artifacts, which can be significant in the IR. Detectors also must be appropriate to the spectral region of interest. The InSb- or HgCdTe-type detectors have been used successfully. In addition, it is often necessary to introduce IR transmission filters to further reduce stray light, but polarization artifacts must be avoided because these are often much larger than the measured signals of interest.

There are two approaches to MVCD measurement: dispersive IR and Fourier transform IR (FTIR). Dispersive spectrometers, which involve diffraction gratings and filters, are very much like those used in the vis–UV as described in Chapter 4. Such spectrometers seem to present fewer polarization artifacts when used over relatively narrow spectral regions ($\sim<200$ cm^{-1}) at medium resolution (5–10 cm^{-1}). The FTIR method can be used for wider spectral regions (~ 1000 cm^{-1}) and higher resolution (0.1 cm^{-1}). In the FTIR method, the IR beam from an interferometer is passed through a linear polarizer and PEM which provides lcp/rcp modulated light. This beam is then passed through the sample in a magnetic field and then focused on the detector. The light modulated at the PEM frequency (usually selected by means of a high-pass filter) is demodulated by a lock-in amplifier tuned to the PEM reference frequency. This signal is combined with the undemodulated signal for the average intensity of the IR light (usually selected by a low-pass filter) and sent to the Fourier transform electronics of the IR spectrometer. The output provides raw MVCD as a ratio of $(A_- - A_+)/(A_- + A_+)$.

The FTIR approach is more prone to artifacts and signal-to-noise problems from phase differences in the interferogram, or determining the zero-path difference for weak signals, than the dispersive measurements, but the greater resolution and wider range possible are often driving forces for careful spectrometer design [refs. 84, 85]. More recently, the development of the step-scan technique [ref. 86] for FTIR has provided some additional improvements for measurements with wider frequency ranges, especially into the near IR region. In the step-scan method (as opposed to rapid-scan, the usual FTIR method) the moving mirror of the interferometer is stopped at different positions while data are collected. All optical frequencies are therefore modulated at the same frequency. This approach removes some of the frequency-dependent attenuation and phase correction problems that are present in rapid-scan FTIR measurements (see refs. 84 and 85 for a description of the step-scan versus rapid-scan methods).

8.1.2. Examples

Several examples of MVCD are described in the following case studies. These examples are not meant to be comprehensive in scope, but rather to give a snapshot of a few of the applications in the literature.

8.1.3. MVCD for Metal Carbonyl Complexes $M(CO)_6$, M = Cr, Mo, and W

The high-symmetry octahedral $M(CO)_6$ complexes present strong ν_6 (T_{1u}) fundamental CO stretching bands near 2000 cm^{-1} in the IR. These bands were found to show strong MVCD signals and present positive A-term-type band shapes for CCl_4 solutions of the $M(CO)_6$ complexes [ref. 87]. As an example, Figure 8.1 presents the absorption and MVCD for $Mo(CO)_6$ in CCl_4 solution. A careful moment analysis gave the relative A_1/D_0 ratio of 1.63 : 1.34 : 1 for the positive A term for M = Cr, Mo, and W, respectively. In addition, a small negative B term was found for each case. Attempts to correlate the positive A term by using an empirical charge model based on the relative motion of charges fixed to the nuclear positions during the ν_6 normal mode gave the correct A-term sign, but were 1–3 orders of magnitude too small compared to the measured values. Instead, a parameterized vibronic coupling model [ref. 88] was applied with some success. In this model, vibronic coupling was assumed between the ν_6 normal mode and the lowest-energy metal-to-ligand (MLCT) electronic excited states of T_{1u} symmetry. These $M(CO)_6$ complexes exhibit two very intense MLCT bands in the UV region as discussed in Chapter 6, Section 6.6.4. Both the T_{1u} [$t_{2g} \rightarrow t_{1u}$] and the T_{1u} [$t_{2g} \rightarrow t_{2u}$] excited states were considered, but the latter, T_{1u} [$t_{2g} \rightarrow t_{2u}$] at higher energy, seemed to provide the greatest contribution. An estimate of the magnetic moment μ for this electronic excited state that would reproduce the observed MVCD A_1/D_0 values was found to be $\mu = -0.48\mu_B$, $-0.35\mu_B$, and $-0.36\mu_B$, (μ_B = Bohr magneton) for M = Cr, Mo, and W, respectively. Thus, the magnetic moment of the CO stretching mode was assumed to be very small, and the MVCD was attributed to significant vibronic coupling with MLCT states.

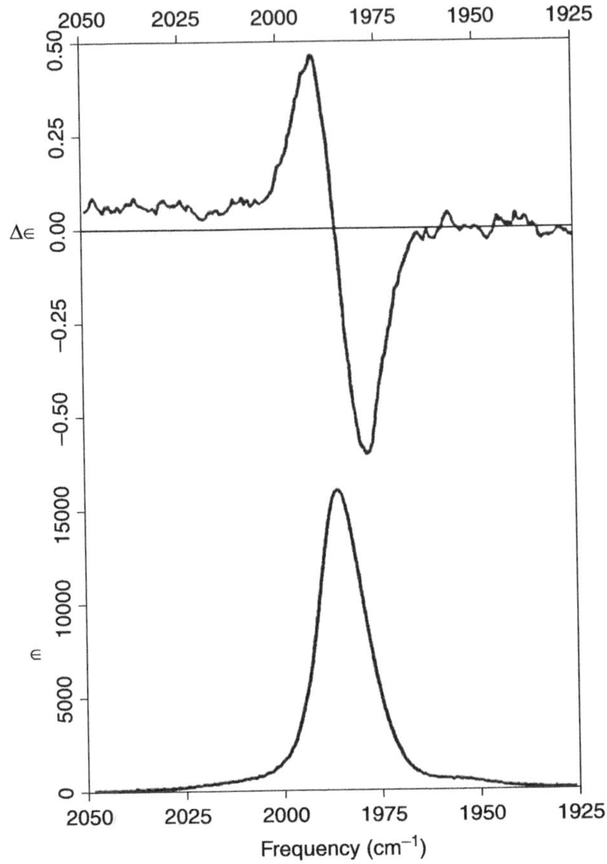

Figure 8.1. Absorption and MVCD for $Mo(CO)_6$ in CCl_4 solution at 4.0 T with a resolution of 10 cm^{-1}, a 10-s time constant, and four scans averaged. Reprinted with permission from ref. 87. Devine, T. R.; Keiderling, T. A. *J. Chem. Phys.* **1985**, *83*, 3749. Copyright 1985 American Institute of Physics.

It may be remarked that these complexes also present combination bands $\nu_1(A_{1g}) + \nu_6(T_{1u})$ and $\nu_3(E_g) + \nu_6(T_{1u})$ in the region 3950–4150 cm^{-1}, but the signals were weaker and had poorer signal-to-noise ratio.

8.1.4. Rotationally Resolved MVCD for Carbon Monoxide

Higher-resolution FTIR methods (0.5 cm^{-1} or better) have enabled MVCD study of gas-phase samples that exhibit rotationally resolved structure (RR MVCD). As an example, Figure 8.2 presents the IR absorption and MVCD spectra for CO gas [ref. 89] which show impressive P and R branches for the CO stretch. The CO spectra represent an interesting test case for RR MVCD because there is only one nondegenerate vibration that cannot give rise to Coriolis coupling and therefore

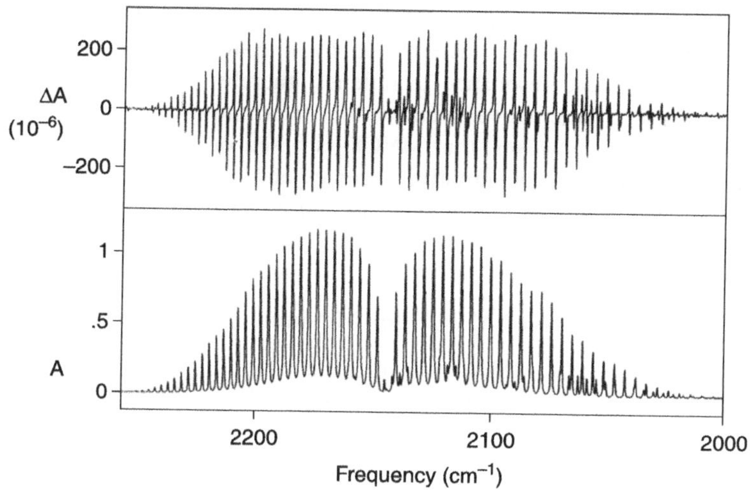

Figure 8.2. Absorption (lower curve) and MVCD (upper curve) for CO in the gas phase; 5-cm path, resolution 0.5 cm^{-1}. The MVCD spectra were determined by the difference between measured spectra at ± 8 T (512 scans each), normalized to 1 T. Reprinted with permission from ref. 89. Wang, B.; Keiderling, T. A. *J. Chem. Phys.* **1993**, *98*, 903. Copyright 1993 American Institute of Physics.

no vibrational angular momentum and no vibrational Zeeman effect. The MVCD is a result entirely of rotational Zeeman spectra. The strong CO bond makes the rigid rotor approximation a good one, and the rotational g value is expected to be the same for the $v = 0$ ground state and $v = 1$ excited vibrational state. In Figure 8.2, each MVCD line has a positive A-term band shape; careful moment analysis of the lines in both absorption and MVCD allow A_1/D_0 ratios to be determined with very good precision. From the A_1/D_0 ratios, rotational molecular g factors were determined. The relationship between the A_1/D_0 ratio and the molecular g_J factor is derived in the standard way by using the rotational Zeeman operator and rovibrational states:

$$H_{\text{Zeeman}} = -\mu_N g_J J_z B \tag{8.1}$$

where μ_N is the nuclear magneton and J_z is the rotational angular momentum projected on the magnetic field B direction. If there is no appreciable vibrational contribution and the $\Delta J = \pm 1$, $\Delta M_J = \pm 1$ selection rules for circularly polarized transitions are assumed, then the result [ref. 89] may be given as

$$g_J = (1/2)(-\mu_B/\mu_N)(A_1/D_0) \tag{8.2}$$

where μ_B is the Bohr magneton. Thus, the positive A terms in the CO MVCD give negative molecular g factors. The average for the R branch was found to be -0.267 ± 0.026, and for the P branch we have -0.250 ± 0.019. These values agree favorably with g values determined by microwave and molecular beam magnetic and electronic resonance techniques, thereby providing validation for MVCD as a useful alternative method for determining molecular Zeeman g factors. It was noted, however, that MVCD is based on absorption intensity and therefore will not have the quantitative accuracy and precision of frequency-based methods, but the sign and magnitudes are fairly easy to measure in a single experiment using MVCD.

8.1.5. Rotationally Resolved MVCD for Methane

Measurements of the RR MVCD spectra for the two allowed T_2 normal modes (ν_3 CH(D) stretch and ν_4 CH(D) deformation) for CH_4 and CD_4 gases at a resolution of 0.5 cm^{-1} by using FTIR methods have produced some very interesting results [ref. 90]. Figure 8.3 shows the absorption and MVCD spectra for CH_4; the spectra for CD_4 show very similar band shapes. There is a marked J dependence at higher J values in these spectra, especially for ν_4 in the P branch. This observation was attributed to Coriolis coupling, which complicates the rovibrational bands. The resolved MVCD features have a negative A-term band shape, and an analysis of A_1/D_0 ratios for the individual bands showed that the g value for ν_3 was $+0.291 \pm 0.023$ (average from 15 transitions) for CH_4 and $+0.167 \pm 0.011$ (average from 16 transitions) for CD_4 by Eq. (8.2). The value for ν_3 for CH_4 agrees within experimental uncertainty ($\pm 10\%$) with the ground-state value of $+0.3133$ determined by molecular beam magnetic resonance and laser saturated level-crossing techniques. However, the values found for ν_4 were 0.44 ± 0.03 (average from five lowest transitions of the R branch) for CH_4 and 0.20 ± 0.03 (average from five lowest transitions of the R branch) for CD_4, which were higher (by 40% for CH_4) than the ground state. These higher values were attributed to a significant contribution from vibrational angular momentum to the molecular magnetic moment. Using a model that takes into account the Coriolis coupling, g values were calculated to be even higher for ν_4, but only negligibly higher for ν_3. Furthermore, a self-consistent field (SCF) quantum mechanical evaluation of the vibrational distortion on the g values was found to be consistent with the dominance of the Coriolis contribution to ν_4 [ref. 90]. Thus in the case of methane, a clear indication of a vibrational contribution to the molecular Zeeman effect is obtained from RR MVCD measurements.

8.1.6. Rotationally Resolved MVCD for Acetylene and Deuterated Isotopomers

An FTIR RR MVCD study of Σ symmetry combination and overtone bands of the linear C_2H_2, C_2HD, and C_2D_2 molecules in the gas phase at 0.1-cm^{-1} resolution have also provided evidence of changes in the molecular g values between the ground and excited vibrational states [ref. 91]. The negative A_1/D_0 values for

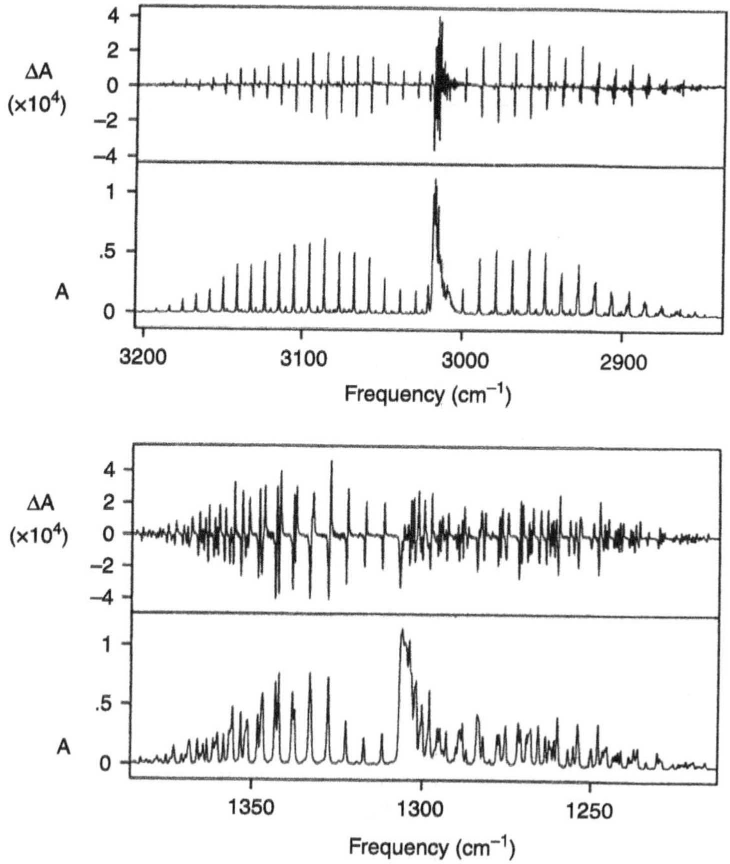

Figure 8.3. Absorption (lower curves) and MVCD (upper curves) for the $\nu_3(T_2)$ CH stretching (top spectra) and the $\nu_4(T_2)$ CH deformation (bottom spectra) bands for CH_4 with a resolution of $0.5\ cm^{-1}$. The spectra were measured by taking the difference between scans at ± 8 T and normalizing to 1 T; each averaged 2048 scans for ν_3 and 1024 scans for ν_4. Reprinted with permission from ref. 90. Wang, B.; Keiderling, T. A. *J. Phys. Chem.* **1994**, *98*, 3957. Copyright 1994 American Chemical Society.

the RR MVCD bands give positive g values for low values of J for $\nu_4+\nu_5$ for C_2H_2, $\nu_4+\nu_5$ and the ν_3 fundamental for C_2HD, and C_2D_2. As an example, the experimental spectra for the $\nu_4+\nu_5$ combination band of C_2H_2 is presented in Figure 8.4. It is apparent that the RR MVCD spectra depend strongly upon the value of J. The analysis of these spectra assumed an effective g value g_{eff} which is given by Eq. (8.3) [analogous to Eq. (8.2) above]. Then if Δg is defined as the

$$g_{eff} = (1/2)(-\mu_B/\mu_N)(A_1/D_0) \tag{8.3}$$

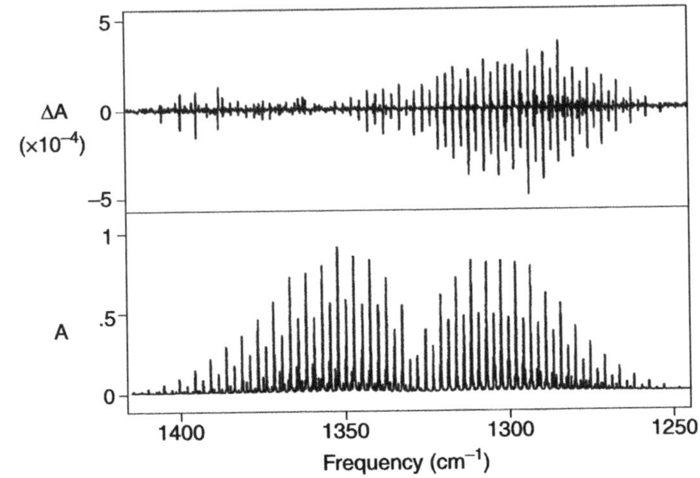

Figure 8.4. Absorption (lower curve) and MVCD (upper curve) for the $\nu_4 + \nu_5$ combination band of C_2H_2 in the gas phase at 0.1-cm^{-1} resolution. The MVCD is the difference of spectra determined at ± 8 T and then normalized to 1 T, each averaged over four blocks of 1024 scans. Reprinted with permission from ref. 91. Tam, C. N.; Bour, P.; Keiderling, T. A. *J. Chem. Phys.* **1996**, *104*, 1813. Copyright 1996 American Institute of Physics.

difference between the excited $g_{v'}$ and ground $g_{v''}$ vibrational states, we obtain

$$\Delta g = g_{v'} - g_{v''} \tag{8.4}$$

g_{eff} is given by the following for the rotational branches of the vibrational spectrum [ref. 91]:

$$g_{\text{eff}} = g_{v'} + \Delta g(J''/2) \quad \text{for } J' = J'' + 1, \text{ the } R \text{ branch} \tag{8.5a}$$

$$g_{\text{eff}} = g_{v'} - \Delta g \quad \text{for } J' = J'', \text{ the } Q \text{ branch} \tag{8.5b}$$

$$g_{\text{eff}} = g_{v'} + \Delta g(J'' + 1)/2 \quad \text{for } J' = J'' - 1, \text{ the } P \text{ branch} \tag{8.5c}$$

From these equations it is seen that if the g values of the ground and excited vibrational states are different (nonzero Δg), then the P and R branches will be linearly dependent upon the J'' value with a slope of $\Delta g/2$. This, in fact, is observed for the $\nu_4 + \nu_5$ bending mode combination band for C_2H_2. Figure 8.5 shows a plot of g_{eff} versus $J'' + 1$ and J'' for the P and R branches of C_2H_2, respectively. The g_J vlaue for the $\nu_4 + \nu_5$ combination band for C_2H_2 was found to be $+0.0535 \pm 0.0033$ and $\Delta g(\nu_4 + \nu_5) = -0.0061 \pm 0.0004$. This analysis predicts a decrease in the value and then a change of sign for the A_1/D_0 values at high J'' for the R branch, just as observed in Figure 8.4. The spectra for the $\nu_4 + \nu_5$ combination band for C_2H_2 were simulated using a $g_{v'}$ value $= 0.86 g_{v''}$ with a high degree of success. The change in g values was interpreted in terms of off-axis deformation of the linear C_2H_2 rotation.

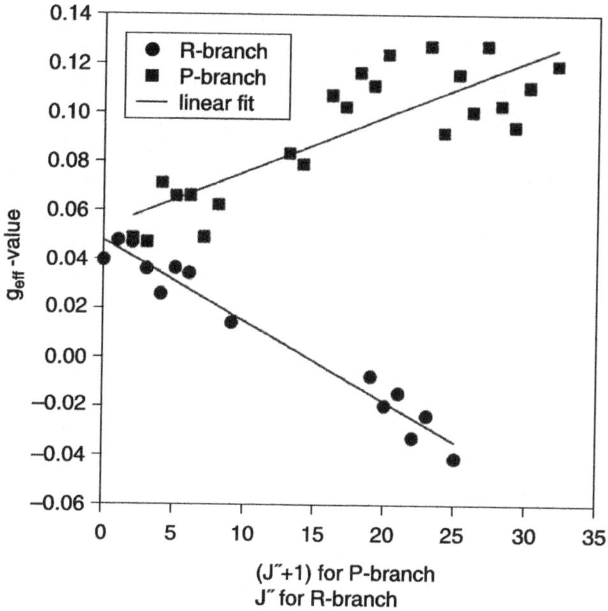

Figure 8.5. Plot of g_{eff} versus $J'' + 1$ for the P branch and J'' for the R branch of the $\nu_4 + \nu_5$ combination band for C_2H_2. Reprinted with permission from ref. 91. Tam, C. N.; Bour, P.; Keiderling, T. A. *J. Chem. Phys.* **1996**, *104*, 1813. Copyright 1996 American Institute of Physics.

8.2. XMCD

MCD measurements in the X-ray region involve excitation of core electrons to empty or partly filled valence orbitals. For example, the *L*-edge absorptions in transition metals involve excitation of the spin–orbit split 2*p* core electrons to the valence 3*d* or 4*p* levels. Interpretation of the MCD of these transitions requires experimental measurements that occur typically at energies of hundreds of electron volts (eV) to several thousand electron volts (keV). An interesting feature of XMCD is that it provides an element-specific spectrum that is a measure of orbital and spin angular momentum and the related magnetic properties which result from such momenta. There are two reviews of XMCD that are worthy of mention, especially for chemists [refs. 92, 93].

8.2.1. XMCD Measurements

XMCD measurements are more challenging than vis–UV or IR MCD experiments (see ref. 92 for a more detailed discussion). A powerful X-ray source is required, which usually means a synchrotron facility. Simpler, laboratory X-ray sources such as X-ray tubes or radioactive nuclei are simply not bright enough to give respectable signal-to-noise ratio. The technology available at a given synchrotron lab will often determine how the XMCD experiment is performed. There are a number of beam configurations and insertion devices that will produce a monochromatic linearly or circularly polarized X-ray beam, and some facilities have more than one type. The most popular approach seems to be to modulate the particle beam with magnets in one of several configurations in order to produce polarized X-radiation in a beamline port. These include the asymmetric wiggler, helical undulator, elliptical wiggler, and elliptically polarizing undulator [ref. 92]. Hard (high-energy) X-ray beamlines often use crystal optics in which the reflection of the beam at the Brewster angle can produce a strong linear polarization. The beam is then converted to circular polarization by means of a quarter-wave plate that exploits the birefringence of a crystal (diamond, Be, or LiF, for example). Detection methods also present special concerns, depending upon the energy region of interest and the type of sample. These include transmission photocurrents (metallic or conducting samples), electron yield measurements (electron multiplier), and fluorescence yield (X-ray fluorescence from the sample of interest). The simplest is probably the channeltron electron multiplier, which provides a photocurrent much the same way as a photomultiplier tube in the vis–UV. However, any method that involves moving electrons to establish a signal current will be affected by the magnetic field of the magnet needed for the MCD effect. Care in the placement of the devices is required.

8.2.2. Sum Rules

A particularly convenient method for the determination of orbital and spin angular momentum and the corresponding magnetic moments from XMCD and XAS (X-ray absorption) spectra has been developed in the form of Sum Rules [refs. 94, 95]. These relations are related to intensities of the XMCD and XAS spectra and are evaluated from integrated areas under the XMCD and XAS curves as illustrated in Figure 8.6. For example, in the case of L-edge ($2p \rightarrow 3d$) transitions for transition metals, the expressions are given in Eq. (8.6), where $\langle L_z \rangle$ and $\langle S_z \rangle$ are expectation values, A represents the area under the L_3 XMCD curve, B the area under the L_2 XMCD curve, C the area of the total XAS absorption curve(s), and n_h the number of holes in the $3d$ level; μ_L and μ_S are orbital and spin magnetic moments, respectively, and μ_B is the Bohr magneton:

$$\langle L_z \rangle = 2(A + B)n_h/(3C) \qquad \mu_L = -\mu_B \langle L_z \rangle \tag{8.6a}$$

$$\langle S_z \rangle = (A - 2B)n_h/(2C) \qquad \mu_S = -g\mu_B \langle S_z \rangle \tag{8.6b}$$

These useful rules allow angular momentum properties and magnetic moments to be determined from integrated spectra without the necessity of laborious spectral simulations.

8.2.3. Magnetic Properties of $GdNi_2$ Laves Phase

The element-specific aspect of XMCD has been useful in the investigation of magnetic properties of alloys and intermetallic compounds. As an example, when

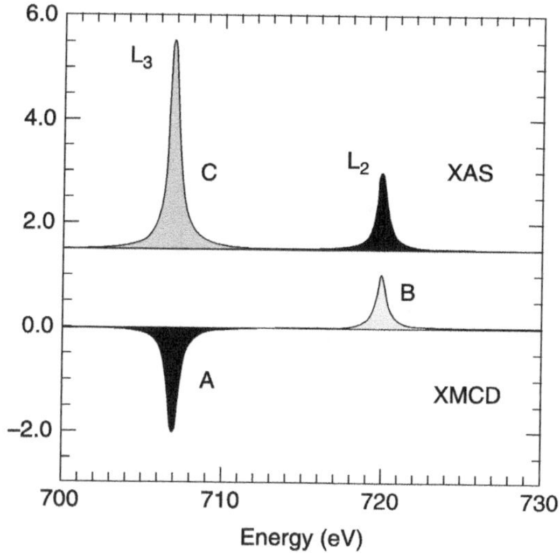

Figure 8.6. Illustration of the band areas used for the Sum Rules of Eq. (8.6). The parameters *A*, *B*, and *C* represent integrated areas under the XAS (upper curves) and XMCD (lower curves) for the L_2 and L_3 transitions for a transition metal. Adapted and reprinted from ref. 92. Funk, T.; Deb, A.; George, S. J.; Wang, H.; Cramer, S. P. *Coord. Chem. Rev.* **2005**, *249*, 3. Copyright 2005, with permission from Elsevier.

Ni is mixed with lanthanide metals, the magnetic properties of Ni tend to decrease in proportion to the relative amount of the lanthanide present. The intermetallic compound $GdNi_2$, however, seems to retain magnetic properties, even up to the Laves phase concentration. Figure 8.7 shows the Ni XMCD and XAS L_2 and L_3 spectra [ref. 96], which, from the Sum Rules of Eq. (8.6), provide evidence for a magnetic moment for Ni in these compounds. The total magnetic moment μ_T was found to be 0.200 μ_B, with orbital and spin components of $\mu_L = 0.060\ \mu_B$ and $\mu_S = 0.140\ \mu_B$, respectively. Thus the 3*d* band of $GdNi_2$ is not completely occupied. The weaker bands in the Ni spectra (*B*, *C*, *E*, and *F* in the XAS, together with B', C', E', and F' in the XMCD) were attributed to eigenstates of $2p^5 3d^9$, possibly from ligand field t_{2g} configurations. Figure 8.8 shows the Gd XMCD and XAS spectra for the M_5 and M_4 $3d \rightarrow 4f$ edges, split by spin–orbit interaction of the 3*d* core hole. The inset shows simulated spectra assuming the Gd 4*f* state is a 7/2 spin state and therefore essentially Gd^{3+}. The opposite signs for the Ni L_3 and L_2 XMCD spectra (positive and negative, respectively) compared to the Gd M_5 and M_4 XMCD spectra (negative and positive, respectively) were interpreted as evidence that the magnetic interaction between Ni and Gd is antiferromagnetic.

8.2.4. XMCD Study of Re 5*d* Magnetism in the Sr_2CrReO_6 Double Perovskite

The ferrimagnetic double perovskite Sr_2CrReO_6 was determined to have an unusually high Curie temperature $T_C = 635$ K. In measurements on related systems,

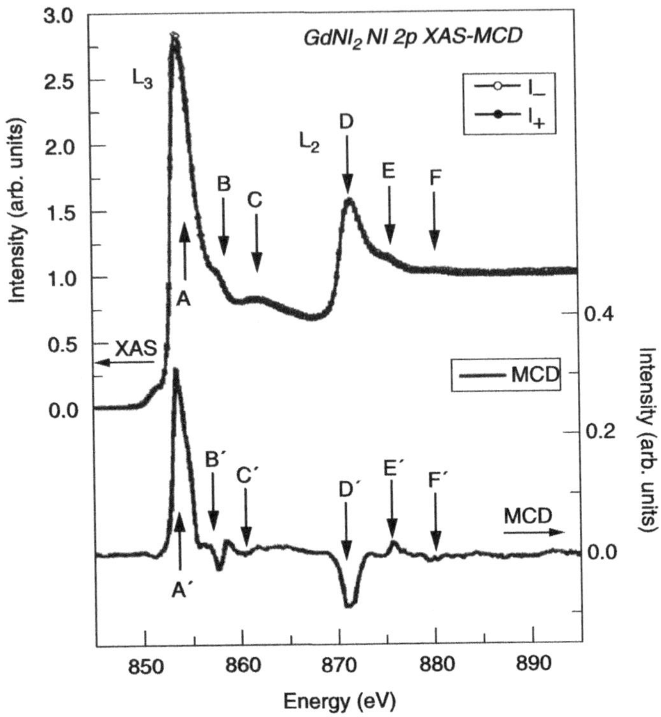

Figure 8.7. Ni L_3 and L_2 ($2p \rightarrow 3d$) XAS (upper curves) and XMCD (lower curve) for $GdNi_2$ at 25 K. Reprinted with permission from ref. 96. Mizumaki, M.; Yano, Y.; Umehara, I.; Ishikawa, F.; Sato, K.; Koizumi, A.; Sakai, N.; Muro, T. *Phys. Rev. B* **2003**, *67*, 132404. Copyright 2003 American Physical Society.

the T_C seems to scale with the magnitude of the spin magnetic moments. In order to investigate the magnetic moment for Sr_2CrReO_6, an XMCD investigation was undertaken [ref. 97], partially motivated by the element-specific nature of the measurement and the prospects of extracting spin and orbital magnetic moments from the experimental spectra by means of the sum rules (see above). Some band structure calculations had predicted the Re magnetic moment to be 0.7–0.85 μ_B, but an experimental neutron scattering experiment had indicated only −0.21 μ_B at 5 K. Therefore the XMCD for both the Re transitions were investigated. Polycrystalline bulk samples of Sr_2CrReO_6 were prepared and checked by macroscopic superconducting quantum interference device (SQUID) magnetization measurements at 5 K and confirmed to have $T_C = 635$ K and a saturation magnetization of 0.89 μ_B per formula unit. The magnetization versus field was also measured by means of the XMCD L_2 signal for Re (see below) and exhibited hysteresis as shown in Figure 8.9, which was consistent with the magnetization hysteresis found in the SQUID measurements. The observed value of the saturation magnetization of 0.89 μ_B per formula unit is quite close to 1 μ_B predicted for antiferromagnetically coupled Cr^{3+} ($S = 3/2$) and Re^{5+} ($S = 5/2$) ions.

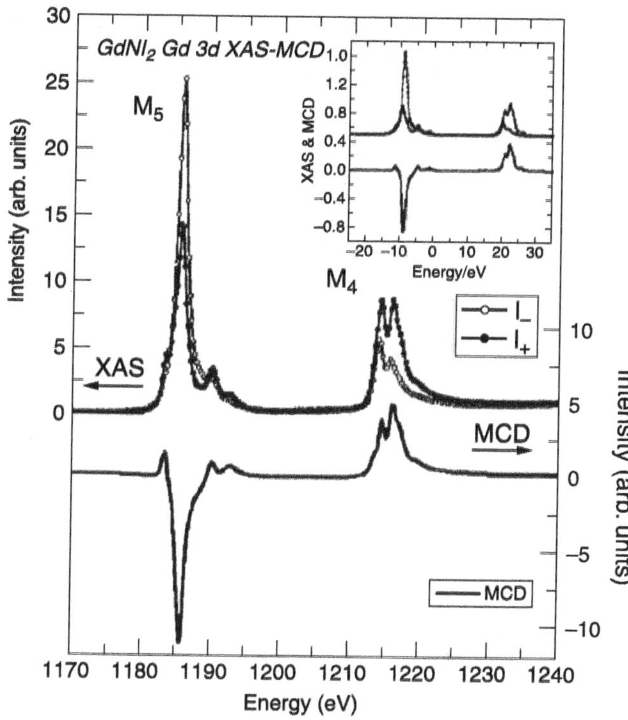

Figure 8.8. Gd M_5 and M_4 ($3d \rightarrow 4f$) XAS (upper curves) and XMCD (lower curve) for $GdNi_2$ at 25 K. The inset shows the simulated spectra. Reprinted with permission from ref. 96. Mizumaki, M.; Yano, Y.; Umehara, I.; Ishikawa, F.; Sato, K.; Koizumi, A.; Sakai, N.; Muro, T. *Phys. Rev. B* **2003**, *67*, 132404. Copyright 2003 American Physical Society.

The X-ray absorption (near-edge absorption, XANES) and XMCD spectra are shown in Figure 8.10. Using the sum rules and data normalized to the number of 5*d* holes in the Re shell ($n_h = 5.3$), the spin and orbital moments, m_S and m_L, were determined to be $-0.68\ \mu_B$ and $+0.25\ \mu_B$ per Re, respectively. These values were larger than calculations that include simple spin–orbit coupling; the larger values were attributed to the importance of relativistic effects due to the heavy Re atom. The moments were also larger than those found for related double perovskite compounds that had lower T_C values, such as Sr_2CrWO_6 ($T_C = 460$ K) and Sr_2FeReO_6 ($T_C = 400$ K). Thus the high T_C observed for Sr_2CrReO_6 parallels the observation of larger m_S moments from the XMCD spectra. A qualitative explanation was given in terms of ferrimagnetism mediated by itinerant minority spin carriers (electrons). It is not clear yet as to whether the large moments are a result of induced moments by pd hybridization with the 3*d* orbitals of Cr or if the moments are intrinsic to the 5*d* orbitals of Re [ref. 97].

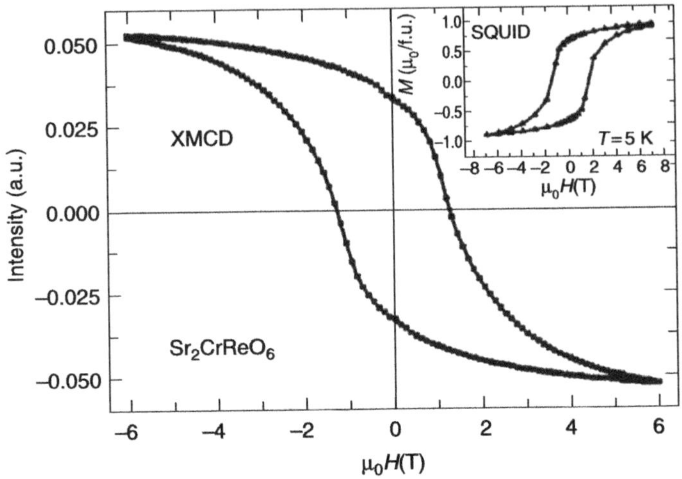

Figure 8.9. Plot of XMCD signal intensity versus field for the Re L_2 edge at 10 K. The inset shows the magnetization versus field for the sample measured by SQUID at 5 K. Reprinted with permission from ref. 97. Majewski, P.; Geprags, S.; Sanganas, O.; Opel, M.; Gross, R.; Wihelm, F.; Rogalev, A.; Alff, L. *Appl. Phys. Letters* **2005**, *87*, 202503. Copyright 2005 American Institute of Physics.

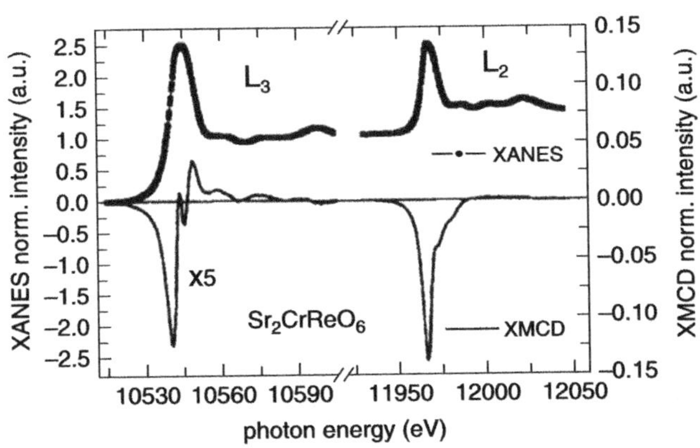

Figure 8.10. X-ray absorption (XANES, upper curves) and XMCD (lower curves) spectra for the Re L_2 (right curves) and L_3 (left curves) edge for Sr_2CrReO_6 at 10 K and 6 T. The L_3 XMCD (left lower curve) was multiplied by 5 before plotting. Reprinted with permission from ref. 97. Majewski, P.; Geprags, S.; Sanganas, O.; Opel, M.; Gross, R.; Wihelm, F.; Rogalev, A.; Alff, L. *Appl. Phys. Letters* **2005**, *87*, 202503. Copyright 2005 American Institute of Physics.

8.2.5. XMCD Study of Mn(III) and Mn(IV) Magnetic Contributions to the SMM $[Mn_{12}O_{12}(CH_3CO_2)_{16}(H_2O)_{24}] \cdot 2CH_3COOH \cdot 4H_2O$

The SMM $Mn_{12}Ac$ studied by MCD C terms (Chapter 7, Section 7.10.1) has also been investigated by XMCD in order to determine the magnetic contributions

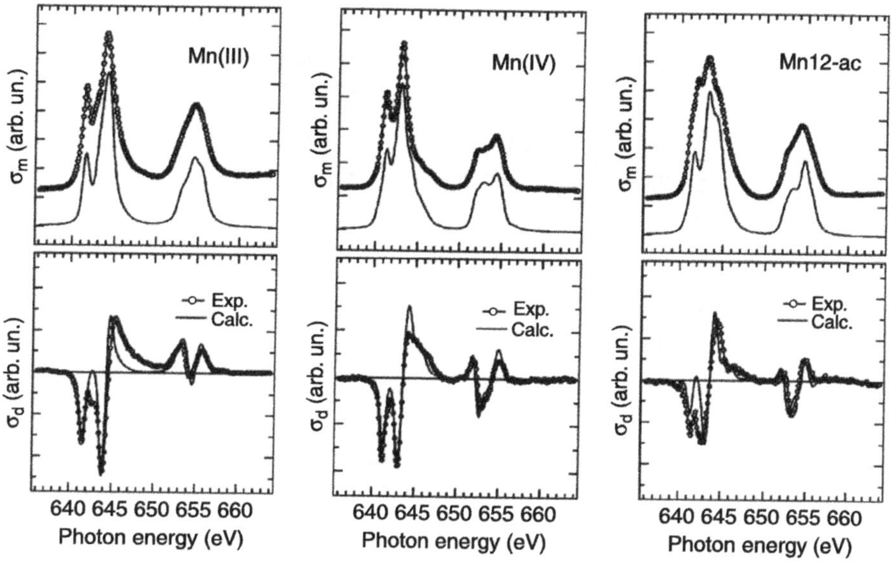

Figure 8.11. X-ray absorption (upper curves) and XMCD (lower curves) L_2 and L_3 spectra for the Mn(III) and Mn(IV) model complexes and the $Mn_{12}Ac$ cluster complex at 1.5 K and 4 T. The solid lines represent the calculated fit to each spectrum. Reprinted with permission from ref. 98. Moroni, R.; dit Moulin, C. C.; Champion, G.; Arrio, M.-A.; Sainctavit, P.; Verdaguer, M.; Gatteshi, D. *Phys. Rev. B* **2003**, *68*, 064407. Copyright 2003 American Physical Society.

from the Mn(III) and Mn(IV) centers in the cluster complex [ref. 98]. XMCD, however, cannot separate the contributions from the Mn ions in different oxidation states effectively, and simple sum rule applications could not be used. Therefore, XMCD spectra were obtained for model Mn(III) and Mn(IV) complexes, their magnetic properties were extracted, and then the properties of the cluster were simulated to provide a fit to the observed XMCD spectra for the cluster. The model Mn(III) complex investigated was $Mn(dbm)_3$, where Hdbm = 1,3,-diphenyl-1,3-propanedione. The complex has a tetragonal structure with some deviations from an idealized D_{4h} symmetry due to Jahn–Teller distortion. The Mn(IV) complex was $[L_2Mn_2(O)(PhBO_2)](ClO_4)_2{\cdot}2(CH_3)_2CO$, where $L =$ 1,4,7-trimethyl-1,4,7-triazacyclononane. The Mn(IV) centers in this complex reside at equivalent octahedral sites and magnetic susceptibility measurements are consistent with two Mn(IV) ions of spin $S = 3/2$ and near-zero ZFS. The XMCD spectra for Mn L_2 ($2p_{1/2} \rightarrow 3d$) and L_3 ($2p_{3/2} \rightarrow 3d$) edges for the Mn(III) and Mn(IV) model complexes and the $Mn_{12}Ac$ cluster complex are shown in Figure 8.11, together with the calculated fit (see ref. 98 for details). From the parameters required for the experimental fit, the magnetic moments for the Mn(III) and Mn(IV) ions in the $Mn_{12}Ac$ cluster complex were determined to be $+3.81\ \mu_B$ and $-2.96\ \mu_B$, respectively. These values extracted from the XMCD spectral results were in good agreement with those found by means of polarized neutron

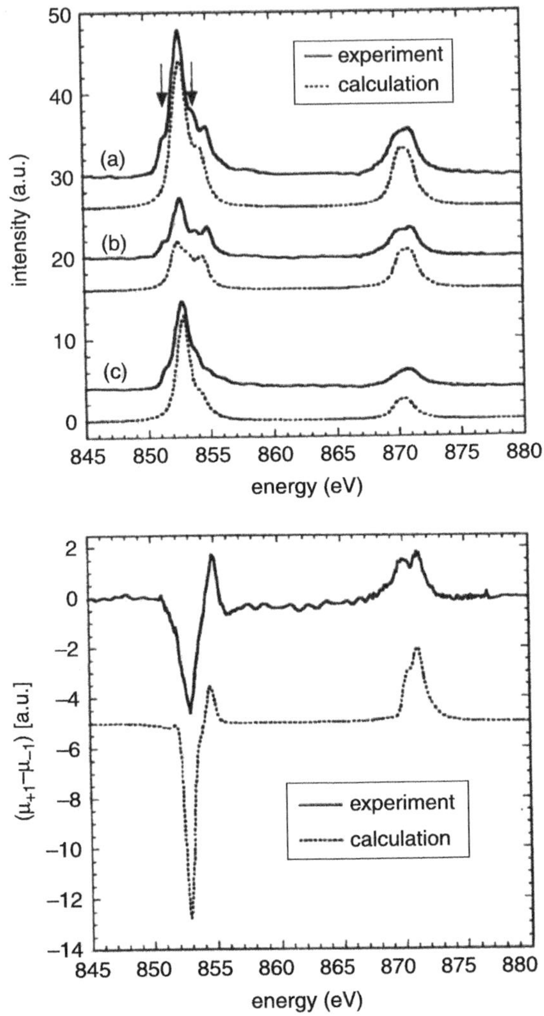

Figure 8.12. Ni *L*-edge XAS (upper curves: a, sum; b, lcp; c, rcp) and 6-T XMCD (lower curves) for thin films of NiAz on a sapphire disk at 2.2 K. In both cases, simulated spectra are presented as dotted curves. Reprinted with permission from ref. 99. Funk, T.; Kennepohl, P.; Di Bilo, A. J.; Wehbi, W. A.; Young, A. T.; Friedrich, S.; Arenholz, E.; Gray, H. B.; Cramer, S. P. *J. Amer. Chem. Soc.* **2004**, *126*, 5859. Copyright 2004 American Chemical Society.

diffraction. The orbital contributions to the moments were found to be negligibly small so that the spin moments in both cases dominate [ref. 98]. Thus the ferrimagnetic coupling between the Mn(III) and Mn(IV) in the $Mn_{12}Ac$ structure was confirmed by the XMCD study.

8.2.6. XMCD for *Pseudomonas aeruginosa* Nickel(II) Azurin (NiAz)

XMCD measurements have been used to investigate the oxidation state and magnetic properties of a small metalloprotein containing Ni(II) recently [ref. 99]. The NiAz structure consists of the Ni trigonally coordinated by side-chain donors Cys112, His46, and His117 at the end of a β barrel, with Met121 thioether and Gly45 peptide carbonyl serving as weak axial ligands. Figure 8.12 shows the XAS and XMCD spectra for NiAz films deposited on sapphire plates. The oxidation state of the Ni in NiAz was determined by the branching ratio of the L-edge spectra defined as the integrated L_3-line intensity at 853.1 eV normalized by the integrated L_3- and L_2-line intensities. The value of $L_3/(L_3 + L_2) = 0.722(4)$ for NiAz was in the range found for a variety of high-spin ($S = 1$) Ni(II) compounds ($L_3/(L_3 + L_2) = 0.71–0.77$). By means of the Sum Rules [Eq. (8.6)], the spin, orbital, and total magnetic moments were determined to be 1.7(3) μ_B, 0.18(7) μ_B, and 1.9 μ_B, respectively, indicating strong covalent bonding to the Ni(II) center. Some density functional theory (DFT) calculations indicate that the $3d_{x^2-y^2}$ and $3d_{z^2}$ provide the two holes in the high-spin Ni(II) ground state, which is in accord with the orbital magnetic moment from XMCD. This example shows that XMCD can serve as a useful probe of the electronic structure of the Ni center in the NiAz metalloprotein.

9 Magnetic Linear Dichroism Spectroscopy

9.1. INTRODUCTION

Magnetic linear dichroism (MLD) spectroscopy [refs. 100–104] is a magneto-optical technique that is closely related to MCD in a complementary way. MLD is based upon the differential absorption of linearly polarized light oriented parallel ($A_{\parallel}$) and perpendicular ($A_{\perp}$) to the direction of an external magnetic field that is *perpendicular* to the direction of light propagation through a sample of interest. The MLD difference absorption (dichroism) is defined as $\Delta A_{\mathrm{MLD}} = A_{\parallel} - A_{\perp}$. For MCD [refs. 1, 105], the dichroism $\Delta A_{\mathrm{MCD}} = A_{\mathrm{lcp}} - A_{\mathrm{rcp}}$ is the differential absorption of left circularly polarized light (lcp) and right circularly polarized light (rcp), respectively, by a sample in a magnetic field that is *parallel* to the direction of light propagation. Thus the difference between the two measurements is the orientation of the magnetic field with respect to the direction of light propagation through the sample of interest: MCD involves a longitudinal orientation of the field, while MLD involves a transverse orientation. They both have their origin in the Zeeman effect: the perturbation of states by an external magnetic field. However, because of the field orientation, the selection rules for optical transitions will be different. While MCD has been developed to a considerable extent and has been used extensively in the investigation of electronic spectra and to probe the properties of electronic states [refs. 1, 105], MLD has not been as widely used for several reasons. The chief reason is that the MLD signal strength is typically an order of magnitude smaller than signals for MCD measurements. Recently, however, several studies have shown that measurements are indeed feasible in some cases and that they produce results that complement those of parallel MCD measurements [refs. 100–104].

9.2. MLD TERMS AND TERM PARAMETERS

The theory for MLD spectroscopy has been developed along the same lines as for MCD [refs. 103, 104]. In particular, the Born–Oppenheimer/Franck–Condon-Rigid Shift (BO-FC-RS) approximations were used to develop a description of MLD

A Practical Guide to Magnetic Circular Dichroism Spectroscopy, by W. Roy Mason

absorption that is parallel to that described for MCD. The perturbation and expansion approximations are similar to those described for MCD and assume the Zeeman energies $-\mu_z B$ are $\ll$ the bandwidth and kT; the expansions were then taken to second order in the field. The expression for $\Delta A/E$ for MLD, however, requires six parameters or terms (compared to three for MCD) as shown in Eq. (9.1):

$$\Delta A/E = \gamma\mu_B B^2\{A_2[(1/2)\partial^2 f(E)/\partial E^2] + [B_1 + C_1/(kT)](-\partial f(E)/\partial E) + [E_0 + F_0/(kT) + G_0/(kT)^2]\}f(E)\} \tag{9.1}$$

where γ is a collection of constants $= 2N_o\pi^3\alpha^2 c\,\ell(\log_{10}e)/(250hcn)$ which include N_0 = Avogadro's number, α = dielectric of the medium, n = index of refraction, c = velocity of light, h = Planck's constant, c = molar concentration, ℓ = path length, μ_B = Bohr magneton, B = magnetic field, $E = h\nu$, and $f(E)$ is the absorption lineshape function. Equation (9.1) shows that MLD depends upon the square of the field. Terms linear in the field, as for MCD, are found to be zero in the approximation. The field dependence represents an important difference between the MCD and MLD measurement. Expressions for the term parameters A_2, B_1, C_1, E_0, F_0, and G_0 are given explicitly in ref. 104; the subscripts, like those of the MCD term parameters, refer to the order of their respective spectral moments. Like those for MCD, the parameters (MLD terms) are dependent upon the nature of the electronic states involved in the transition. The A_2 and B_1 parameters require ground- or excited-state degeneracy, similar to the MCD A term. The C_1, F_0, and G_0 parameters require degeneracy in the ground state and are temperature-dependent, like the MCD C term. The E_0 parameters are present for all transitions as a result of field-induced intermixing of the states involved in the transitions with all other states of the system, analogous to MCD B terms. Equation (9.1) also predicts that the MLD A term should appear as a second derivative of the lineshape function (Figure 9.1). The B and C terms should be first derivatives, similar to the MCD A

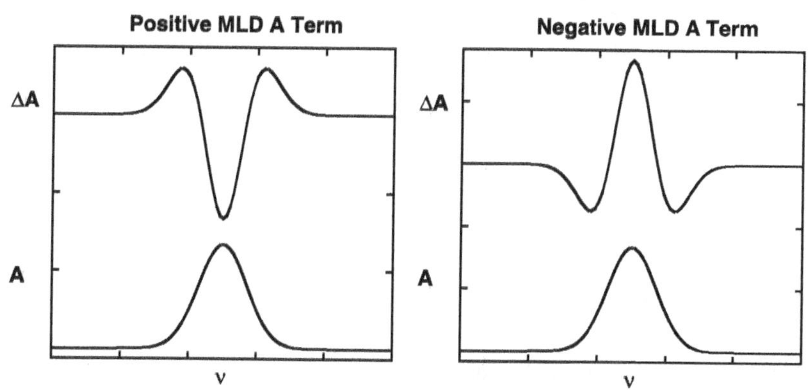

Figure 9.1. Positive and negative MLD A terms.

term (see Chapter 3, Figure 3.4), and the E, F, and G terms should follow the line-shape function itself. The relative magnitude of the A–G MLD terms is given approximately as [ref. 104]

$$A:B:C:E:F:G \approx 5/\Gamma^2 : 1/(\Delta W\Gamma) : 1/(kT\Gamma) : 1/(\Delta W)^2 : 1/(kT\Delta W) : 1/(kT)^2$$

where Γ = bandwidth at half-maximum, ΔW = energy difference between intermixing states, and kT = Boltzmann's constant times Kelvin temperature. With typical values of $kT = 200\ \text{cm}^{-1}$, $\Gamma = 10^3\ \text{cm}^{-1}$, and $\Delta W = 10^4\ \text{cm}^{-1}$, the ratio of magnitudes are [ref. 104]

$$\text{A:B:C:E:F:G} = 500 : 10 : 500 : 1 : 50 : 2500$$

Thus for a diamagnetic ground state and a degenerate excited state, the A term dominates, but if the ground state is degenerate, then the G term is likely to be the largest, especially at low temperature. The E term is small for most cases except when ΔW is $\leq 10^2\ \text{cm}^{-1}$.

As an example, consider the atomic $s^2 \rightarrow sp$ transition discussed earlier in Chapter 3 to illustrate the MCD A term; the ground state is nondegenerate and the excited state is threefold degenerate. The allowed Zeeman transitions are shown in Figure 9.2 for both MCD and MLD in the presence of the respective oriented applied field. Two transitions are expected for the MCD measurement which lead to an A term as discussed in Chapter 3, while three transitions are possible for the MLD measurement. In this case the diamagnetic ground state dictates that the MLD parameters will be $C_1 = F_0 = G_0 = 0$. Furthermore, if

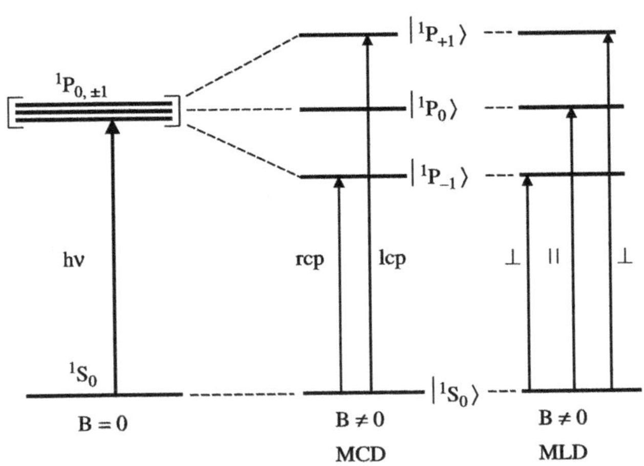

Figure 9.2. Zeeman energy levels for the $^1S_0$ (s^2) $\rightarrow$ $^1P_{1,0,-1}$ (sp) showing the allowed MCD and MLD transitions upon application of a longitudinal- or transverse-oriented external magnetic field, respectively.

field-induced mixing with other states is small by virtue of large energy differences between the other states and either of the S_0 or $P_{0,\pm 1}$ states, then B_1 and E_0 also will be ≈ 0. Therefore the MLD A_2 parameter will be the only one of significance so that from Eq. (9.1) the MLD A term will be given by

$$\Delta A/E = \gamma B^2 A_2 (1/2)\partial^2 f(E)/\partial E^2$$

The spectrum will thus appear as a *second* derivative of the lineshape function. Similarly, for the MCD, the A term will be the only term of significance since C terms will be zero and B terms will be small, and as shown in Chapter 3, the MCD A term will appear as a *first* derivative of the line shape function. The expression for the MLD A_2 parameter for a transition $A \rightarrow J$ is given in Eq. (9.2) [ref. 104],

$$\begin{aligned} A_2 = (1/d_A) \sum\nolimits_{\alpha\beta} (|\langle A\alpha|m_{\|}|J_{\beta}\rangle|^2 - |\langle A_{\alpha}|m_{\perp}|J_{\beta}\rangle|^2) \\ \times (\langle J_B|\mu_z|J_{\beta}\rangle - \langle A_{\alpha}|\mu_z|A_{\alpha}\rangle)^2 \end{aligned} \tag{9.2}$$

where the light propagation is assumed to be in the $+y$ direction and the transverse magnetic field is oriented in the $+z$ direction. In Eq. (9.2), d_A = degeneracy of the ground-state A, $m_{\|} = m_z$, $m_{\perp} = m_x = (1/\sqrt{2})(m_+ + m_-)$, and $\mu_z = -\mu_B(L_z + 2S_z)$. When applied to the $^1S_0(s^2) \rightarrow {}^1P_{1,0,-1}$ (sp) transition, where $d_A = 1$, $\alpha = 0$, and $g = +1$, the A_2 parameter is given by

$$\begin{aligned} A_2 &= \sum\nolimits_{\beta=0,\,\pm 1} \{|\langle ^1S_0|m_z|^1P_{\beta}\rangle|^2 - [(1/2)|\langle ^1S_0|m_+|^1P_{\beta}\rangle|^2 \\ &\quad + (1/2)|\langle ^1S_0|m_-|^1P_{\beta}\rangle|^2]\}^2 (\langle ^1P_{\beta}|\mu_z|^1P_{\beta}\rangle - \langle ^1S_0|\mu_z|^1S_0\rangle)^2 \\ &= -(1/2)|\langle ^1S_0|m_+|^1P_{-1}\rangle|^2(\langle ^1P_{-1}|\mu_z|^1P_{-1}\rangle)^2 + |\langle ^1S_0|m_z|^1P_0\rangle|^2 \\ &\quad \times (|\langle ^1P_0|\mu_z|^1P_0\rangle)^2 - (1/2)|\langle ^1S_0|m_-|^1P_{+1}\rangle|^2(\langle ^1P_{+1}|\mu_z|^1P_{+1}\rangle)^2 \\ &= -(1/2)|\langle ^1S_0|m_+\rangle ^1P_{-1}\rangle|^2[(-1)(-\mu_B)]^2 + 0 - (1/2)|\langle ^1S_0|m_-|^1P_{+1}\rangle|^2 \\ &\quad \times [(+1)(-\mu_B)]^2 \\ &= -|\langle ^1S_0|m_{\pm}|^1P_{\mp 1}\rangle|^2\mu_B^2 \end{aligned}$$

Therefore

$$\Delta A/E = -\gamma B^2[|\langle ^1S_0|m_{\pm}|^1P_{\mp 1}\rangle|^2\mu_B^2](1/2)\partial^2 f(E)/\partial E^2$$

and thus the MLD should consist of a negative second derivative of the lineshape function and therefore a negative MLD A term as shown in Figure 9.1.

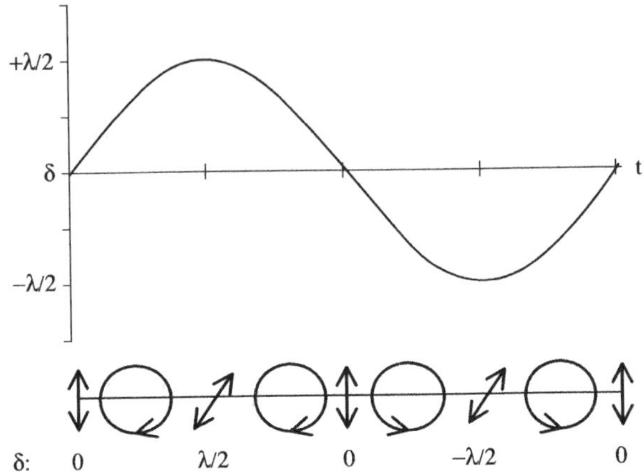

Figure 9.3. Polarization of light emergent from the PEM over the course of the sine-wave modulation. Two cycles of linear polarization ($\pm\lambda/2$) are produced during a single period.

9.3. MEASUREMENT OF MLD SPECTRA

The MLD experimental setup is essentially the same as for MCD as described in Chapter 4. The field is oriented at right angles to the direction of light propagation, but otherwise the methodology is similar. In order to produce a modulated light beam that provides linearly polarized light alternately parallel and perpendicular to the transverse field, the PEM axis and the input plane polarized light must be at a 45° angle to the transverse field. The PEM amplitude is then adjusted to give half-wave ($\lambda/2$) retardation. A phase shift by $\pm\lambda/2$ has the effect of rotating the emergent light polarization by $\pm 45°$, thereby giving linear polarization along the $\parallel$ and $\perp$ directions required. This is illustrated in Figure 9.3, where the polarization of the light emergent from the PEM is shown for a single sine-wave modulation frequency period. Note that linear polarization ($\pm\lambda/2$) occurs at twice the modulation frequency of the PEM, and therefore the detection frequency by the lock-in amplifier must be tuned to twice that used for circularly polarized ($\pm\lambda/4$) modulation for MCD as described in Chapter 4. The lock-in amplifier gives a dc signal proportional to $A_{\parallel} - A_{\perp}$, provided that the dc signal from PMT 1 is held constant, just as for MCD measurements.

9.4. SOME EXAMPLES OF MLD SPECTRA

The number of examples of MLD spectra in the literature is considerably more limited than for MCD. In this section a few such examples will be presented in an illustrative way. In most cases, conditions are set so that only one of the six MLD terms dominates the spectrum, which allows for interpretation within the theory. Often, MCD measurements are obtained at the same time and under the same conditions for comparison.

9.4.1. Atomic Mercury Vapor

The MCD spectra for atomic Hg vapor were presented in Chapter 6 (Section 6.2, Figure 6.1), and showed clearly the two Zeeman transitions for the excited $|P\pm_1\rangle$ energy levels in the $^1S_0(6s^2) \rightarrow {}^3P_1(6s6p)$ 253.65-nm absorption, which are allowed by the strong Hg spin–orbit coupling with the $^1S_0(6s^2) \rightarrow {}^1P_1(6s6p)$ transition that occurs at shorter wavelength (184.957 nm). The analogous MLD measurement (Figure 9.4, [ref. 106]) for the 253.65-nm band shows the three

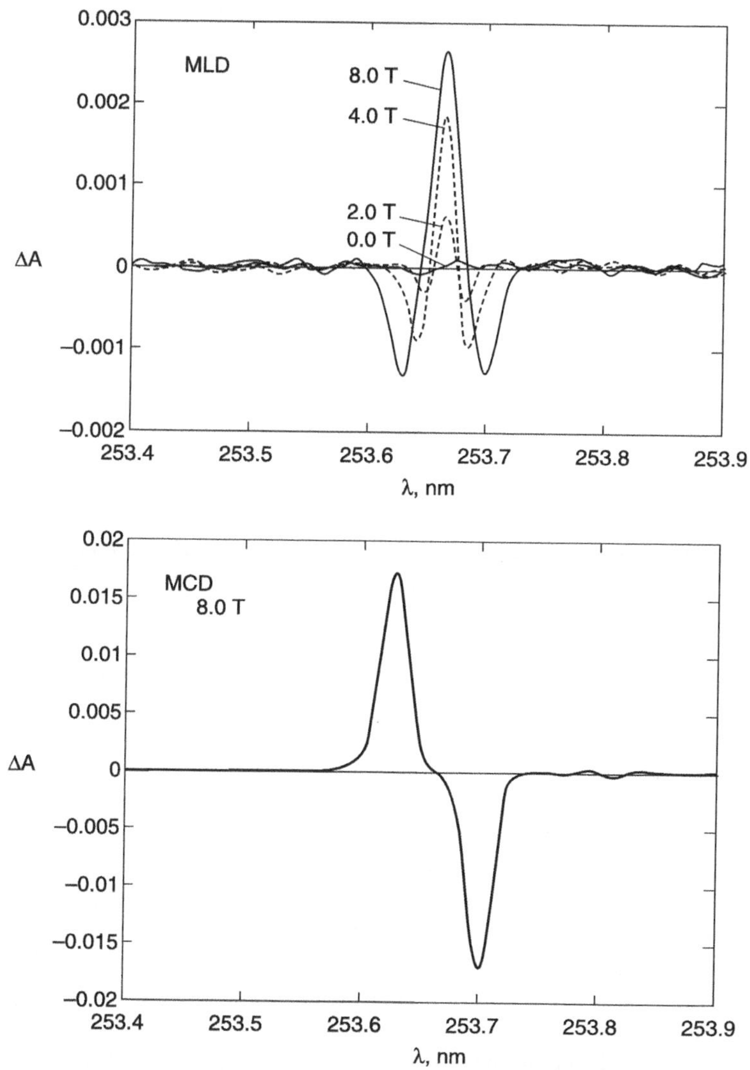

Figure 9.4. MLD at 0.0, 2.0, 4.0, and 8.0 T (top curves) and MCD at 8.0 T (bottom curve) for Hg vapor in air in a 1-cm path at 295 K. (Adapted from ref. 106.)

allowed MLD transitions predicted for the Zeeman sublevels $|P_{0,\pm 1}\rangle$ (see Figure 9.2, but with spin multiplicities removed because of the spin–orbit intermixing of the "triplet" and "singlet" $|P_{0,\pm 1}\rangle$ states as explained in Chapter 6, Section 6.2), Figure 9.4 includes MLD spectra at 8.0, 4.0, and 2.0 T, along with MCD spectra at 8.0 T for comparison. Note the nearly 10-fold smaller MLD signal compared to the MCD signal at 8.0 T. The MCD spectra show that at low field the unresolved Zeeman transitions give the appearance of a positive A term, while at higher field, as in Figure 9.4, the appearance is that of two oppositely signed B terms and shows the resolved transitions to the Zeeman sublevels. The observed MCD and MLD spectra are in accord with the selection rule expectations shown in Figure 9.2, apart from the spin multiplicity labels. At fields where the Zeeman sublevel energy splitting is greater than the absorption bandwidths, the transition probabilities in an electric dipole approximation can be written for ΔA_{MCD} and ΔA_{MLD} in Eqs. (9.3) and (9.4). The light

$$\Delta A_{\text{MCD}}/E = (A_{\text{lcp}} - A_{\text{rcp}})/E = \gamma(N_S/N)\{(|\langle S_0|m_-|P_{+1}\rangle|^2)f(E - E_{+1}) - (|\langle S_0|m_+|P_{-1}\rangle|^2)f(E - E_{-1})\} \tag{9.3}$$

$$\Delta A_{\text{MLD}}/E = (A_\parallel - A_\perp)/E = \gamma(N_S/N)\{|\langle S_0|m_z|P_0\rangle|^2 f(E - E_0) - (1/2)[|\langle S_0|m_x|P_{+1}\rangle|^2 f(E - E_{+1}) + |\langle S_0|m_x|P_{-1}\rangle|^2 f(E - E_{-1})]\} \tag{9.4}$$

propagation is assumed to be along the z direction for MCD and along the y direction for MLD, with the field along the z direction in both cases. In Eqs. (9.3) and (9.4), $\gamma = 2N_0\,\pi^3\,\alpha^2\,c\,\ell\,(\log_{10}e)/(250hcn)$; $E = h\nu$; m_γ is the electric dipole operator, with + or − subscripts for MCD referring to rcp or lcp light, respectively, and for MLD $m_\parallel = m_z$ and $m_\perp = m_x$; N_S/N = fractional population in the $|S_0\rangle$ ground state; and $f(E - E_\gamma)$ = the absorption lineshape function centered about the respective Zeeman transition energies E_{-1}, E_0, and E_{+1}. Since all of the electric dipole matrix elements in Eqs. (9.3) and (9.4) are squared, and therefore positive, the signs of ΔA_{MCD} and ΔA_{MLD} can be predicted for the levels of Figure 9.2. For MCD spectra, two ΔA_{MCD} bands of equal intensity should be resolved that differ in sign with the lower-energy band (rcp) being negative and the higher-energy band (lcp) positive, while for MLD spectra, three ΔA_{MLD} bands are expected with the center band positive and twice the intensity of the two negative outside bands. Furthermore, the energy difference between the two MCD bands and the two outside MLD bands should be the same for the same applied field strength. Furthermore, the g value for the P_1 state is positive so that the central MLD band is predicted to be positive (negative A_2 term parameter as in the example given above).

The Zeeman energy splitting between the P_{+1} and the P_{-1} levels is given by

$$\Delta E = 2g\mu_B B \tag{9.5}$$

where g is the excited-state g factor, $\mu_B = 0.46680\ \mathrm{cm^{-1}T^{-1}}$ is the Bohr magneton, and B is the magnetic field. The g factor is given by the Landé formula with $J = L = S = 1$ in Eq. (9.6).

$$\begin{aligned} g &= 1 + \frac{J(J+1) + S(S+1) - L(L+1)}{2J(J+1)} \\ &= 1.5 \quad \text{for } J = L = S = 1 \end{aligned} \tag{9.6}$$

Experimentally, from atomic emission spectral data [ref. 107] the g value for the Hg(0) $^3P_1$ ($6s6p$) state was found to be 1.479, close to the value of 1.5 predicted by Eq. (9.6); the experimental g value thus gives 11.05 $\mathrm{cm^{-1}}$ at 8.00 T for ΔE from Eq. (9.5). From the experimental data, the ΔE values from the two outside negative MLD transitions measured at 8.00 T (where the spectra are substantially resolved) and the two oppositely signed MCD bands at 8.00 T are found to be $11.1 \pm 0.2\ \mathrm{cm^{-1}}$ and $11.3 \pm 0.2\ \mathrm{cm^{-1}}$, respectively, and are thus within experimental uncertainty of the splitting predicted from the atomic spectral g value. Furthermore, the relative peak magnitudes for the ΔA_{MLD} bands at 8.00 T ($-1.30 \times 10^{-3} : +2.68 \times 10^{-3} : -1.32 \times 10^{-3}$) are close to the relative intensity prediction of $-1:+2:-1$ from Eq. (9.4), assuming the bandwidths of the three bands are the same. At lower fields, it is no surprise that the splitting of the negative MLD bands and the relative peak magnitudes depart from expectations as a consequence of incomplete resolution of the Zeeman multiplet.

The results of this study provide a clear, simple example of how MLD and MCD measurements complement each other. The two spectra together fully characterize the triplet Zeeman multiplet. The relative signs of the individual transitions serve to establish the energy ordering of the multiplet. Experimentally, the weaker MLD signals unfortunately make their measurement more difficult than for MCD spectra, but MLD has the advantage of showing the $\Delta M_J = 0$ transition, which is forbidden for MCD.

9.4.2. Metal Atoms Isolated in Noble Gas Matrices

A combination of MCD and MLD measurements have been used to interpret spectra for metal atoms isolated in noble gas matrices [refs. 108–110]. As an example, Figures 9.5 and 9.6 show the absorption and MCD spectra for Ni atoms isolated in an argon matrix at 4.2 K and the MCD and MLD spectra for Ni atoms in an annealed Ar matrix, respectively [ref. 109]. The interpretation of the spectra is complicated by the identification of two different ground states for Ni atoms in the matrix: $^3D_3$ and $^3F_4$. These states are separated by

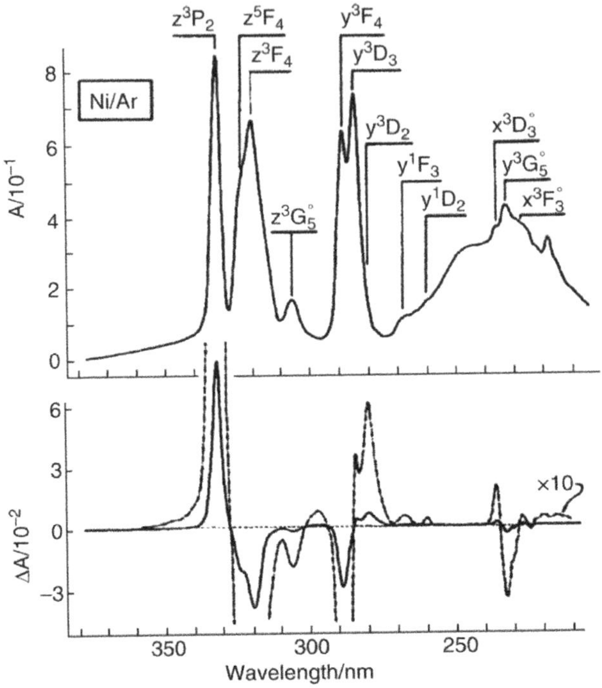

Figure 9.5. Absorption (upper curve) and MCD (lower curve) at 0.29 T for Ni atoms in an Ar matrix at 4.2 K (unannealed). The excited states indicated by small open circles are attributed to transitions from an $^3F_4$ ground state, while the others are believed to be from a $^3D_3$ ground state. (Reprinted with permission from ref. 109. Vala, M.; Eyring, M.; Pyka, J.; Rivoal, J.-C.; Grisolia, C. *J. Chem. Phys.* **1985**, *83*, 969. Copyright 1985 American Institute of Physics.)

only 205 cm^{-1}, and both are found when Ni atoms are isolated in unannealed Ar matrices. Upon thermal annealing, however, the $^3F_4$ state reverts to the lower-energy $^3D_3$ state. The complementary use of MCD and MLD to assign the spectra is demonstrated and rests on the relative signs of the features. Thus the $\Delta J = J_{ex} - J_{gs}$ values of $+1$, 0, and -1 were predicted to give MCD/MLD signs of $-/-$, $+/+$, and $+/-$, respectively [ref. 109]. For example, the band labeled z^3P_2 in Figure 9.6 has a $\Delta J = -1$ and shows positive MCD and negative MLD signals, respectively. Thus knowing the ground-state J value and the signs of the MCD and MLD features, the excited state J value can be determined and supported by spectroscopic experiment.

Finally, it may be remarked that MLD (and MCD) field and temperature saturation studies for iron atoms isolated in Xe and Kr matrices have been used to determine the Fe site-symmetry and crystal-field parameters [ref. 110]. Saturation effects are observed for MLD spectra in much the same way as described earlier in Chapter 6 for MCD spectra. MLD spectra present terms that depend upon T^{-2} and B^2, so in principle the changes will be larger than for MCD spectra with the

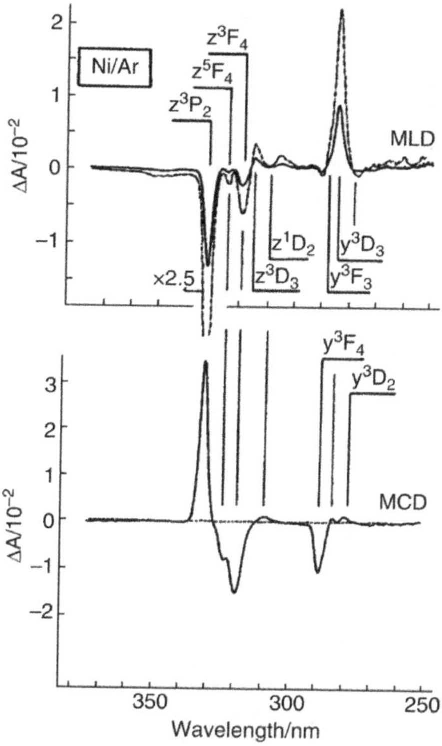

Figure 9.6. MCD (lower curve) at 1.58. T and MLD (upper curve) at 3.0 T for Ni atoms isolated in an Ar matrix at 4.2 K. The MLD spectrum was obtained after annealing for 10 min at 38 K and cooling to 4.2 K. (Reprinted with permission from ref. 109. Vala, M.; Eyring, M.; Pyka, J.; Rivoal, J.-C.; Grisolia, C. *J. Chem. Phys.* **1985**, *83*, 969. Copyright 1985 American Institute of Physics.)

same difference in field or temperature. In the case of Fe atoms in Kr or Xe matrices, an octahedral site symmetry with a crystal-field parameter of $0.06 \pm 0.05\ \text{cm}^{-1}$ and a ground-state splitting of $3.2 \pm 2.5\ \text{cm}^{-1}$ are consistent with measured MCD and MLD saturation curves [ref. 110].

9.4.3. Lanthanide Metal Ions in Solution: Ho^{3+}

The MLD signals are generally weak compared to MCD; however, if the sample has a narrow band, a paramagnetic ground state, and the magnetic field is high, then as Eq. (9.1) indicates, the MLD will be enhanced. The spectra of the lanthanide ions in solution provide an example of paramagnetic ions with narrow bands for the $4f \rightarrow 4f$ transitions. As an example, a portion of the absorption, MCD, and MLD spectra for Ho^{3+} in aqueous $HClO_4$ solution is shown in Figure 9.7 [ref. 111]. It is clear from the $4f^{10}$ ground state ($^5I_8$) and the various excited terms indicated in the figure that the spectra are complicated and exhibit

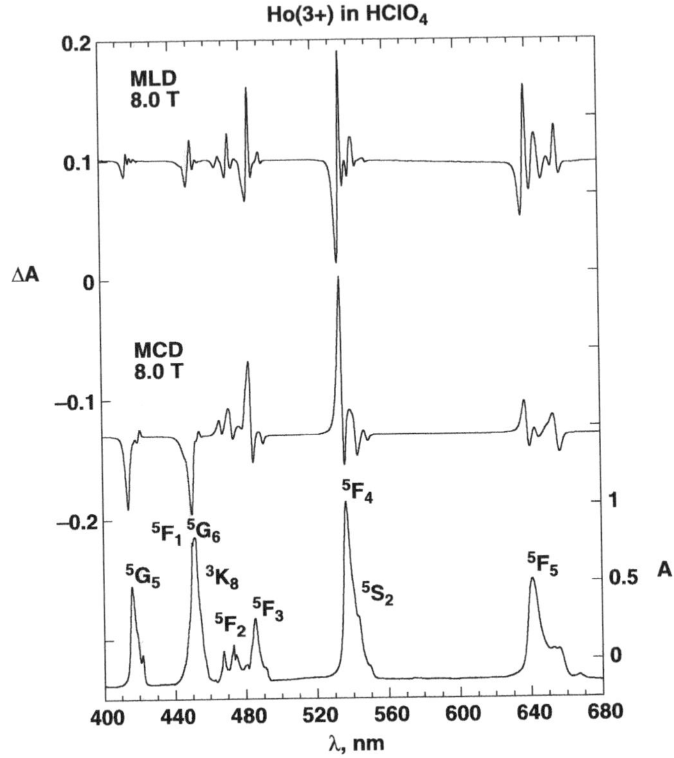

Figure 9.7. Absorption, MCD, and MLD spectra for Ho^{3+} in 0.10 M $HClO_4$. The MLD and MCD spectra at 8.0 T and 295 K have been scaled by factors of 10 and 0.40 and ΔA has been offset by +0.1 and −0.13, respectively, before plotting. The ground-state term for $4f^{10}$ Ho^{3+} is $^5I_8$; excited-state terms are indicated on the absorption spectrum. (Adapted from ref. 111.)

many closely spaced overlapping transitions that have degeneracies. There are both spin-allowed and spin-forbidden transitions present which depend upon spin–orbit coupling mechanisms for intensity. These overlapping transitions present a challenge for interpretation. Since the MLD and MCD spectra differ from each other, and each of them shows a higher degree of resolution of individual spectral features than the absorption spectrum by virtue of the positive and negative signs in their respective spectral curves, having all three spectra can provide an enormous aid in interpretation. It is highly likely that the MLD terms that dominate are *G* terms, so that low temperature measurements would also be beneficial, however, the ion must be embedded in a suitable transparent, non-birefringent and non-linearly dichroic matrix to avoid spurious signals. Such measurements have not been reported.

Figure 9.8 shows an expanded portion of the spectrum near 535 nm. The MLD spectra were measured at different fields from 0 to 8.0 T. It is clear that the

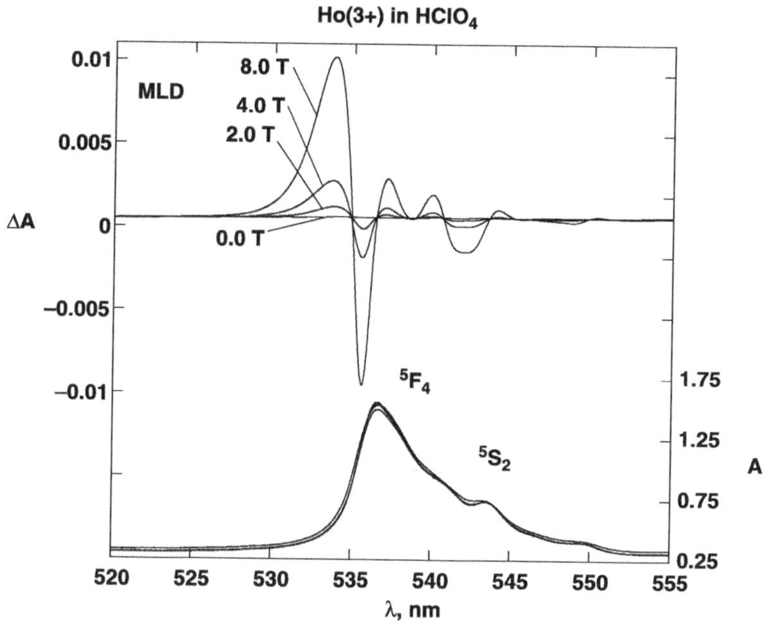

Figure 9.8. An expanded portion of the 535-nm band for Ho^{3+} in 0.010 M $HClO_4$ showing the field dependence of the MLD spectra. (Adapted from ref. 111.)

quadratic dependence on field enhances the signals dramatically at high field. If the field could be high enough, MLD could easily compete with MCD signals, but such fields are not feasible experimentally at the present time. For example, in Figure 9.7, the MCD and MLD signals are about 25 : 1, respectively. Therefore, the field would have to be increased by a factor of 25 to yield comparable signals; such a field is not possible with conventional equipment.

9.4.4. Ferrocytochrome *c* and Deoxymyoglobin

An application of MLD spectroscopy to metalloproteins was demonstrated by spectral measurements for the diamagnetic ($S = 0$) ferrocytochrome *c* and the paramagnetic ($S = 2$) deoxymyoglobin [ref. 100]. In both cases, MLD and comparison MCD spectra were obtained in the visible region for frozen glass samples. An experimental protocol was used whereby the MLD was measured for the transverse field in both the $+z$ and $-z$ directions, and then the resulting spectra were added. This protocol essentially removed any nonmagnetic residual linear dichroism (LD) exhibited by the frozen samples. The diamagnetic heme porphyrin chromophore ferrocytochrome *c* showed measurable MLD for the Q band near 550 nm (see Figure 9.9), which appear in the form of a second derivative curve; the MCD for the Q band showed a positive A term consistent with the assignment as ${}^1A_{1g} \rightarrow {}^1E_u$ based on a nominal D_{4h} symmetry. Since the ground state is

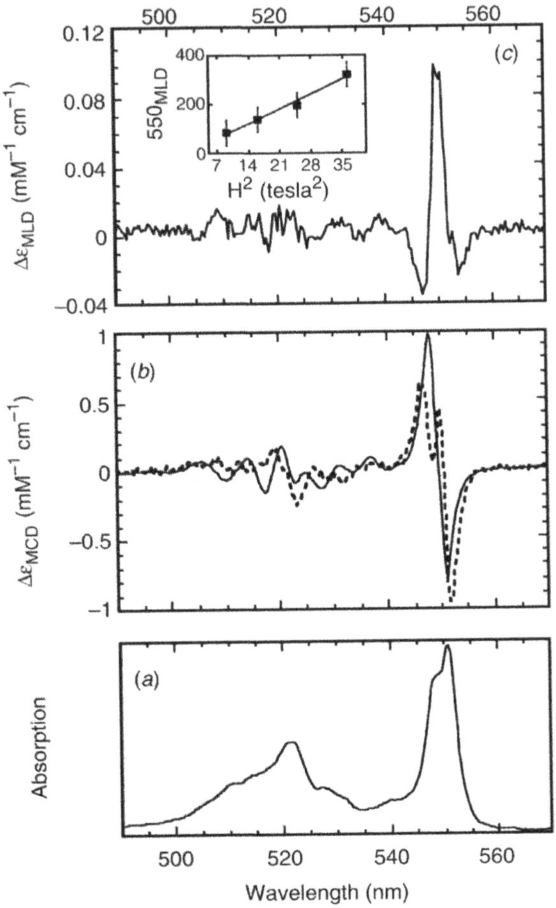

Figure 9.9. Absorption (*a*), MCD (*b*), and MLD (*c*) spectra for ferrocytochrome *c* ($S = 0$) at 4.2 K in a 50%(v/v) glycerol glass. MCD measured at 1.0 T; MLD measured at 4.0 T. The inset shows the MLD field dependence. (Reprinted with permission from ref. 100. Peterson, J.; Pearce, L. L.; Bominaar, E. L. *J. Am Chem Soc.* **1999**, *121*, 5972. Copyright 1999 American Chemical Society.)

diamagnetic, the MLD terms in Eq. (9.1) are $C = F = G = 0$. The interpretation included allowance for a crystal-field splittling, Δ, of the ^1E_u excited state, which was assumed to be smaller than the bandwidth. The detailed interpretation of the MLD and MCD spectra was given in terms of a pseudo A_2 parameter (two closely spaced B_1 parameters) for the MLD and a pseudo A term for the MCD. The MLD and MCD together were able to provide an estimate of Δ (≤ 93 cm^{-1}) and the excited-state g_z factor (9.2 ± 0.8). The latter value showed that the angular momentum for the ^1E_u excited state ($g_z/2 = 4.9 \pm 0.4 = M_L$) was smaller than earlier estimates had assumed ($M_L = 7.8$–7.9) from MCD and absorption data.

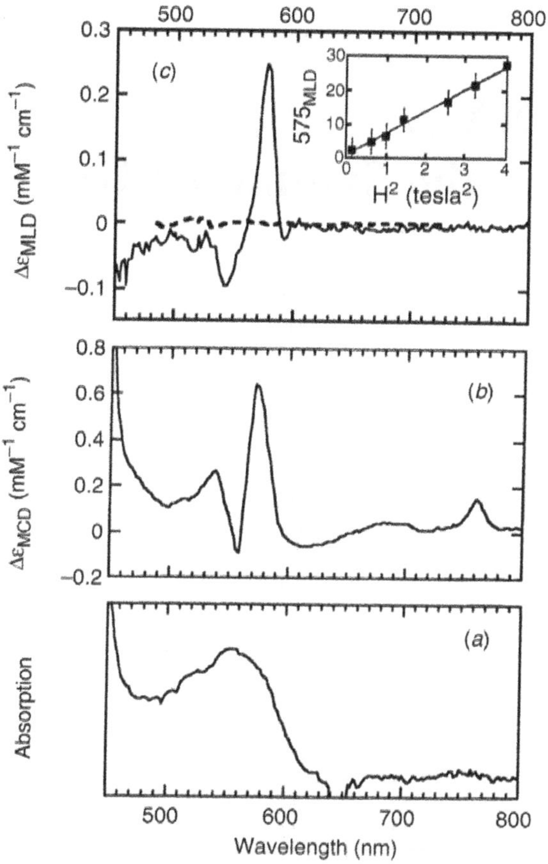

Figure 9.10. Absorption (*a*), MCD (*b*), and MLD (*c*) spectra for deoxymyogloben at 4.2 K in 50% (v/v) glycerol glass. MCD at 5.0 T and MLD at 7.0 T. In (*c*) the broken line represents the MLD for ferricytochrome *c* ($S = 1/2$) and the inset shows the MLD field dependence. (Reprinted with permission from ref. 100. Peterson, J.; Pearce, L. L.; Bominaar, E. L. *J. Am Chem Soc.* **1999**, *121*, 5972. Copyright 1999 American Chemical Society.)

Figure 9.10 shows the MLD and MCD spectra for deoxymyoglobin, a heme protein with a paramagnetic ground state ($S = 2$) [ref. 100]. Also shown in the figure is the MLD measured for the paramagnetic ferricytochrome *c* ($S = 1/2$), which shows no signal in the same region as the strong signal for deoxymyoglobin. Thus the MLD spectra can detect the deoxymyoglobin in the presence of ferricytochrome *c*, whereas the MCD spectra can detect the ferricytochrome *c* in the presence of deoxymyoglobin [see ref. 54, Figure 3]. The authors note the potential for measurement of paramagnetic metalloproteins that differ in spin states. The MLD terms contributing to the deoxymyoglobin spectrum are temperature-dependent with *G* terms likely dominant.

APPENDIX A
Tables for the Symmetry Groups *O* and T_d

TABLE A.1. Function ψ and Operator Op Transformation Coefficients for Groups *O* and T_d

These coefficients enable the conversion of functions ψ and Operators Op to standard basis functions $|\psi a\alpha\rangle$ and operators Op^f_ϕ by means of the relations

$$|\psi_{a\alpha}\rangle = \sum_i \langle a\alpha|\psi_i\rangle * \psi_i$$

$$\mathrm{Op}^f_\phi = \sum_i \langle f_\phi|\mathrm{Op}_i\rangle * \mathrm{Op}_i$$

Bases are usually chosen so the sums in these relations are unnecessary [ref. 1, Section C.7].

ψ or Op	$\langle a\alpha\|\psi_i\rangle$ or $\langle f\phi\|\mathrm{Op}_i\rangle$	Standard Basis $\|\psi a\alpha\rangle$ or Op^f_ϕ
	O	
x, p_x, V_x, R_x	-1	T_1x
y, p_y, V_y, R_y	$-i$	T_1y
z, p_z, V_z, R_z	1	T_1z
d_{z^2}	-1	$E\theta$
$d_{x^2-y^2}$	1	$E\varepsilon$
dyz	i	$T_2\xi$
dzx	1	$T_2\eta$
dxy	$-i$	$T_2\zeta$
$\|1-1\rangle, V_{-1}, R_{-1}$	-1	T_1-1
$\|1\ 0\rangle, V_0, R_0$	1	T_10
$\|1+1\rangle, V_{+1}, R_{+1}$	-1	T_11

(*Continued*)

TABLE A.1. (*Continued*)

ψ or Op	$\langle a\alpha\vert\psi_i\rangle$ or $\langle f\phi\vert \mathrm{Op}_i\rangle$	Standard Basis $\vert\psi a\alpha\rangle$ or Op^f_ϕ
	T_d	
x, p_x, V_x	-1	$T_2\xi(x)$
y, p_y, V_y	i	$T_2\eta(y)$
z, p_z, V_z	1	$T_2\zeta(z)$
$\vert 1\ -1\rangle, V_{-1}$	1	$T_2 1$
$\vert 1\ 0\rangle, V_0$	1	$T_2 0$
$\vert 1\ +1\rangle, V_{+1}$	1	$T_2 - 1$

TABLE A.2. *3j*, *2j*, and *2jm* Phases in the *O* and T_d Bases for Single-Valued Irreps. See Ref. 1, Section C.11, for Phases for Double-Valued Irreps E', E'', and U', Which are Used in Double Groups Involving Half-Integral Angular Momentum

3*j*	2*j*
$\{abc\} = (-1)^a(-1)^b(-1)^c$ where $(-1)^{A_1} = (-1)^E = (-1)^{T_2} = 1$, $(-1)^{A_2} = (-1)^{T_1} = -1$	$\{a\} = 1$ for A_1, A_2, E, T_1, and T_2

2*jm*

$\vert a\alpha\rangle$	$\begin{pmatrix} a \\ \alpha \end{pmatrix}$	$\vert a\alpha\rangle$	$\begin{pmatrix} a \\ \alpha \end{pmatrix}$
A_1a_1	1	T_1x	-1
A_2a_2	1	T_1y	1
$E\theta$	1	T_1z	-1
$E\varepsilon$	1	$T_2\xi$	-1
		$T_2\eta$	1
		$T_2\zeta$	-1

TABLE A.3. 3*jm* for *O* and T_d Bases for Single-Valued Irreps. The Cases Indicated by 3*jm*† Change Sign for an Odd Permutation of Columns Within the $\begin{pmatrix} a & b & c \\ \alpha & \beta & \gamma \end{pmatrix}$ 3*jm* Symbol. See Ref. 1, Section C.12 and Table C.12.1, for 3*jm* Involving Double-Valued Irreps E', E'', and U'

A_1	A_1	A_1	3*jm*
a_1	a_1	a_1	1

A_2	A_2	A_1	3*jm*
a_2	a_2	a_1	1

E	E	A_1	3*jm*
θ	θ	a_1	$(2)^{-1/2}$
ε	ε	a_1	$(2)^{-1/2}$

E	E	A_2	3*jm*†
θ	ε	a_2	$-(2)^{-1/2}$

E	E	E	3*jm*
θ	θ	θ	$-1/2$
θ	ε	ε	$1/2$

T_1	T_1	A_1	3*jm*
x	x	a_1	$-(3)^{-1/2}$
y	y	a_1	$(3)^{-1/2}$
z	z	a_1	$-(3)^{-1/2}$
-1	1	a_1	$(3)^{-1/2}$
0	0	a_1	$-(3)^{-1/2}$

T_1	T_1	E	3*jm*
x	x	θ	$1/2(3)^{1/2}$
x	x	ε	$1/2$
y	y	θ	$-1/2(3)^{1/2}$
y	y	ε	$1/2$
z	z	θ	$-(3)^{-1/2}$
-1	-1	ε	$-1/2$
-1	1	θ	$-1/2(3)^{1/2}$
0	0	θ	$-(3)^{-1/2}$
1	1	ε	$-1/2$

T_1	T_1	T_1	3*jm*†
x	y	z	$(6)^{-1/2}$
-1	0	1	$-(6)^{-1/2}$

T_2	T_1	A_2	3*jm*
ξ	x	a_2	$(3)^{-1/2}$
η	y	a_2	$(3)^{-1/2}$
ζ	z	a_2	$(3)^{-1/2}$
-1	-1	a_2	$-(3)^{-1/2}$
0	0	a_2	$-(3)^{-1/2}$
1	1	a_2	$-(3)^{-1/2}$

T_2	T_1	E	3*jm*†
ξ	x	θ	$1/2$
ξ	x	ε	$-1/2(3)^{1/2}$
η	y	θ	$-1/2$
η	y	ε	$-1/2(3)^{1/2}$
ζ	z	ε	$(3)^{-1/2}$
-1	-1	ε	$1/2(3)^{1/2}$
-1	1	θ	$-1/2$
0	0	ε	$-(3)^{-1/2}$
1	-1	θ	$-1/2$
1	1	ε	$1/2(3)^{1/2}$

T_2	T_1	T_1	3*jm*
ξ	y	z	$(6)^{-1/2}$
η	z	x	$-(6)^{-1/2}$
ζ	x	y	$(6)^{-1/2}$
-1	1	0	$(6)^{-1/2}$
0	-1	-1	$(6)^{-1/2}$
0	1	1	$-(6)^{-1/2}$
1	0	-1	$-(6)^{-1/2}$

(*Continued*)

TABLE A.3. (*Continued*)

T_2	T_2	A_1	$3jm$
ξ	ξ	a_1	$-(3)^{-1/2}$
η	η	a_1	$(3)^{-1/2}$
ζ	ζ	a_1	$-(3)^{-1/2}$
-1	1	a_1	$(3)^{-1/2}$
0	0	a_1	$-(3)^{-1/2}$

T_2	T_2	E	$3jm$
ξ	ξ	θ	$1/2(3)^{1/2}$
ξ	ξ	ε	$1/2$
η	η	θ	$-1/2(3)^{1/2}$
η	η	ε	$1/2$
ζ	ζ	θ	$-(3)^{-1/2}$
-1	-1	ε	$-1/2$
-1	1	θ	$-1/2(3)^{1/2}$
0	0	θ	$-(3)^{-1/2}$
1	1	ε	$-1/2$

T_2	T_2	T_1	$3jm^{\dagger}$
ξ	η	z	$(6)^{-1/2}$
η	ζ	x	$(6)^{-1/2}$
ζ	ξ	y	$-(6)^{-1/2}$
-1	0	-1	$(6)^{-1/2}$
-1	1	0	$(6)^{-1/2}$
0	1	1	$(6)^{-1/2}$

T_2	T_2	T_2	$3jm$
ξ	η	ζ	$(6)^{-1/2}$
-1	-1	0	$(6)^{-1/2}$
0	1	1	$-(6)^{-1/2}$

TABLE A.4. ***6j* for *O* and T_d Bases for Single-Valued Irreps**

All $6j$ containing one or more A_1 irreps are determined by

$$\begin{Bmatrix} A_1 & b & c \\ d & e & f \end{Bmatrix} = \delta_{bc}\delta_{ef}\delta(dbf)|b|^{-1/2}|e|^{-1/2}\{dbf\}$$

and are not given in Table A.4. All other single-valued $6j$ not related by symmetry rules to those given below are zero. See ref. 1, Section C.13 and Table C.13.1, for $6j$ involving double-valued irreps E', E'', and U'.

A_2	E	E	$6j$
A_2	E	E	$1/2$
E	E	E	$1/2$
T_1	T_1	T_2	$(6)^{-1/2}$
T_2	T_1	T_2	$-(6)^{-1/2}$

E	E	E	$6j$
E	E	E	0
T_1	T_1	T_1	$-1/2(3)^{1/2}$
T_1	T_1	T_2	$-1/2(3)^{1/2}$
T_1	T_2	T_2	$-1/2(3)^{1/2}$
T_2	T_2	T_2	$-1/2(3)^{1/2}$

(*Continued*)

TABLE A.4. (*Continued*)

A_2	T_1	T_2	$6j$
A_2	T_1	T_2	$1/3$
E	T_1	T_2	$1/3$
E	T_2	T_1	$1/3$
T_1	T_1	T_2	$1/3$
T_1	T_2	T_1	$1/3$
T_2	T_1	T_2	$-1/3$
T_2	T_2	T_1	$-1/3$

E	T_1	T_1	$6j$
E	T_1	T_1	$1/3$
E	T_1	T_2	0
E	T_2	T_2	$-1/3$
T_1	T_1	T_1	$1/6$
T_1	T_1	T_2	$-1/2(3)^{1/2}$
T_1	T_2	T_2	$-1/6$
T_2	T_1	T_1	$-1/6$
T_2	T_1	T_2	$-1/2(3)^{1/2}$
T_2	T_2	T_2	$1/6$

E	T_2	T_2	$6j$
E	T_2	T_2	$1/3$
T_1	T_2	T_2	$1/6$
T_2	T_2	T_2	$-1/6$

E	T_1	T_2	$6j$
E	T_1	T_2	$1/3$
E	T_2	T_2	0
T_1	T_1	T_2	$-1/6$
T_1	T_2	T_1	$1/6$
T_1	T_2	T_2	$-1/2(3)^{1/2}$
T_2	T_1	T_2	$1/6$
T_2	T_2	T_1	$-1/6$
T_2	T_2	T_2	$-1/2(3)^{1/2}$

T_1	T_1	T_1	$6j$
T_1	T_1	T_1	$1/6$
T_1	T_1	T_2	$1/6$
T_1	T_2	T_2	$-1/6$
T_2	T_2	T_2	$-1/6$

T_1	T_1	T_2	$6j$
T_1	T_1	T_2	$1/6$
T_1	T_2	T_2	$1/6$
T_2	T_2	T_1	$-1/6$
T_2	T_2	T_2	$-1/6$

T_1	T_2	T_2	$6j$
T_1	T_2	T_2	$1/6$
T_2	T_2	T_2	$1/6$

T_2	T_2	T_2	$6j$
T_2	T_2	T_2	$1/6$

APPENDIX B
Tables for the Fourfold Symmetry Group D_4

TABLE B.1. Function ψ and Operator Op Transformation Coefficients for Group D_4. See Table A.1 for Definitions

ψ or Op	$\langle a\alpha\|\psi_i\rangle$ or $\langle f\phi\|\mathrm{Op}_i\rangle$	Standard Basis $\|\psi a\alpha\rangle$ or Op_ϕ^f
x, p_x, V_x, R_x	-1	Ex
y, p_y, V_y, R_y	$-i$	Ey
z, p_z, V_z, R_z	1	A_2a_2
d_{z^2}	-1	A_1
$d_{x^2-y^2}$	1	B_1b_1
d_{yz}	i	Ex
d_{zx}	1	Ey
d_{xy}	$-i$	B_2b_2
$\|1\ -1\rangle, V_{-1}, R_{-1}$	-1	$E-1$
$\|1\ 0\rangle, V_0, R_0$	1	A_2a_2
$\|1\ +1\rangle, V_{+1}, R_{+1}$	-1	$E1$

TABLE B.2. $3j$, $2j$, and $2jm$ Phases in D_4 Bases for Single-Valued Irreps. See Ref. 1, Section D.8, for Phases for Double-Valued Irreps E' and E''

$3j$	$2j$
$\{abc\} = (-1)^a(-1)^b(-1)^c$ where $(-1)^{A_1} = (-1)^{B_1} = (-1)^{B_2} = 1$; $(-1)^{A_2} = (-1)^E = -1$	$\{a\} = 1$ for A_1, A_2, B_1, B_2, and E

$2jm$

$\|a\alpha\rangle$	$\begin{pmatrix} a \\ \alpha \end{pmatrix}$	$\|a\alpha\rangle$	$\begin{pmatrix} a \\ \alpha \end{pmatrix}$
A_1a_1	1	Ex	-1
A_2a_2	-1	Ey	1
B_1b_1	1	$E1$	1
B_2b_2	-1	$E-1$	1

TABLE B.3. $3jm$ in D_4 Bases for Single-Valued Irreps. The Cases Indicated by $3jm^\dagger$ Change Sign for an Odd Permutation of Columns Within the $\begin{pmatrix} a & b & c \\ \alpha & \beta & \gamma \end{pmatrix}$ $3jm$ Symbol. See Ref. 1, Section D.9 and Table D.9.2, for $3jm$ Involving Double-Valued Irreps E' and E''

A_1	A_1	A_1	$3jm$
a_1	a_1	a_1	1

A_2	A_2	A_1	$3jm$
a_2	a_2	a_1	-1

B_1	B_1	A_1	$3jm$
b_1	b_1	a_1	1

B_2	B_2	A_1	$3jm$
b_2	b_2	a_1	-1

B_2	B_1	A_2	$3jm^\dagger$
b_2	b_1	a_2	1

E	E	A_2	$3jm^\dagger$
x	y	a_2	$-(2)^{-1/2}$
-1	1	a_2	$-(2)^{-1/2}$

E	E	B_2	$3jm$
x	y	b_2	$(2)^{-1/2}$
-1	-1	b_2	$(2)^{-1/2}$
1	1	b_2	$-(2)^{-1/2}$

E	E	A_1	$3jm$
x	x	a_1	$-(2)^{-1/2}$
y	y	a_1	$(2)^{-1/2}$
-1	1	a_1	$(2)^{-1/2}$

E	E	B_1	$3jm$
x	x	b_1	$(2)^{-1/2}$
y	y	b_1	$(2)^{-1/2}$
-1	-1	b_1	$-(2)^{-1/2}$
1	1	b_1	$-(2)^{-1/2}$

TABLE B.4. $6j$ in D_4 Bases for Single-Valued Irreps. See Ref. 1, Section D.10 and Table D.10.1, for $6j$ Involving Double-Valued Irreps E' and E''

All $6j$ containing one or more A_1 irreps are determined by

$$\begin{Bmatrix} A_1 & b & c \\ d & e & f \end{Bmatrix} = \delta_{bc}\delta_{ef}\delta(dbf)|b|^{-1/2}|e|^{-1/2}\{dbf\}$$

and are not given in the table below. All other single-valued $6j$ not related by symmetry rules to those given below are zero.

A_2	B_1	B_2	$6j$
A_2	B_1	B_2	1
E	E	E	$-(2)^{-1/2}$

B_1	E	E	$6j$
B_1	E	E	1/2
B_2	E	E	$-1/2$

A_2	E	E	$6j$
A_2	E	E	1/2
B_1	E	E	1/2
B_2	E	E	1/2

B_2	E	E	$6j$
B_2	E	E	1/2

APPENDIX C
Tables for the Threefold Symmetry Group D_3

TABLE C.1. Function ψ and Operator Op Transformation Coefficients for Group D_3. See Table A.1 for Definitions

ψ or Op	$\langle a\alpha \vert \psi_i \rangle$ or $\langle f\phi \vert \mathrm{Op}_i \rangle$	Standard Basis $\vert \psi a\alpha \rangle$ or Op^f_ϕ
x, p_x, V_x, R_x	-1	$E\theta$
y, p_y, V_y, R_y	$-i$	A_2a_2
z, p_z, V_z, R_z	1	$E\varepsilon$
$\vert 1\ -1\rangle, V_{-1}, R_{-1}$	-1	$E\ -1$
$\vert 1\ 0\rangle, V_0, R_0$	1	A_2a_2
$\vert 1\ +1\rangle, V_{+1}, R_{+1}$	-1	$E\ +1$

TABLE C.2. 3*j*, 2*j*, and 2*jm* Phases in D_3 Bases for Single-Valued Irreps. See Ref. 1, Section E.8, for Phases for Double-Valued Irreps E' P_1 and P_2

3*j*

$\{abc\} = (-1)^a(-1)^b(-1)^c$

where

$(-1)^{A_1} = 1$

$(-1)^{A_2} = (-1)^E = -1$

The law breaking 3*j* phase $\{EEE\} = 1$

2*j*

$\{a\} = 1$ for A_1, A_2, and E

2*jm*

$\vert a\alpha\rangle$	$\begin{pmatrix} a \\ \alpha \end{pmatrix}$	$\vert a\alpha\rangle$	$\begin{pmatrix} a \\ \alpha \end{pmatrix}$
A_1a_1	1	$E1$	1
A_2a_2	1	$E-1$	1
$E\theta$	-1		
$E\varepsilon$	-1		

TABLE C.3. $3jm$ in D_3 Bases for Single-Valued Irreps. The Cases Indicated by $3jm^\dagger$ Change Sign for an Odd Permutation of Columns within the $\begin{pmatrix} a & b & c \\ \alpha & \beta & \gamma \end{pmatrix}$ $3jm$ Symbol. See Ref. 1, Section E.9 and Table E.9.2 for $3jm$ Involving Double-Valued Irreps E' and E''

A_1	A_1	A_1	$3jm$	A_2	A_2	A_1	$3jm$
a_1	a_1	a_1	1	a_2	a_2	a_1	1

E	E	A_1	$3jm$	E	E	E	$3jm$
θ	θ	a_1	$-(2)^{-1/2}$	θ	θ	ε	$-i/2$
ε	ε	a_1	$-(2)^{-1/2}$	ε	ε	ε	$i/2$
-1	1	a_1	$(2)^{-1/2}$	-1	-1	-1	$(2)^{-1/2}$
				1	1	1	$(2)^{-1/2}$

E	E	A_2	$3jm^\dagger$
θ	ε	a_2	$(2)^{-1/2}$
-1	1	a_2	$-(2)^{-1/2}$

TABLE C.4. $6j$ in D_3 Bases for Single-Valued Irreps. See Ref. 1, Table E.10.1, for $6j$ Involving Double-Valued Irreps E', P_1, and P_2

All $6j$ containing one or more A_1 irreps are determined by

$$\begin{Bmatrix} A_1 & b & c \\ d & e & f \end{Bmatrix} = \delta_{bc}\delta_{ef}\delta(dbf)|b|^{-1/2}|e|^{-1/2}\{dbf\}$$

and are not given in the table below. All other single-valued $6j$ not related by symmetry rules to those given below are zero. See ref. 1, Table E.10.1, for $6j$ involving double-valued irreps E', P_1, and P_2.

A_2	E	E	$6j$	E	E	E	$6j$
A_2	E	E	1/2	E	E	E	0
E	E	E	1/2				

APPENDIX D
3*jm* Factors for Single-Valued Irreps of the $SO_3 \supset O$ and $O \supset D_4$ Chains

TABLE D.1. Partial List of $SO_3 \supset O$ 3*jm* Factors $\begin{pmatrix} S & f & S \\ a_1 & f_1 & b_1 \end{pmatrix}^{SO_3}_{O}$ Involving Single-Valued Irreps. A More Extensive List that Includes E', E'', and U' Double Group Irreps can be Found in Ref. 1, Table B.5.1

0	0	0	3*jm*
A_1	A_1	A_1	1

1	1	0	3*jm*
T_1	T_1	A_1	1

1	1	1	3*jm*
T_1	T_1	T_1	-1

2	1	1	3*jm*
E	T_1	T_1	$(2/5)^{1/2}$
T_2	T_1	T_1	$(3/5)^{1/2}$

2	2	0	3*jm*
E	E	A_1	$(2/5)^{1/2}$
T_2	T_2	A_1	$(3/5)^{1/2}$

2	2	1	3*jm*
T_2	E	T_1	$-(2/5)^{1/2}$
T_2	T_2	T_1	$-(5)^{-1/2}$

2	2	2	3*jm*
E	E	E	$-2(2/35)^{1/2}$
T_2	T_2	E	$-(6/35)^{1/2}$
T_2	T_2	T_2	$-3(35)^{-1/2}$

TABLE D.2. Partial List of $O \supset D_4$ 3jm Factors $\begin{pmatrix} a_1 & f_1 & b_1 \\ a_2 & f_2 & b_2 \end{pmatrix}\begin{matrix} O \\ D_4 \end{matrix}$ Involving Single-Valued Irreps. Factors that Include E', E'', or U′ in O and E' or E'' in D_4 can be Found in Ref. 1, Table D.12.1. Here, the Cases that Include Phase Values of ±1 Indicate that the 3jm Factor Value Either Changes Sign (−1) or Remains Unchanged (+1) for an Odd Permutation of Columns in the Factor; if no Phase is Indicated, +1 is Assumed

A_1	A_1	A_1	3jm
A_1	A_1	A_1	1

A_1	E	E	3jm
A_1	A_1	A_1	$(2)^{-1/2}$
A_1	B_1	B_1	$(2)^{-1/2}$

A_1	T_2	T_2	3jm
A_1	B_2	B_2	$(3)^{-1/2}$
A_1	E	E	$(2/3)^{1/2}$

A_2	T_1	T_2	3jm	Phase
B_1	A_2	B_2	$(3)^{-1/2}$	−1
B_1	E	E	$(2/3)^{1/2}$	+1

E	T_1	T_1	3jm
A_1	A_2	A_2	$(3)^{-1/2}$
A_1	E	E	$-(6)^{-1/2}$
B_1	E	E	$(2)^{-1/2}$

E	T_2	T_2	3jm
A_1	B_2	B_2	$(3)^{-1/2}$
A_1	E	E	$-(6)^{-1/2}$
B_1	E	E	$(2)^{-1/2}$

T_1	T_1	T_2	3jm	Phase
A_2	E	E	$(3)^{-1/2}$	−1
E	E	B_2	$(3)^{-1/2}$	+1

T_2	T_2	T_2	3jm
B_2	E	E	$(3)^{-1/2}$

A_1	A_2	A_2	3jm
A_1	B_1	B_1	1

A_1	T_1	T_1	3jm
A_1	A_2	A_2	$(3)^{-1/2}$
A_1	E	E	$(2/3)^{1/2}$

A_2	E	E	3jm	Phase
B_1	A_1	B_1	$-(2)^{-1/2}$	−1

E	E	E	3jm
A_1	A_1	A_1	$-1/2$
A_1	B_1	B_1	$1/2$

E	T_1	T_2	3jm	Phase
A_1	E	E	$(2)^{-1/2}$	−1
B_1	A_2	B_2	$-(3)^{-1/2}$	+1
B_1	E	E	$(6)^{-1/2}$	−1

T_1	T_1	T_1	3jm
A_2	E	E	$-(3)^{-1/2}$

T_1	T_2	T_2	3jm	Phase
A_2	E	E	$-(3)^{-1/2}$	+1
E	B_2	E	$-(3)^{-1/2}$	−1

Reviews and References

REVIEWS OF MCD

1. Buckingham, A. D.; Stephens, P. J. *Annu. Rev. Phys. Chem.* **1966**, *17*, 399.
2. Schatz, P. N.; McCaffery, A. J. *Quart. Rev.* **1969**, *23*, 552.
3. Stephens, P. J. *Adv. Chem. Phys.* **1976**, *35*, 197.
4. Michl, J. *Tetrahedron* **1984**, *40*, 3845.
5. Ball, D. W. *Spectroscopy* **1991**, *6*, 18.
6. Hollebone, B. R. *Spectrochim. Acta Reviews* **1993**, *15*, 493.
7. Pavel, E. G.; Solomon, E. I. *Am. Chem. Soc. Symp. Ser.* **1998**, *692*, 119.
8. Andersson, L. A. *Encyclopedia of Spectroscopy and Spectrometry*; Lindon, J. C., Ed; Academic Press, San Diego, 2000, p. 1217.
9. Johnson, M. K. *Physical Methods in Bioinorganic Chemistry*; Que, L. Ed; University Science Books, Sausalito, CA, 2000, p. 233.
10. Mason, W. R. *Comprehensive Coordination Chemistry II*; Vol. 2; Lever, A. B. P., Vol. Ed; Elsevier, Oxford, 2004, ch 2.25, p. 327.
11. Solomon, E. I.; Neidig, M. L.; Schenk, G. *Comprehensive Coordination Chemistry II*; Vol. 2; Lever, A. B. P., Vol. Ed; Elsevier, Oxford, 2004, ch 2.26, p. 339.

REFERENCES

1. Piepho, S. B.; Schatz, P. N. *Group Theory in Spectroscopy with Applications to Magnetic Circular Dichroism*; Wiley-Interscience: New York, 1983.
2. Born, M.; Wolf, E. *Principles of Optics*, 7th ed., Cambridge University Press: Cambridge, 1999.
3. Condon, E. U. *Rev. Mod. Phys.*, **1937**, *9*, 432.
4. Eyring, H.; Walter, J.; Kimball, G. E. *Quantum Chemistry*, Wiley: New York, 1965.
5. Mason, W. R. *Anal. Chem.* **1982**, *54*, 646–48.
6. Butler, P. H. *Point Group Symmetry Applications: Methods and Tables*, Plenum Press: New York, 1981.
7. Mason, W. R. *Inorg. Chem.* **1988**, *27*, 437.
8. Smith, D.; Williamson, B. E.; Schatz, P. N. *Chem. Phys. Lett.* **1986**, *131*, 457.

9. Martin, D. S., Jr.; Foss, J. G.; McCarville, M. E.; Tucker, M. A.; Kassman, A. J. *Inorg. Chem.* **1966**, *5*, 491.
10. McCaffery, A. J.; Schatz, P. N.; Stephens, P. J. *J. Am. Chem. Soc.* **1968**, *90*, 5730.
11. Nakayama, K.; Komorita, T.; Shimura, Y. *Bull. Chem. Soc. Jpn.* **1984**, *57*, 972.
12. Martin, D. S., Jr.; Tucker, M. A.; Kassman, A. J. *Inorg. Chem.* **1965**, *4*, 1682.
13. Bosnich, B. *J. Am. Chem. Soc.* **1966**, *88*, 2606.
14. Isci, H.; Mason, W. R. *Inorg. Chem.* **1983**, *22*, 2266.
15. Isci, H.; Mason, W. R. *Inorg. Chem.* **1984**, *23*, 1565.
16. Olsson, L.-F. *Inorg. Chem.* **1986**, *25*, 1697.
17. Piepho, S. B.; Schatz, P. N.; McCaffery, A. J. *J. Am. Chem. Soc.* **1969**, *91*, 5994; Piepho, S. B. *PhD Dissertation*, University of Virginia, Charlottesville, VA, 1970.
18. Isci, H.; Mason, W. R. *Inorg. Chem.* **1975**, *14*, 905.
19. Mason, W. R. *Inorg. Chem.* **1986**, *25*, 2925; Isci, H.; Mason, W. R. *Inorg. Nucl. Chem. Lett.* **1972**, *8*, 885.
20. Savas, M. M.; Mason, W. R. *Inorg. Chem.* **1987**, *26*, 301.
21. Koutek, M. E.; Mason, W. R. *Inorg. Chem.* **1980**, *19*, 648.
22. Savas, M. M.; Mason, W. R. *Inorg. Chem.* **1988**, *27*, 658.
23. Jaw, H.-R. C.; Mason, W. R. *Inorg. Chem.* **1989**, *28*, 4370.
24. Mason, W. R. *Inorg. Chem.* **2001**, *40*, 6316.
25. Henning, G. N.; Dobosh, P. A.; McCaffery, A. J.; Schatz, P. N. *J. Am. Chem. Soc.* **1970**, *92*, 5377.
26. Schatz, P. N.; Stephens, P. J.; Henning, G. N.; McCaffery, A. J. *Inorg. Chem.* **1968**, *7*, 1246.
27. Mason, W. R. *Inorg. Chem.* **1999**, *38*, 2742.
28. Chastain, S. K.; Mason, W. R. *Inorg. Chem.* **1981**, *20*, 1395.
29. Jaw, H.-R. C.; Mason, W. R. *Inorg. Chem.* **1990**, *29*, 3452.
30. Mason, W. R. *Inorg. Chem.* **1997**, *36*, 1164.
31. Jaw, H.-R. C.; Mason, W. R. *Inorg. Chem.* **1991**, *30*, 275.
32. Adrowski, M. J.; Mason, W. R. *Inorg. Chem.* **1997**, *36*, 1443.
33. Mason, W. R. *Inorg. Chem.* **2000**, *39*, 370.
34. Isci, H.; Mason, W. R. *Inorg. Chem.* **1985**, *24*, 271.
35. Gedanken, A.; Rowe, M. D. *Chem. Phys. Lett.* **1975**, *34*, 39.
36. Johnson, B. R.; Kinsey, J. L. *J. Phys. Chem.* **1996**, *100*, 18937.
37. Scott, J. D.; Felps, W. S.; Findley, G. L.; McGlynn, S. P. *J. Chem. Phys.* **1978**, *68*, 4678; McGlynn, S. P.; Scott, J. D.; Felps, W. S. *J. Phys.* **1982**, *43*, C2305.
38. Felps, W. S.; Scott, J. D.; McGlynn, S. P. *J. Chem. Phys.* **1996**, *104*, 419.
39. Douglas, I. N.; Grinter, R.; Thomson, A. J. *Molec. Phys.* **1973**, *26*, 1257.
40. Allen, S. D.; Mason, M. G.; Schnepp, O.; Stephens, P. J. *Chem. Phys. Lett.* **1975**, *30*, 140.
41. Barton, T. J.; Douglas, I. N.; Grinter, R.; Thomson, A. J. *Molec. Phys.* **1975**, *30*, 1677.
42. Boyle, M. E.; Williamson, B. E.; Schatz, P. N.; Marks, J. P.; Snyder, P. A. *Chem. Phys. Lett.* **1986**, *130*, 33.

43. Snyder, P. A.; Lund, P. A.; Schatz, P. N.; Rowe, E. M. *Chem. Phys. Lett.* **1981**, *82*, 546.
44. Ishii, K.; Kobayashi, N.; Matsuo, T.; Tanaka, M.; Sekiguchi, A. *J. Am. Chem. Soc.* **2001**, *123*, 5356.
45. Michl, J. *J. Am. Chem. Soc.* **1978**, *100*, 6801.
46. Michl, J. *Tetrahedron* **1984**, *40*, 3845.
47. Mack, J.; Stillman, M. J. *J. Phys. Chem.* **1995**, *99*, 7935.
48. Mack, J.; Stillman, M. J. *Inorg. Chem.* **2001**, *40*, 812.
49. VanCott, T. C.; Rose, J. L.; Misener, G. C.; Williamson, B. E.; Schrimpf, A. E.; Boyle, M. E.; Schatz, P. N. *J. Phys. Chem.* **1989**, *93*, 2999.
50. Houssier, C.; Sauer, K. *J. Am. Chem. Soc.* **1970**, *92*, 779.
51. Frackowiak, D.; Bauman, D.; Manikowski, H.; Browett, W. R.; Stillman, M. J. *Biophys. Chem.* **1987**, *28*, 101.
52. Umetsu, U.; Seki, R.; Wang, Z.-Y.; Kumagai, I.; Nozawa, T. *J. Phys. Chem. B* **2002**, *106*, 3987.
53. Hughes, J. L.; Razeghifard, R.; Logue, M.; Oakley, A.; Wydrzynski, T.; Krausz, E. *J. Am. Chem. Soc.* **2006**, *128*, 3649.
54. Zaitoun, M. A.; Mason, W. R.; Lin, C. T. *J. Phys. Chem. B* **2001**, *105*, 6780.
55. Gale, R.; McCaffery, A. J. *J. Chem. Soc. Chem. Commun.* **1972**, 832.
56. Upton, A. H. P.; Williamson, B. E. *J. Phys. Chem.* **1994**, *98*, 71.
57. Isci, H.; Mason, W. R. *Inorg. Chim. Acta* **2004**, *357*, 4065.
58. Graham, R. G.; Grinter, R.; Perutz, R. N. *J. Am. Chem. Soc.* **1988**, *110*, 7036.
59. Williamson, B. E.; VanCott, T. C.; Boyle, M. E.; Misener, G. C.; Stillman, M. J.; Schatz, P. N. *J. Am. Chem. Soc.* **1992**, *114*, 2412.
60. Samet, C.; Rose, J. L.; Piepho, S. B.; Laurito, J.; Andrews, L. A.; Schatz, P. N. *J. Am. Chem. Soc.* **1994**, *116*, 11109; as amended in *J. Am. Chem. Soc.* **1995**, *117*, 9381.
61. Solomon, E. I.; Brunold, T. C.; Davis, M. I.; Kemsley, J. N.; Lee, S.-K.; Lehnert, N.; Neese, F.; Skulan, A. J.; Yang, Y.-S.; Zhou, J. *Chem. Rev.* **2000**, *100*, 235.
62. Solomon, E. I.; Pavel, E. G.; Loeb, K. E.; Campochiaro, C. *Coord. Chem. Rev.* **1995**, *144*, 369.
63. Solomon, E. I. *Inorg. Chem.* **2001**, *40*, 3656.
64. Solomon, E. I.; Neidig, M. L.; Schenk, G. *Comprehensive Coordination Chemistry II*; Vol. 2, Lever, A. B. P., Ed.; Elsevier, Ltd.: Amsterdam, 2004, p. 339.
65. Lee, S.-K.; George, S. D.; Antholine, W. E.; Hedman, B.; Hodgson, K. O.; Solomon, E. I. *J. Am. Chem. Soc.* **2002**, *124*, 6180.
66. Yoon, J.; Mirica, L. M.; Stack, T. D. P.; Solomon, E. I. *J. Am. Chem. Soc.* **2005**, *127*, 13680; and references cited therein.
67. Oganesyan, V. S.; George, S. J.; Cheesman, M. R.; Thomson, A. J. *J. Chem. Phys.* **1999**, *110*, 762.
68. Oganesyan, V. S.; Thomson, A. J. *J. Chem. Phys.* **2000**, *113*, 5003.
69. Larrabee, J. A.; Leung, C. H.; Moore, R. L.; Thamrong-nawasawat, T.; Wessler, B. S. H. *J. Am. Chem. Soc.* **2004**, *126*, 12316.
70. Lowther, W. T.; Orville, A. M.; Madden, D. T.; Lim, S.; Rich, D. H.; Matthews, B. W. *Biochemistry* **1999**, *38*, 7678.

71. Bingham, S. J.; Suter, D.; Schweiger, A.; Thomson, A. J. *Chem. Phys. Lett.* **1997**, *266*, 543.

72. Borger, B.; Bingham, S. J.; Gutschank, J.; Schweika, M. O.; Suter, D.; Thomson, A. J. *J. Chem. Phys.* **1999**, *111*, 8565.

73. Bingham, S. J.; Borger, B.; Suter, D.; Thomson, A. J. *Rev. Sci. Instr.* **1998**, *69*, 3403

74. Borger, B.; Gutschank, J.; Suter, D.; Thomson, A. J.; Bingham, S. J. *J. Am. Chem. Soc.* **2001**, *123*, 2334.

75. McInnes, E. J. L.; Pidcock, E.; Oganesyan, V. S.; Cheesman, M. R.; Powell, A. K.; Thomson, A. J. *J. Am. Chem. Soc.* **2002**, *124*, 9219.

76. Collison, D.; Murrie, M.; Oganesyan, V. S.; Piligkos, S.; Poolton, N. R. J.; Rajaraman, G.; Smith, G. M.; Thomson, A. J.; Timko, G. A.; Wernsdorfer, W.; Winpenny, R. E. P.; McInnes, E. J. L. *Inorg. Chem.* **2003**, *42*, 5293.

77. Macfarlane, R. M. *J. Chem. Phys.* **1963**, *39*, 3118.

78. Görller-Walrand, C.; Verhoeven, P.; D'Olieslager, J.; Fluyt, L.; Binnemans, K.; *J. Chem. Phys.* **1994**, *100*, 815; as amended in *J. Chem. Phys.* **1994**, *101*, 7189.

79. Fluyt, L.; Couwenberg, I.; Lambaerts, H.; Binnemans, K.; Görller-Walrand, C.; Reid, M. F. *J. Chem. Phys.* **1996**, *105*, 6117.

80. Heyde, K.; Binnemans, K.; Görller-Walrand, C. *J. Chem Soc. Faraday Trans.* **1998**, *94*, 843.

81. Görller-Walrand, C.; Behets, M.; Porcher, P.; Laursen, I. *J. Chem. Phys.* **1985**, *83*, 4329.

82. Reid, M. F.; Richardson, F. S. *J. Chem. Phys.* **1983**, *79*, 5735; Richardson, F. S.; Saxe, J. D.; Davis, S. A.; Faulkner, T. R. *Mol. Phys.* **1981**, *42*, 1401; Reid, M. F.; Richardson, F. S. *Chem. Phys. Lett.* **1983**, *95*, 501; Reid, M. F.; Richardson, F. S. *J. Phys. Chem.* **1984**, *88*, 3579.

83. Keiderling, T. A. *J. Chem Phys.* **1981**, *75*, 3639.

84. Wang, B. *Am. Lab.* **1996**, 36C, and references therein.

85. Yoo, R. K.; Croatto, P. V.; Wang, B.; Keiderling, T. A. *Appl. Spectrosc.* **1991**, *45*, 231.

86. Wang, B.; Keiderling, T. A. *Appl. Spectrosc.* **1995**, *49*, 1347.

87. Devine, T. R.; Keiderling, T. A. *J. Chem. Phys.* **1985**, *83*, 3749.

88. Pawlikowski, M.; Keiderling, T. A. *J. Chem. Phys.* **1984**, *81*, 4765; Pawlikowski, M.; Devine, T. R. *J. Chem. Phys.* **1985**, *83*, 950.

89. Wang, B.; Keiderling, T. A. *J. Chem. Phys.* **1993**, *98*, 903.

90. Wang, B.; Keiderling, T. A. *J. Phys. Chem.* **1994**, *98*, 3957; Wang, B.; Yoo, R. K.; Croatto, P. V.; Keiderling, T. A. *Chem. Phys. Lett.* **1991**, *180*, 339.

91. Tam, C. N.; Bour, P.; Keiderling, T. A. *J. Chem. Phys.* **1996**, *104*, 1813; Tam, C. N.; Keiderling, T. A. *Chem. Phys. Lett.* **1995**, *243*, 55.

92. Funk, T.; Deb, A.; George, S. J.; Wang, H.; Cramer, S. P. *Coord. Chem. Rev.* **2005**, *249*, 3.

93. Cramer, S. P. *ACS Symposium Series 858 Paramagnetic Resonance of Metallobiomolecules* **2003**, 358.

94. Carra, P.; Thole, B. T.; Altarelli, M.; Wang, X. *Phys. Rev. Lett.* **1993**, *70*, 694.

95. Thole, B. T.; Carra, P.; Sette, F.; van der Laan, G. *Phys. Rev. Lett.* **1992**, *68*, 1943.

96. Mizumaki, M.; Yano, Y.; Umehara, I.; Ishikawa, F.; Sato, K.; Koizumi, A.; Sakai, N.; Muro, T. *Phys. Rev. B* **2003**, *67*, 132404.

97. Majewski, P.; Gepragss, S.; Sanganas, O.; Opel, M.; Gross, R.; Wilhelm, F.; Rogalev, A.; Alff, L. *Appl. Phys. Lett.* **2005**, *87*, 202503.

98. Moroni, R.; dit Moulin, C. C.; Champion, G.; Arrio, M.-A.; Sainctavit, P.; Verdaguer, M.; Gatteshi, D. *Phys. Rev. B* **2003**, *68*, 064407.

99. Funk, T.; Kennepohl, P.; Di Bilo, A. J.; Wehbi, W. A.; Young, A. T.; Friedrich, S.; Arenholz, E.; Gray, H. B.; Cramer, S. P. *J. Am. Chem. Soc.* **2004**, *126*, 5859.

100. Peterson, J.; Pearce, L. L.; Bominaar, E. L. *J. Am. Chem. Soc.* **1999**, *121*, 5972; and references cited therein.

101. Peterson, J.; Bominaar, E. L. *Anal. Chem.* **2002**, *74*, 527A.

102. Bominaar, E. L.; Peterson, J. *J. Chem. Phys.* **1999**, *111*, 7512.

103. Bominaar, E. L.; Achim, C.; Peterson, J. *J. Chem. Phys.* **1998**, *109*, 942.

104. Kreglewski, M.; Vala, M. *J. Chem. Phys.* **1981**, *74*, 5411; and references therein.

105. Mason, W. R., "*Comprehensive Coordination Chemistry II*, McCleverty, J. A.; Meyer, T. J., Eds, Vol. 2, Section 2.25, pp. 327–337, Lever, A. P. B., Vol. ed., Elsevier Ltd, Oxford, 2004.

106. Mason, W. R. Unpublished results 2004.

107. Moore, C. E. *Natl. Bur. Stand. Circ. (U.S.)* **1958**, No. 467, vol. III.

108. Vala, M.; Rivoal, J.-C.; Grisolia, C.; Pyka, J. *J. Chem. Phys.* **1985**, *82*, 4376.

109. Vala, M.; Eyring, M.; Pyka, J.; Rivoal, J.-C.; Grisolia, C. *J. Chem. Phys.* **1985**, *83*, 969.

110. Rivoal, J.-C.; Zoueu, J.; Blanchard, X.; Nahoum, R.; Eyring, M.; Pyka, M.; Vala, M. *J. Chem. Phys.* **1994**, *101*, 2684.

111. Mason, W. R. Unpublished results 2003.

INDEX

A Practical Guide to Magnetic Circular Dichroism Spectroscopy, by W. Roy Mason

Printed and bound by CPI Group (UK) Ltd, Croydon, CR0 4YY

16/09/2026

14963458-0004